is boo' is ⸱ - 'efore

⅟

MECHANICAL FILTERS
IN ELECTRONICS

WILEY SERIES ON FILTERS:
Design, Manufacturing, and Applications

Editors: Robert A. Johnson and George Szentirmai

Mechanical Filters in Electronics
  Robert A. Johnson

LC - Filters: Design, Testing, and Manufacturing
  Erich Christian

MECHANICAL FILTERS IN ELECTRONICS

ROBERT A. JOHNSON
Rockwell International

A Wiley-Interscience Publication

JOHN WILEY & SONS

New York   Chichester   Brisbane   Toronto   Singapore

Library of Congress Cataloging in Publication Data:

Johnson, Robert A., 1932–
    Mechanical filters in electronics.

    (Wiley series on filters)
    Includes index.
    1. Mechanical filters (Electronic engineering)
I. Title. II. Series.
TK7872.F5J66 1982   621.3815'32   82-10922
ISBN 0-471-08919-2

Printed in the United States of America
10 9 8 7 6 5 4 3 2 1

*To my wife Lois, whose support was essential;
and to those who use their knowledge to bring peace,
dignity, and a better life to the world's people*

SERIES PREFACE

The primary objective of the Wiley Series on Filters is to bring together theory and industrial practice in a series of volumes written for filter users as well as those involved in filter design and manufacturing. Although this is a difficult task, the authors in this series are well qualified for the job. They bring both strong academic credentials and many years of industrial experience to their books. They have all designed filters, have been involved in manufacturing, and have had experience in interacting with the filter user.

Each of the books covers a wide range of subjects including filter specifications, design, theory, parts and materials, manufacturing, tuning, testing, specific applications, and help in using the filter in a circuit. The books also provide a broad view of each subject based on the authors' own work and involvement with filter experts from around the world.

The most outstanding feature of this series is the broad audience of filter builders and users, which it addresses. This includes filter research and development engineers, filter designers, and material specialists, as well as industrial, quality control, and sales engineers. On the filter user's side, the books are of help to the circuit designer, the system engineer, as well as applications, reliability, and component test experts, and specifications and standards engineers.

ROBERT A. JOHNSON
GEORGE SZENTIRMAI

PREFACE

The purpose of this book is to describe the design, manufacturing, and use of mechanical filters to practicing engineers. The audience I am addressing includes people involved in filter research, design, and manufacturing, in addition to those involved in the use of filters: systems engineers, circuit designers, applications engineers, and reliability and test engineers. Because this is the first comprehensive book written on mechanical filters in the western world, I have given it an international slant to make it helpful to those working outside of the United States.

This book was written for both the skilled filter designer and the person who has a basic knowledge of electrical engineering. There is enough theory developed to enable the person who is new to mechanical filters to understand how mechanical filters work, as well as how they are manufactured, specified, tested, and used. For the person knowledgeable in mechanical filters, the book provides access to a considerable amount of new and difficult-to-obtain information and also can be used as a quick reference.

In this book I use filters that are presently being manufactured to illustrate basic concepts. In this way, while learning the fundamentals, the reader is also made aware of the state of the art in mechanical filter technology. Also, because this is a "real world" book I have taken some risks by expressing value judgments based on 25 years of electromechanical filter design, but I have usually warned the reader when I have done so.

Because the book's wide audience is composed of filter builders and users, I have worked toward making each chapter as independent of the others as possible. For example, this allows the filter user who is writing a specification to start with Chapter 8 and avoid having to read the chapters on designing a mechanical filter. At the same time, the chapters that are heavy with design are not compromised. I would suggest, though, that Chapter 1 be read by everyone.

The main thrust of Chapter 1 is to provide answers to such questions as, what are mechanical filters, and what are their characteristics? Chapter 2, which deals with transducers, is important in that it bridges the electrical/mechanical gap through the use of analogies and equivalent circuits. For a filter designer the entire chapter is essential, whereas the nonspecialist can skip the equations and read to the section on magnetostrictive transducers.

At that point, the nonspecialist can continue or can jump to the section on piezoelectric ceramics, which parallels the magnetostrictive section.

Chapter 3, on resonators and coupling elements, is important for the filter designer (in particular the concept of equivalent mass), but only the first few pages and the section on coupled resonators need to be read by the filter user. Chapter 4, on circuit design methods, is of interest to the filter user because it deals with the question of the conditions for electrical tuning. Also the figures illustrate the various transducer/resonator/coupling combinations used by mechanical filter manufacturers. Since the material of Chapter 4 is long and detailed, the nonspecialist who reads all of it deserves a great deal of praise.

Chapters 5 and 6 are the nuts and bolts of the mechanical filter technology and are of value to both the filter designer and user. These chapters are of particular value to applications and reliability engineers, who have responsibility for seeing that filter components selected are reliable and will withstand expected environmental stresses.

Chapter 7, applications of mechanical filters, discusses the various systems in which mechanical filters are used, and therefore is of interest to applications and systems engineers and to equipment designers. Chapter 8, on specifying and testing mechanical filters, is of help to everyone, from the filter designer to the filter user's incoming-inspection test technician.

The chapter on using mechanical filters, Chapter 9, was written for those who employ mechanical filters in their equipment. This chapter completes the cycle that started in Chapter 1 with the filter user's question of what filter to choose, and proceeded through subsequent chapters to cover subjects such as specifications, design, manufacturing, and testing.

ROBERT A. JOHNSON

Tustin, California
December 1982

ACKNOWLEDGMENTS

This book was made possible by those people who were willing to share their time and their ideas with me. They have become my friends and so the distinction between our friendship and our professional relationship has blurred. Each of these people has also shared my dream of a free exchange of ideas among technical people throughout the world. This has been true of my supervisors, Rich Muret, Jack Graham, and Wes Peterson, who created a work climate that allowed for this open exchange of ideas. I also thank my co-workers whose help started 25 years ago when Herb Lewis spent his lunch hours teaching me filter theory. At the present time I have been getting technical help and encouragement from Bill Domino, Pete Ysais, Fred Fanthorpe, Lee Cornet, and Don Havens.

I also want to express my appreciation to Gabor Temes, who taught me about filter design with transformed variables and started me in my first book writing; to Carl Kurth, who involved me in IEEE technical activities; to Desmond Sheahan, who worked with me in building bridges to other parts of the world; and to George Szentirmai, my Wiley Filter Series co-editor, who has been a teacher and a support to me for many years.

Also, I want to thank my international friends for their contributions to this book. Masashi Konno was my first contact outside of the United States and he has been a wonderful friend and source of help over the past two decades. I have been privileged to write papers with Manfred Börner and Kazuo Yakuwa and discuss both design and factory concepts with Takeshi Yano and Tasuku Yuki. A little more than ten years ago I met Alf Günther, who has become a friend, history teacher, and always a willing source of technical help. I want to thank also Hans Schüssler, Josef Deckert, Francesco Molo, and Satoru Fujishima for helpful discussions, correspondence, and factory visits. It has been a privilege to know Josef Trnka, Herbert Ernyei, and Franciszek Kamiński whose work, due to space limitations, has been somewhat neglected in this book. Thanks also to Tetsuro Takaku, Izumi Kawakami, Yasukaza Kawamura, and Yasuo Koh for their help; and to Yoshi Tomikawa who taught me about rigid-body modes and mechanical equivalent circuits while he was in the United States.

Finally, thanks to Dot Hopkins and Jan Chantland for their help in typing the manuscript; our factory and marketing people; and again to Wes Peterson, who, because of the special person he is, made this book possible.

R.A.J.

CONTENTS

MECHANICAL FILTERS
IN ELECTRONICS

*Chapter One*_____

INTRODUCTION

This is a book about mechanical filters—or more precisely, electromechanical bandpass filters. The term *electromechanical* comes from the transformation, within the filter, of electrical signals into mechanical energy, and after the filtering takes place, the transformation of the remaining signals back into the original electrical form. The term *bandpass* means that the filter will pass a band (a spectrum) of frequencies and attenuate frequencies outside of that band. This description of electromechanical filters is not unique to mechanical filters. There are other electromechanical bandpass filters: crystal filters, ceramic filters, and surface acoustic wave (SAW) filters, so we must further define what we mean by the words *mechanical filter*. We will do this by comparing the mechanical filter to the other electromechanical filters.

A mechanical filter is composed of acoustically (mechanically) coupled mechanical resonators. This differentiates a mechanical filter from lumped-component quartz-crystal filters or ceramic filters that are composed of electrically coupled resonators. A mechanical filter has bulk or lumped resonators that are usually wire coupled, which distinguishes it from acoustically coupled monolithic crystal filters and ceramic filters, which use two or more resonators on a single substrate. A mechanical filter is differentiated from a SAW filter by the fact that signals in a SAW filter are propagated in only one direction, whereas signals in a mechanical filter are allowed to propagate back and forth between the input and the output.

A mechanical filter also is a passive analog component; it requires no external power sources or external clocks, and it processes analog rather than digital data. This differentiates the mechanical filter from active, switched capacitor (SC), charged coupled device (CCD), and digital filters.

BASIC PRINCIPLES

Understanding the operation of a mechanical filter involves an understanding of basic principles of physics, circuit theory, and vibration theory, along with a familiarity of electromechanical transducer concepts. It is this breadth of necessary knowledge that frightens people when confronted with understanding the mechanical filter. Therefore, when talking to an electronics engineer, the mechanical filter designer usually talks in terms of electrical analogies and equivalent circuits, and when discussing mechanical filters with a mechanical engineer, they will probably talk about springs and masses.

A transition between the two disciplines can be made with motor and loudspeaker examples (electrical-to-mechanical) and generator and phonograph pick-up examples (mechanical-to-electrical). The motor-generator concepts are helpful in understanding how impedances are reflected through electromechanical transducers. For example, when the electrical output terminals of a generator are shorted, the mechanical impedance (ratio of torque to angular velocity) across the armature suddenly increases. Loudspeaker and phonograph pick-up concepts are helpful in understanding the transduction of many signals having various amplitudes and frequencies. Because of frequency response limitations of electromechanical transducers, loudspeakers and phonograph pick-ups act as electrical-to-mechanical and mechanical-to-electrical bandpass filters. The broader the passband and the flatter the amplitude response in the passband, the better the fidelity.

Elements of a Mechanical Filter

The mechanical filter is a device composed of electrical, electromechanical, and mechanical elements combined in such a way that bandpass filtering takes

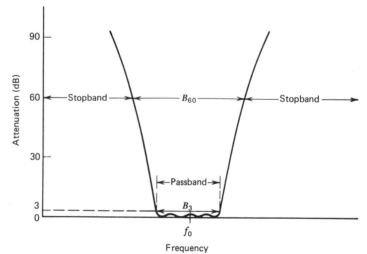

FIGURE 1.1.   Attenuation versus frequency characteristics of a bandpass filter.

place. Figure 1.1 shows the frequency response of a mechanical filter. The curve corresponds to the voltage, expressed in dB, at the output terminals of the filter. The input signal, which drives the source resistance and filter, is a constant voltage, variable-frequency sinusoidal wave. The attenuation of the input signal is measured with respect to the maximum voltage output. The difference between the two frequencies corresponding to 3 dB attenuation is the 3 dB bandwidth B_3. The region between the two 3 dB frequencies is the filter passband. The difference between the 60 dB attenuation frequencies is the 60 dB bandwidth B_{60}. The region outside the 60 dB points is the stopband. The ratio of B_{60} to B_3 is the shape factor of the filter. The center frequency f_0 is often defined as the frequency midway between the 3 dB attenuation frequencies. The ratio of B_3 to f_0 is the fractional bandwidth of the filter. Having defined the frequency characteristics of the mechanical filter, let's now look at the elements, which when properly combined achieve a desired response.

Figure 1.2 shows the electrical, electromechanical, and mechanical elements used in a mechanical filter. The electrical tuning element resonates with the electrical reactance of the electromechanical transducer but is only used in so-called intermediate-band and wideband filters. The transducer is either magnetostrictive or piezoelectric, and when alone or when attached to a metal rod or bar, it resonates at a frequency within the filter passband. The transducer drives a system of wire-coupled resonators which, in turn, drives the output transducer and electrical tuning coil or capacitor. The terminating resistors R represent the resistance of the input (source) and output (load) circuits. It is necessary to terminate the filter with resistance in order to achieve a flat or moderate ripple passband response. All of the filter components have linear and bilateral characteristics, so different amplitude and frequency signals can be treated independently or simply added as they propagate in both directions between the filter input and output resistors.

So far, we have treated the mechanical filter in very general terms. Let's next look at a specific filter and trace a signal from the generator to the load resistor. Figure 1.3 shows a disk-wire mechanical filter that employs magnetostrictive ferrite transducers. Looking at Figure 1.3, electrical current I from the voltage generator flows into the resonating capacitor C_R and the transducer coil. Current through the coil produces a magnetic field which passes through the ferrite rod causing the ferrite to vibrate at the frequency of the generator

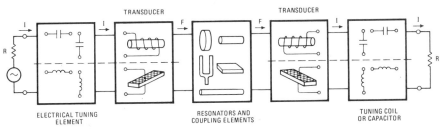

FIGURE 1.2. A mechanical filter circuit showing a variety of its primary elements. (Reprinted from *ELECTRONICS*, Oct. 13, 1977, copyright © McGraw-Hill, Inc., 1977. All rights reserved.)

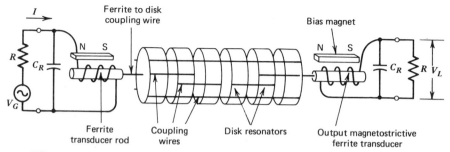

FIGURE 1.3. Essential elements of the six-disk mechanical filter shown in Figure 1.4.

signal. The change in dimensions and resultant vibration due to a changing magnetic field is called *magnetostriction*. This vibration is coupled to the first disk resonator by means of a small wire. Mechanical energy is coupled between disks by wires welded to the circumference of each disk. Vibrations from the end disk and coupling wire excite the output transducer. Strains in the output transducer produce an alternating magnetic field which induces a voltage V_L across the coil and load resistor R. Although we described the signal as flowing from the input to the output, it should be remembered that under steady-state conditions, energy is passing in both directions. Figure 1.4 shows the actual disk-wire mechanical filter.

Turning to Figure 1.5, we see an electrical equivalent circuit of the disk-wire mechanical filter. The mechanical components—the springs, masses, and damping elements—have been transformed to their electrical equivalents by means of the direct, or mobility, analogy. If we were to make a digital-computer frequency-response analysis of the inductor-coupled network in Figure 1.5, we would obtain a curve similar to that in Figure 1.1. That is evident to a person skilled in basic filter theory but far from obvious to the novice.

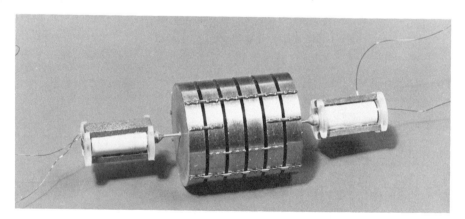

FIGURE 1.4. Mechanical filter used in FDM telephone systems. (Courtesy of Rockwell International, USA.)

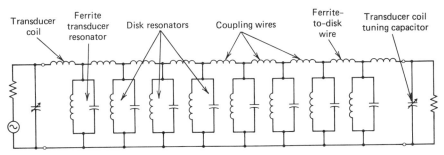

FIGURE 1.5. Inductor-coupled electrical equivalent circuit of the mechanical filter of Figures 1.3 and 1.4.

For the person who has some training in electronics, but is not skilled in filter theory, the following discussion will be helpful. Figure 1.6 shows a section of the inductor-coupled mechanical filter equivalent circuit. Both tuned circuits are resonant at frequency f_1. Because the section has two resonators, it has two natural resonances. In other words, when excited, the circuit will vibrate at two frequencies. One frequency is where the currents i in the resonators are in phase causing the potential across the inductor L_{12} to be zero at all times. The second natural frequency is where the currents in the tuned circuits are out of phase with respect to one another, causing a potential to be developed across the coupling inductor L_{12}. Because the potentials on opposite sides of L_{12} are equal in magnitude but opposite in sign, the potential at the midpoint of the coupling inductor is always equal to zero. The midpoint could, therefore, be grounded and each half of the section would vibrate indepen-

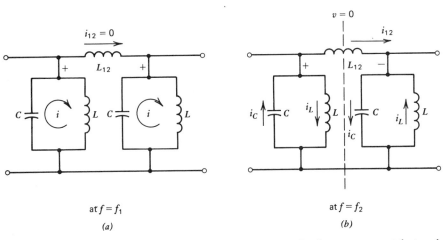

FIGURE 1.6. Resonance conditions for a two-resonator filter. (a) In-phase resonance at the tuned circuit's frequency f_1, and (b) the out-of-phase resonance at the higher frequency f_2.

dently but at the same frequency,

$$f = f_2 = 1/2\pi\sqrt{L'C} \quad \text{where } L_1' = (L_{12}/2)L \Big/ \big[(L_{12}/2) + L\big].$$

If we were to drive one end of the circuit of Figure 1.6 with a current source and measure the voltage at the opposite end, we would see large voltage peaks at the frequencies f_1 and f_2. If we replaced the current source with a voltage source in series with a resistor and also resistively terminated the output, then the output voltage response, as a function of frequency, would be that of a bandpass filter (see Figure 4.9) where the frequencies f_1 and f_2 define the ends of the passband. This type of analysis can be extended to a larger number of resonators, as is done in the spring-mass analysis of Figures 3.33 and 3.34. If your background is strong in mechanical systems, the analysis of Figure 3.33 may be more helpful than Figure 1.6. General relationships between the mechanical filter elements and the filter characteristics (bandwidth, center frequency, and shape factor) are discussed in the first few pages of Chapter 3.

Types of Mechanical Filters

The disk-wire filter discussed in the previous section is only one of a large number of mechanical filter types. Figures 1.7 to 1.10 show mechanical filters that use flexural, extensional, and torsional modes of vibration of the resonators and the coupling wires. Mechanical filter transducers range from simple rods or bars to composite ceramic/metal resonators like those shown in Figure 2.18. Methods of coupling the transducer to the first metal resonator are shown in Figure 4.27. Coupling methods for metal resonators in the

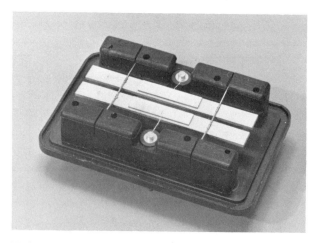

FIGURE 1.7.   Mechanical filter for telephone signaling applications, $f_0 = 3.825$ kHz. (Courtesy of Rockwell International, USA.)

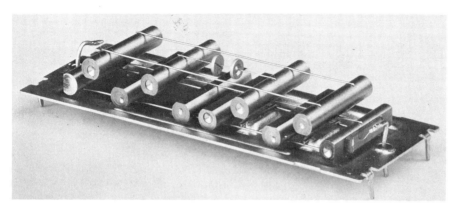

FIGURE 1.8. 48 kHz telephone channel filter with mass-loaded wire bridging. (Courtesy of Telettra, Italy.)

interior of the mechanical filter are shown in Figure 4.21, and methods of realizing finite-frequency attenuation poles are shown in Figure 4.29. Electrical tuning and termination methods are described in Figure 4.34.

WHY MECHANICAL FILTERS?

Mechanical filters are used in systems that demand narrow bandwidth, low loss, and good stability. These requirements can be achieved with mechanical filters because of the high Q and excellent temperature and aging characteris-

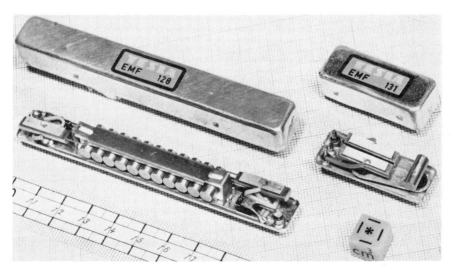

FIGURE 1.9. Telephone channel and signaling filters. Carrier frequency $f_c = 128$ kHz. (Courtesy of Tesla, C.S.R.)

FIGURE 1.10.   48 kHz (top) and 128 kHz telephone channel filters. (Courtesy of Siemens, F.R.G.)

tics of the mechanical resonators. Iron-nickel alloy resonators have Q values that range from 10,000 to 25,000. This allows mechanical filters to be built with bandwidths as narrow as 0.05 percent without excessive loss or passband rounding.

Equally important as high Q is the temperature coefficient of frequency and the frequency aging of mechanical resonators. Typical resonator frequency shifts are parabolic and are on the order of 2 ppm/°C, which at 455 kHz and a temperature range of ± 50°C means a shift in the filter passband of approximately 45 Hz. This shift is in either the positive or negative direction, depending on the material. This degree of stability is important because in single-sideband equipment, frequency shifts greater than 100 Hz result in either poor carrier rejection or a loss of the audio response. Regarding aging, typical mechanical filter metal resonators are stable to within 50 ppm over the lifetime of the equipment. At 455 kHz the passband drift can be held to less than 25 Hz.

Characteristics of Mechanical Filters

Attempting to characterize mechanical filters in a single table is risky because of the wide range of configurations, materials, and manufacturing processes that are used throughout the world. Therefore, the data shown in Table 1.1 should be used with caution. Also, the values are not independent. For instance, the narrowest bandwidth filters will not have high shock resistance and low loss. Wide bandwidth filters and one-and two-resonator filters tend to have the worst temperature stability. High-frequency filters and wide-bandwidth filters are inclined to have low terminating resistance, and so on. Table 1.1 covers most mechanical filter designs. Low-frequency filters with tuning coils were omitted because they are only manufactured in small quantities.

TABLE 1.1. Characteristics of Narrowband Tuning-Fork and Flexure-Mode Filters, and Intermediate-Band Communication Filters

Filter Characteristic	Filter Type					
	Narrowband Tuning-Fork		Narrowband Flexure-Mode		Intermediate-Band Communication	
	(Min.)	(Max.)	(Min.)	(Max.)	(Min.)	(Max.)
Center frequency (f_0)	200 Hz	25 kHz	2 kHz	75 kHz	50 kHz	600 kHz
Fractional bandwidth (B/f_0)	0.3%	1.0%	0.15%	10%	0.05%	10%
Number of poles (resonators)	1	2	2	4	2	15
Insertion loss (IL)	2 dB	20 dB	1.5 dB	10 dB	2 dB	15 dB
Temperature coefficient of f_0 (TC_f)	±20 ppm/°C	±60 ppm/°C	±3 ppm/°C	±25 ppm/°C	±1.5 ppm/°C	±5 ppm/°C
Terminating resistance (R_T)	10 kΩ	300 kΩ	2 kΩ	50 kΩ	100 Ω	30 kΩ
Aging (10 years)	200 ppm	750 ppm	100 ppm	750 ppm	50 ppm	250 ppm
Shock endurance	10 g	35 g	15 g	200 g	15 g	250 g

Comparisons with Other Filters

In this section we make some broad comparisons between mechanical filters and other filter types. This is, of course, risky because of changing technologies and the exceptions to the generalizations. Therefore, use the following with caution!

The two most important filter user specifications are center frequency and bandwidth. Figure 1.11 shows a comparison between various electromechanical filters; Figure 1.12 shows the bandwidth and frequency ranges of other filter types. It is important to note that the allowable center frequency shift, chosen for these curves, over a temperature range of $\pm 50°C$, is equal to $0.2B$, where B is the filter's 3 dB bandwidth. Also, the length of a resonator cannot exceed 5 cm or its volume exceed 10 cm³. In addition, resonator contributions to the filter loss are restricted to 1.5 dB per resonator, which corresponds to a normalized Q (QB/f_0) of 5. Only filters that meet the above criteria are included in Figures 1.11 and 1.12. Also these criteria are somewhat arbitrary, in that they do not take into account the application. For instance, a voice-bandwidth single-sideband filter requires a stability of more than 10 times the $0.2B$ frequency shift, whereas a simple tone-receiving filter may only need $0.5B$ frequency stability. The bandwidth limitation of switched-capacitor filters is due to increased noise and reduced dynamic range as the bandwidth narrows.

In Figure 1.11, we see that mechanical filters overlap crystal and ceramic filters up to the mechanical-filter frequency limit of about 600 kHz. Below 2 kHz, tuning-fork mechanical filters stand alone. Looking at Figure 1.12, we see that active filters and switched-capacitor filters overlap with the tuning-fork resonator filters, as well as some of the flexural-mode bar and multiresonator mechanical filters. Because of the extensive overlap, we must look at other factors in judging whether a mechanical filter is the right filter for an application.

Some of these additional factors include price, design cost, power dissipation, the need for an external clock, dynamic range, and reliability, as well as a more detailed analysis of stability, size, and Q-rounding. Let's next look at some comparisons between filters in the overlap regions of Figures 1.11 and 1.12.

Comparing a mechanical filter to a switched-capacitor filter, the SC filter has the advantage of size, and in large production quantities, a possible advantage in cost. The negative aspects of using SC filters are the high design cost, long design time, moderate to poor dynamic range, and the fact that both power and an external clock are needed.

Active filters require external power, but not a clock, and have a somewhat better dynamic range than the SC filters. Prices, design costs, and time-to-design are comparable to those of mechanical filters. An advantage of active filters is that they can be designed in-house, but this sometimes leads to production problems and surprisingly high costs. Active filters are comparable in size with mechanical filters, but in general, are not as stable.

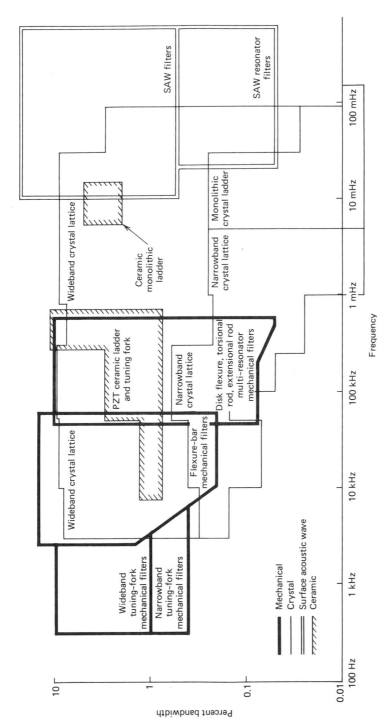

FIGURE 1.11. Application range of electromechanical filters.

11

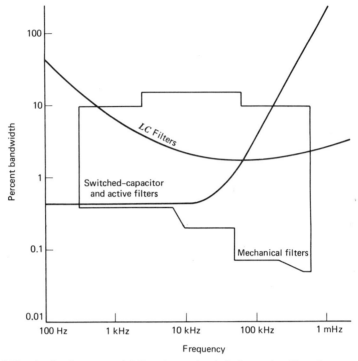

FIGURE 1.12. Application range of *LC*, active, and switched-capacitor filters in comparison to mechanical filters.

In the overlap regions, *LC* filters are generally larger and less stable then mechanical filters. Like mechanical filters, they do not require external power sources or clocks. Also, on the positive side, they can be designed and manufactured in-house and large production volumes are not needed to make them cost-effective.

In the case of multiple-resonator filters, there is considerable competition between crystal and mechanical filters. Generally, mechanical filters have a cost advantage and crystal filters have a performance advantage, but for some applications these advantages are reversed. Design costs and times are comparable.

Ceramic-resonator filters in the 10 percent fractional bandwidth region have stopband performance advantages over mechanical filters, but in the narrow bandwidth regions, ceramic filters suffer greatly from stability and *Q* problems. Prices for multielement ceramic filters are similar to mechanical filters, but simple one- and two-element ceramic roofing ladders for spurious signal rejection are quite inexpensive. Design costs for multielement ladders at nonstandard frequencies and bandwidths are high.

Having compared various filters, let us now look at applications of mechanical filters that stretch from watches to telephone systems.

Uses of Mechanical Filters

Mechanical filters are used in communication systems as IF (intermediate frequency) filters in radios; they are used as channel filters, signaling filters, and pilot tone filters in FDM (frequency-division multiplex) telephone systems, and are used in FSK (frequency-shift keying) modems. Radio applications range from inexpensive AM car radios to high performance HF (high frequency) transceivers. Mechanical filters have set the world's standards for telephone voice filtering and are the primary selectivity components in many, if not most, navigation receivers and train control systems. In addition, one- and two-resonator tuning forks are used in large numbers in tone generators and decoders for long-distance monitoring and control systems, as well as in paging and alarm systems. Finally, millions of small three-terminal zinc-oxide tuning-fork resonators are used in oscillator circuits for wrist watches.

   In summary, mechanical filters are found where stability and narrowband selectivity are needed. Details regarding most of the above applications are discussed in Chapter 7.

A BRIEF HISTORY OF MECHANICAL FILTERS

Leon Brillouin, in his excellent book *Wave Propagation in Periodic Structures* [1], begins his discussion of lattices with Newton's use of a spring-mass model to calculate the velocity of sound in air. This may also be a good starting point for the history of mechanical filters. Newton's model was composed of identical masses separated by identical springs, in other words, a ladder network having a lowpass filter response. Following the work of Newton, studies of resonators and wave propagation were carried out by many giants of the physical sciences in the seventeenth, eighteenth, and nineteenth centuries: Hook, Taylor, the Bernoullis, Euler, D'Alembert, Fourier, Lagrange, Cauchy, Kelvin, and finally Rayleigh, who in 1877 published his still very useful book *The Theory of Sound* [2]. By the end of the nineteenth century, the study of acoustics and mechanical vibrating systems was a mature science.

   Still missing, before mechanical filters could be realized, were circuit theory concepts relating to bandpass filtering and electromechanical transducers that could be used at IF frequencies. In 1915 the major breakthrough was made when Campbell and Wagner developed the first electrical bandpass filters [3]. People like Edward Norton, who had a strong mechanical background, applied the electrical theory to electromechanical systems such as loudspeakers and phonograph recording equipment [4]. During this time, transducers that could be applied to higher-than-audio frequencies, and constant-modulus resonator materials were developed. Because of the work done by Cady and people like Warren Mason at Bell Telephone Laboratories, most early narrowband filter development was on crystal filters, but in 1946, Robert Adler of Zenith developed the first IF mechanical filter [5]. Adler's work was closely followed by developments at Collins Radio and RCA, and within a decade, by work in Europe and Japan.

As shown in the following historical outline, the mechanical filter has become a vital part of the electronics industry through the efforts of people throughout the world. New concepts such as the trapped-energy structure of Strashilov et al. have not yet been applied to practical applications but may be forerunners for the mechanical filters of the future.

Key Events in Mechanical Filter History

1914 Wagner in Germany and Campbell in the USA conceived the first electrical bandpass filters.

1918 The Langevin-type composite resonator was developed in France [7].

1920 Guillaume and Chevenard, in France, developed constant-modulus metal alloys [6].

1923 Harrison and Norton at Bell Telephone Laboratories (BTL) developed sound equipment using mechanical filter concepts [8].

1927 Pierce in the USA invented the magnetostriction resonator [9].

1941 Mason at BTL developed a transmission-line mechanical filter with piezoelectric transducers, electrical termination, and attenuation poles [10].

1942 Mason's book *Electromechanical Transducers and Wave Filters* was published [11].

1944 Barium titanate transducers were developed in Japan, the USSR, and the USA [12].

1946 Robert Adler designed the first IF mechanical filter for radios. The filter employed plate resonators and coupling wires [5].

1948 Doelz at Collins Radio developed the first high production-volume mechanical filter [13]. Production began in 1952.

1954 Bernard Jaffe and U.S. co-workers developed the first PZT-ceramic transducer materials [12].

1956 Börner in West Germany and Tanaka in Japan independently developed the widely used torsional-resonator, coupling-wire filter [14].

1956 Konno in Japan invented the flexural-resonator, torsional-coupling wire filter [15].

1957 Collins Radio developed and manufactured the first FDM telephone channel filters [16].

1961 Börner and Johnson independently developed the first bridging-wire filters [14].

1970 Onoe and Yano in Japan developed the mechanical wave-separating filter [17].

1971   Siemens engineers in West Germany developed the 1/20 CCITT telephone channel filter [18].

1974   Yano and NEC team developed the parallel ladder telephone channel and signaling filters with common transducer input [19]. Yakuwa and Okuda at Fujitsu developed a simplified phase-inversion structure for pole-type channel filters [20].

1977   Fujishima and Murata co-workers developed the 0.003 cm^3 sputtered ZnO composite tuning-fork two-port resonator [21].

1980   Strashilov, Hinkov, and Branzalov developed a trapped-energy mechanical filter structure [22].

REFERENCES

1. L. Brillouin, *Wave Propagation in Periodic Structures*. New York: Dover, 1946.

2.. J. W. S. Rayleigh, *The Theory of Sound*. New York: Dover, 1945.

3.. G. A. Campbell, "Physical theory of the electric wave-filter," *Bell Syst. Tech. J.*, **1**, 1–32 (Nov. 1922).

4. E. L. Norton, "Wave filter," U.S. Patent 1,681,554 (Aug. 1928).

5. R. Adler, "Compact electromechanical filters," *Electronics*, **20**, 100–105 (Apr. 1947).

6. M. P. Chevenard, "Etude de l'elastice de torsion des aciers au nickel a'haute teneur en chrome," *Compt. Rend.*, **171**, 93–96 (July 1920).

7. P. Langevin, French Patent No. 505,703.

8. F. V. Hunt, *Electroacoustics: The Analysis of Transduction and Its Historical Background*. New York: Wiley, 1954.

9. G. W. Pierce, "Magnetostrictive vibrator," U.S. Patent 1,882,397 (Oct. 1932).

10. W. P. Mason, "Wave transmission network," U.S. Patent 2,345,491 (Mar. 1944).

11. W. P. Mason, *Electromechanical Transducers and Wave Filters*. New York: Van Nostrand, 1942.

12. B. Jaffe, W. R. Cook, and H. Jaffe, *Piezoelectric Ceramics*. New York: Academic Press, 1971.

13. M. L. Doelz, "Electromechanical filter," U.S. Patent 2,615,981 (Oct. 1952).

14. R. A. Johnson, M. Börner, and M. Konno, "Mechanical filters—A review of progress," *IEEE Trans. Sonics Ultrason.*, **SU-18**, 155–170 (July 1971).

15. M. Konno, "Theoretical considerations of mechanical filters" (in Japanese), *J. Inst. Elec. Commun. Eng. Jap.*, **40**, 44–51 (1957).

16. J. C. Hathaway and D. F. Babcock, "Survey of mechanical filters and their applications," *Proc. IRE*, **45**, 5–16 (Jan. 1957).

17. M. Onoe and T. Yano, "Electromechanical wave separating filters," in *Proc. 20th IEEE Electronic Component Conf.* (May 1970).

18. F. Künemund, "Channel filters with longitudinally coupled flexural mode resonators," *Siemens Forsch. -u. Entwickl.-Ber.*, **1** (4), 325–328 (1972).

19. T. Yano, T. Futami, and S. Kanazawa, "New torsional mode electromechanical channel filter," in *Proc. 1974 European Conf. on Circuit Theory and Design*, London, 121–126, (July 1974).

20. K. Yakuwa, S. Okuda, and M. Yanagi, "Development of new channel bandpass filters," in *Proc. 1974 IEEE ISCAS*, San Francisco, 100–105 (Apr. 1974).

21. S. Fujishima, H. Nonaka, T. Nakamura, and H. Nishiyama, "Tuning fork resonators for electronic wrist watches using ZnO sputtered film," *1st Meeting on Applications of Ferroelectric Materials in Japan* (1977).

22. V. L. Strashilov, V. P. Hinkov, and K. P. Branzalov, "Trapped energy thickness-flexural modes of elongated rectangular plates," *IEEE Trans. Sonics Ultrason.*, **SU-28**, 349–355 (Sept. 1981).

ELECTROMECHANICAL TRANSDUCERS

Because mechanical filters are used in electronic circuits, it is necessary to convert electrical energy to mechanical energy and then reverse this process at the output of the filter. This is one of the functions of the electromechanical transducer. To design the transducer into a filter, it is necessary to be able to describe it with a physical and mathematical model. In this chapter we look at equivalent circuit models and associated equations for magnetostrictive and piezoelectric transducers. Transducer material characteristics are also discussed as they relate to filter design and to the understanding of the performance limitations of mechanical filters in general. In the last section of the chapter, we look at methods of measuring transducer characteristics.

INTRODUCTION

Electromechanical transducers used in mechanical filters must perform tasks other than energy conversion. For instance, the transducer must be designed to impedance-match the electrical and mechanical circuits. Also, the transducer will play a part in the input and output mechanical resonance circuits, and if tuning is used, the transducer inductance will be an element in the electrical tuned circuit. Although the transducer has a number of functions, its behavior under various conditions can be described by a few important circuit parameters. These parameters, for example resonance frequency and Q, are in turn affected by environmental conditions and material and manufacturing variations. Table 2.1 shows some of the more important interrelated parameters. The circuit parameters could have been expressed in other ways and the list of external variables is not complete; but under most conditions these lists are

TABLE 2.1.   Interrelated Electromechanical Transducer Parameters

Transducer Functions as Applied to a Mechanical Filter	Circuit Parameters Governing Transducer Performance	External Variables and Conditions Affecting Transducer Circuit Parameters
Energy conversion	Electromechanical coupling coefficient	Temperature
Impedance matching		Aging
Providing selectivity through mechanical resonance	Electrical (static) inductance or capacitance	Material property variations
		Fabrication tolerances
Being part of an electrical tuned circuit	Mechanical (motional) equivalent compliance and mass (or analogous L and C).	Assembly variations
		Electrical signal level
	Electrical and mechanical resonance frequencies	
	Electrical and mechanical Q values	

sufficient and form the basis for the study of transducers in this book. If we had expressed Table 2.1 as a three-dimensional matrix showing interrelationships, we would have seen that most of the elements of the matrix were non-zero. As an example, impedance matching is affected by coupling, static inductance or capacitance, resonance frequency, and Q. In turn, all of these are affected by all of the external variables such as temperature, aging, and so on. Fortunately, most mechanical filters use only magnetostrictive or piezoelectric transducers, which reduces the amount of subject material that we must consider. Furthermore, most of the piezoelectric transducers use PZT ceramics and the magnetostrictive transducers use Fe-Ni-Co oxides. This makes our study of transducers even more manageable. As a means of further introduction to the subject of transducers, let us look at the transducer of Figure 2.1
The transducer of Figure 2.1 can be viewed as an input transducer with electrical signals V and I, which are driving the device, or viewed as an output transducer being driven by the torque τ and angular velocity $\dot{\theta}$. The term *transducer* has been used in the broad sense to describe this particular resonator, which is composed of an alloy metal bar and a piezoelectric ceramic.

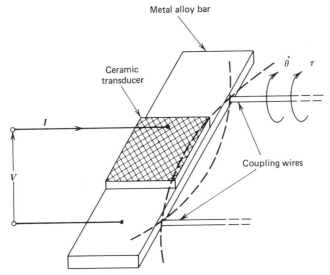

FIGURE 2.1.   Composite piezoelectric-ceramic, metal-alloy flexural-mode transducer.

The ceramic is a transducer material, and alone would be a transducer vibrating in another mode, such as extension in the direction of its length. When bonded to the metal bar, the extension is converted to flexure of the entire resonator. The dashed lines in Figure 2.1 show the motion of the bar in the fundamental flexural mode. We will usually use the term *transducer* to describe a device when concerned about energy conversion or impedance matching, and will use the term *resonator* when describing selectivity or a mode of vibration.

To describe the operation of the transducer in more detail, let us look at the function of the piezoelectric ceramic. The fact that the material is *piezoelectric* means that a voltage impressed across the two major surfaces results in the material expanding or contracting in the length, width, and thickness directions. When the impressed frequency coincides with a resonance of the entire structure, the displacement (motion) of the ceramic will coincide with the shape of that mode. In the flexural-mode case, the ceramic bends, changes length, and, to a lesser extent, changes dimensions in the thickness and width directions (due to the Poisson effect). The ceramic bar is plated on its upper and lower major surfaces providing an electric field through the thickness of the bar across its entire length and width. One terminal is connected to the upper plated surface and the other is attached to the metal bar, which makes contact with the bottom electrode. An applied voltage causes the ceramic to expand and contract resulting in bending of the entire resonator. Maximum displacement occurs when the applied signal frequency is the same as the frequency of the resonator.

We see from the dashed lines in Figure 2.1 that there are two points of zero flexural motion at the edge of the bar. These nodal points are nodes only in the sense of vertical and horizontal motion but not torsion. Wires attached at these two points will be driven in torsion, or conversely, torque applied to the wires will drive the resonator in a flexural mode resulting in a voltage being generated across the transducer terminals.

To this point, we have barely touched on the parameters of Table 2.1. In the next sections, we examine their relationships in detail starting with a discussion of electrical and mechanical equivalent circuits.

EQUIVALENT CIRCUITS

The types of equivalent circuits we deal with in this section are those which simplify the design process or make more clear our understanding of mechanical filters. As an example, a distributed mechanical network like the transducer described in the previous section will often be converted to a mechanical lumped-element circuit composed of springs and masses. The mechanical circuit is then converted to its electrical analogy. Converting from the bar resonator to a spring-mass system is a network simplification but is not an exact conversion. The bar resonator has an infinite number of natural resonances, whereas the number of lumped-element equivalent-circuit resonances are governed by the number of springs and masses in the equivalent circuit. Because mechanical filters are narrow-bandwidth devices, this is not a strong limitation and the lumped-element equivalent circuits are very useful.

The second equivalent circuit mentioned is the electrical analogy of the mechanical network. This equivalence is of most help in understanding and analyzing not only the transducer but the entire filter. Most often the mechanical filter designer or user understands electrical networks better than dynamics or acoustics. Also, most computer programs for filter design and most of the literature on filter design is expressed in electrical terms. Therefore, rather than attempt to use both electrical and mechanical variables, the mechanical networks are usually transformed to their electrical analogy. The fact that the electromechanical filter can be converted, on paper, to an "all-electrical" network is very important to the filter user. In the initial system-planning phases, the filter user can treat the mechanical filter as if it is an *LC* filter, or more generally, an *n*-resonator ladder network. Tables or curves or even a computer analysis can be consulted to find the frequency and time responses of an *n*-resonator filter. There may be a need to consult with the filter designer with regard to resonator *Q*'s or if a certain percent bandwidth flat-delay response can be realized; but at least the filter user has a good start toward a system plan and a preliminary filter specification. But let us first look at mechanical schematic diagrams and then look at the subject of analogies and equivalent circuits.

Schematic Diagrams

Both schematic diagrams and mathematical equations can be used to describe physical systems. The diagrams may contain lumped elements such as masses and springs, or capacitors and inductors, or may be transmission lines characterized by an impedance and a propagation constant. The mathematical representation may be in terms of differential equations or steady-state equations. In all cases, our starting point is the physical system, which in turn, can be represented by a pictorial diagram [1].

The pictorial diagram is simply a representation, in drawing form, of the essentials of an actual device. It is most often used to describe a mechanical system and acts as a bridge between an actual device and its schematic diagram. The pair of coupled bar-resonators of Figure 2.2(a) can be represented by the pictorial diagram shown in 2.2(b). The spring-mass system could also represent other types of coupled systems, for instance two ceramic/alloy composite resonators coupled in torsion, as was shown in Figure 2.1, or coupled tuning forks, or disks, or any number of other acoustically coupled mechanical resonators.

Comparing the actual device in Figure 2.2 to the pictorial diagram we see that the bar resonators are represented by spring constants K, masses M, damping elements D, and have the same resonance frequencies ω_1 and ω_2. We will assume the coupling wire is short compared to a wavelength and has little mass, and therefore it can be represented by a simple spring having stiffness K_{12}. The stiffness (spring constant) and mass of each resonator is an "equivalent" value and is a function of the point at which the wire is attached. If the coupling wire is near a nodal point, the equivalent mass and stiffness increase, although the resonance frequency, which is governed by the equation

$$\omega_i = 2\pi f_i = \sqrt{\frac{K_i}{M_i}}, \tag{2.1}$$

remains constant. The equivalent mass is the apparent mass of a resonator measured at a point on the resonator and in a specific direction. In other words, we have replaced the resonator with a spring-mass combination that has the same impedance characteristics near the resonance frequency. In terms of a force F and a velocity $\dot{x}$ we can write the resonator steady-state equation

$$F = j\omega M\dot{x} + \frac{K}{j\omega}\dot{x} + D\dot{x}, \tag{2.2}$$

or in differential equation form

$$F\sin\omega t = \frac{M\,d\dot{x}}{dt} + K\int \dot{x}\,dt + D\dot{x}. \tag{2.3}$$

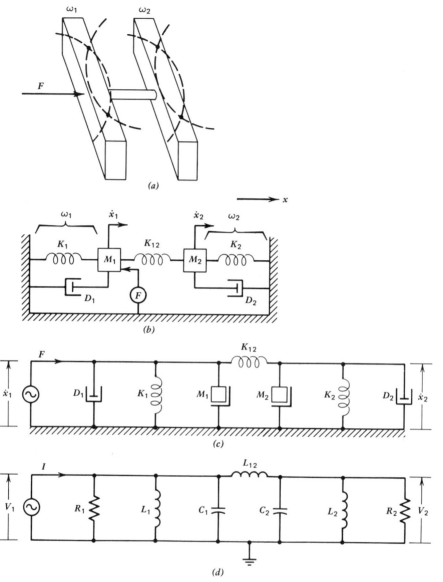

FIGURE 2.2. A two-resonator mechanical filter and its equivalent pictorial and schematic forms; (*a*) two wire-coupled, flexural-mode bars, (*b*) the lumped spring-mass pictorial diagram, (*c*) the mechanical schematic diagram, and (*d*) the electrical analogy.

Rather than write equations for the entire system shown in Figure 2.2(b), let's find a way to convert the pictorial diagram to a schematic diagram [1]. Having done that, conventional methods such as nodal or loop equations or matrices can be used to analyze the network.

A way to construct the schematic diagram is to first identify the two terminals of each element. In the case of a spring and damping element this is not a problem, but you may ask where the two terminals of the mass are. Because the velocity of a mass element is always referenced to an inertial reference frame, we simply add a terminal to the mass and connect it to the frame, which we will call "reference ground." Next, all terminals that move together are tied together. All terminals that do not move, but are clamped, are connected to the reference ground. Apply this procedure to the two-resonator filter in Figure 2.2, we obtain the schematic diagram of 2.2(c).

Note in Figure 2.2(c) that the velocities $\dot{x}_1$ and $\dot{x}_2$ are measured "across" the elements. In the case of a spring, for example, we are interested in the displacement or velocity between (i.e., across) the two ends of the spring, the force being the quantity that passes "through" the element. It is this concept of "through" and "across" variables that is basic to the understanding of electromechanical analogies.

Electromechanical Analogies

The discussion in this section is centered on one type of electromechanical analogy, which is the "mobility" analogy where across-variables are equated to across-variables and through-variables to through-variables. This means that velocity is analogous to voltage (V) and force to current (I). In this analogy, the network topologies of the mechanical and electrical circuits are also the same, as is the form of the differential equations. Writing Equations (2.2) and (2.3) in electrical terms we obtain

$$I = j\omega CV + \frac{1}{j\omega L}V + GV \qquad (2.4)$$

and

$$I \sin \omega t = C\frac{dV}{dt} + \frac{1}{L}\int V\,dt + GV. \qquad (2.5)$$

In summary, the mobility analogy has the following characteristics:

1. Differential equations have the same form.
2. Across variables are analogous to across variables, through variables are analogous to through variables.
3. Network topologies are the same.

Applying the preceding to the mechanical network of Figure 2.2(c), we obtain the electrical network of Figure 2.2(d). Also, we can construct Table 2.2 which

TABLE 2.2. Electromechanical Mobility Analogy

		Mechanical	Electrical
Variables	Across	Velocity ($\dot{x}$ or V) Angular Velocity ($\dot{\theta}$)	Voltage (V)
	Through	Force (F) Torque (τ)	Current (I)
Lumped network elements		Damping (D)	Conductance (G) Resistance^{-1} ($1/R$)
		Compliance (C)	Inductance (L)
		Stiffness (K)	Inductance^{-1}
		Mass (M) Mass Moment of Inertia (J)	Capacitance (C)
Transmission lines		Compliance/Unit Length (C_l)	Inductance/Unit Length (L_l)
		Mass/Unit Length (M_l)	Capacitance/Unit Length (C_l)
		Characteristic Mobility (Z_0)	Characteristic Impedance (Z_0)
		Propagation Constants (β)	Propagation Constants (β)
		Phase Velocity (v_p)	Phase Velocity (v_p)
Immittances		Mobility (Z)	Impedance (Z)
		Impedance (Y)	Admittance (Y)
		Clamped Point ($Z = 0$)	Short Circuit ($Z = 0$)
		Free Point ($Y = 0$)	Open Circuit ($Y = 0$)
Source immittances		Force ($Z = \infty$) Velocity ($Y = \infty$)	Current ($Z = \infty$) Voltage ($Y = \infty$)
Topology		Loop	Loop
		Node	Node
		Series Connection	Series Connection
		Parallel Connection	Parallel Connection

relates analogous electrical and mechanical quantities. Through use of this table, we can easily convert equations or schematic diagrams by simply replacing F by I, M by C, and so on.

Note that in Table 2.2 some new parameters have been added to those we have already discussed, for instance, compliance (C) which is the reciprocal of stiffness, and mass moment of inertia (J). In a rotational system, the steady-state equation equivalent to Equation (2.2) for the linear-motion system is

$$\tau = j\omega J\dot{\theta} + \frac{K}{j\omega}\theta + D\dot{\theta}, \qquad (2.6)$$

where τ is torque, $\dot{\theta}$ is angular velocity, K is torsional stiffness, and D is the torsional damping. Also included in Table 2.2 are parameters that are used to describe transmission lines.

Wires and bars vibrating in longitudinal or torsional modes can be represented by transmission-line equations of the same form as electrical networks. Also, they can be represented schematically in a similar way, as shown in Figure 2.3.

Since mechanical filters are generally cascade structures, such as ladder networks, we will most often make use of the transmission (or cascade) $ABCD$ matrix for writing equations describing two-port networks. The equations

$$x_1 = Ax_2 + By_2$$

$$y_1 = Cx_2 + Dy_2$$

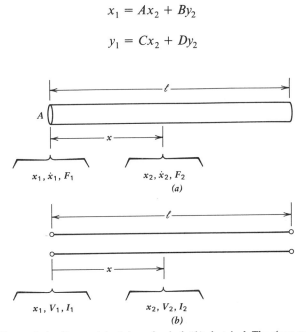

FIGURE 2.3.   Transmission line models, (a) mechanical, (b) electrical. The characteristics of these lines are described by Equations (2.8) and (2.9).

can be expressed in matrix form as

$$\begin{bmatrix} x_1 \\ y_1 \end{bmatrix} = \begin{bmatrix} A & B \\ C & D \end{bmatrix}\begin{bmatrix} x_2 \\ y_2 \end{bmatrix}. \tag{2.7}$$

The transmission line equations, in terms of $ABCD$'s for the mechanical line between two points a distance x apart, are (for lossless lines)

$$\begin{bmatrix} \dot{x}_1 \\ F_1 \end{bmatrix} = \begin{bmatrix} \cos \beta x & jZ_0 \sin \beta x \\ \dfrac{j}{Z_0} \sin \beta x & \cos \beta x \end{bmatrix}\begin{bmatrix} \dot{x}_2 \\ F_2 \end{bmatrix}, \tag{2.8}$$

where

$$Z_0 = \frac{1}{A\sqrt{\rho E}}\sqrt{\frac{C_l}{M_l}}, \qquad \beta = \frac{\omega}{v_p}, \qquad v_p = \sqrt{\frac{E}{\rho}} = \frac{1}{\sqrt{C_l M_l}}.$$

The analogous electrical network can be described by the equations

$$\begin{bmatrix} V_1 \\ I_1 \end{bmatrix} = \begin{bmatrix} \cos \beta x & jZ_0 \sin \beta x \\ \dfrac{j \sin \beta x}{Z_0} & \cos \beta x \end{bmatrix}\begin{bmatrix} V_2 \\ I_2 \end{bmatrix}, \tag{2.9}$$

where

$$Z_0 = \sqrt{\frac{\mu}{\varepsilon}} = \sqrt{\frac{L_l}{C_l}}, \qquad \beta = \frac{\omega}{v_p}, \qquad v_p = \frac{1}{\sqrt{\mu\varepsilon}} = \frac{1}{\sqrt{L_l C_l}}.$$

In the Equations (2.8) and (2.9), A represents the cross-sectional area of the mechanical line, ρ is the density, E is the elastic (Young's) modulus, μ is permeability, and ε is permittivity. The remaining terms are described in the table of analogies, 2.2. In equations where electrical and mechanical terms are both included, reciprocal stiffness is substituted for compliance C.

In this section a number of concepts have been introduced. The most important of these is the mobility analogy which is used throughout this book. Also, the $ABCD$ matrix was applied to electrical and mechanical transmission lines and is used again later in describing lumped-element networks.

Throughout this book we define alpha-numeric symbols according to the context in which they are found. If the term Z_0 is found in an equation relating mechanical terms, then Z_0 is the mechanical characteristic mobility, as shown in Table 2.2. In a mechanical system, μ corresponds to Poisson's ratio and V to velocity, and so on. Also we use springs to identify stiffness or compliance in a mechanical system, and standard inductor symbols to represent inductance in

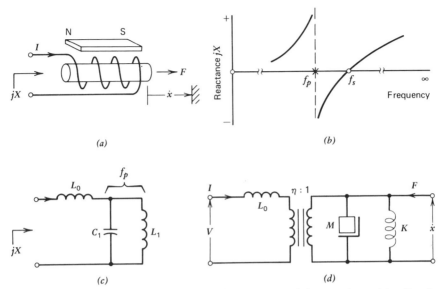

FIGURE 2.4. Development of equivalent circuits for a magnetostrictive transducer; (a) coil, rod, and biasing magnet, (b) pole-zero reactance diagram, (c) electrical equivalent circuit, and (d) electromechanical equivalent circuit.

an electrical circuit. Therefore an electrical analogy (equivalent circuit) will contain only electrical symbols. By having the symbols be consistent with the system context we can use symbols that are most commonly used in each discipline.

Transducer Equivalent Circuits

In this section, some generalized concepts regarding transducer equivalent circuits are developed by examining the frequency response characteristics of the magnetostrictive-rod resonator shown in Figure 2.4(a).

The Magnetostrictive Transducer

Magnetostrictive transducers are composed of three main elements: the magnetostrictive ferrite or metal alloy, a coil, and a means of providing a constant (DC) flux field. The resonator of Figure 2.4 uses a permanent magnet to "bias" the alternating (AC) field generated by the coil. The coil makes little or no contact with the bar, allowing it to be free to vibrate in a half-wave length longitudinal mode. Measuring the reactance of the transducer, under the condition of no external acoustic loading of the bar, we obtain the reactance curve of Figure 2.4(b) [2]. The pole at frequency f_p and the zero at frequency f_s correspond to parallel and series resonances of a network containing L's and

C's. Two networks satisfy these poles and zeros; one is shown in Figure 2.4(c). The choice of this network, rather than its dual equivalent, is on the basis that (1) the frequency f_p is the mechanical resonance frequency of the transducer bar under electrical open-circuit conditions, and (2) the use of parallel resonant circuits to represent mechanical resonators is consistent with the mobility analogy. Note that we have idealized the transducer by not including electrical or mechanical losses. Losses are discussed in a subsequent section.

At this point, we have derived an all-electrical equivalent circuit. But how does the capacitor C_1 relate to the equivalent mass M of the resonator? What is the relationship of the "actual" inductance L_0 and the "equivalent" inductance L_1, to the compliance $C = 1/K$ of the mechanical circuit? The answer to these questions is that there is, in general, not a one-to-one correspondence. The mass of the resonator may increase and the capacitance C_1 remain constant. For instance, this is true when the resonance frequency and electromechanical coupling do not change. A better question to ask is "how do the equivalent electrical element values vary in relationship to the external electrical element values?" This question is answered through a knowledge of the mechanical resonance frequency and the electromechanical coupling. As the mechanical resonance frequency and the coupling between the electrical and mechanical networks is varied, the equivalent circuit values, as seen from the electrical terminals, will vary.

If a network composed of electrical and mechanical network element-values is desired, we can relate these values through an electromechanical transformer whose "turns ratio" η varies with coupling. Being consistent with the mobility analogy where the electrical and mechanical circuit topologies remain the same, the equivalent circuit of Figure 2.4(d) can be drawn. Viewing the mechanical circuit through the transformer terminals we should see C_1 and L_1 where $C_1 = M/\eta^2$ and $L_1 = \eta^2/K$. Although η can be found for each specific transducer by making electrical and mechanical measurements (or calculations of equivalent mass, for instance), it is seldom used in practical design work. A more useful parameter is the electromechanical coupling coefficient which relates the electrical equivalent circuit values of Figure 2.4(c) and the reactance plot of Figure 2.4(b).

The electromechanical coupling coefficient k_{em} is the parameter that links the electrical and mechanical sides of the transducer network by relating the electrical element values to the electrical equivalent mechanical values. In other words, it expresses the relationship between L_0 and L_1 of Figure 2.4(c). Knowing this one electromechanical relationship as well as the relationships between the equivalent masses and compliances of the transducer and other mechanical elements, an electrical equivalent circuit representation of the entire filter can be found.

The electromechanical coupling coefficient is most often defined as the ratio of the energy stored in the mechanical circuit to the total input energy. From this definition and the fact that for temperature-stable magnetostrictive trans-

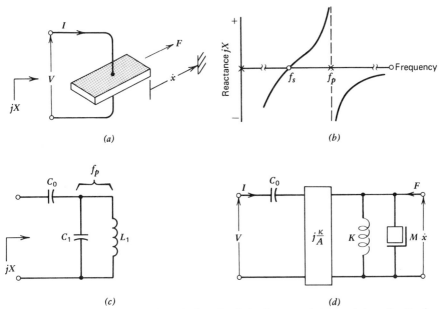

FIGURE 2.5. Piezoelectric ceramic equivalent circuit development; (a) ceramic bar vibrating in a "31" mode, (b) driving-point reactance diagram, (c) electrical equivalent circuit, and (d) electromechanical equivalent circuit.

ducers the coupling coefficient is low (typically 0.10 to 0.15),

$$ k^2_{em} \simeq \frac{L_1}{L_0} \simeq \frac{2(f_s - f_p)}{f_p}. \tag{2.10} $$

By measuring the pole and zero frequencies of the transducer input imped-ance, the coupling coefficient and therefore the ratio L_1 to L_0 can be found. L_0 can be measured with a reactance bridge and then L_1 can be calculated from the L_1/L_0 ratio. The remaining element of the electrical equivalent circuit is C_1, which can be obtained from L_1 and the mechanical resonance frequency f_p.

The Piezoelectric Transducer

Figure 2.5(a) shows a piezoelectric-ceramic transducer bar like that shown in Figure 2.1. The applied voltage and resultant electric field causes the bar to vibrate in a half-wavelength longitudinal mode in the direction shown by velocity $\dot{x}$, and force F. By ignoring losses and higher frequency (order) modes we can obtain the curve of Figure 2.5(b) from reactance measurements at the electrical terminals. In contrast to the magnetostrictive case, the series reso-nance frequency f_s is lower than the parallel resonance f_p. There are two circuits having three reactance elements that have this reactance curve. One is a capacitor in parallel with a series-tuned circuit and the other is the network shown in Figure 2.5(c) [2]. In the latter case, the resonant circuit $L_1 C_1$ corresponds to the open-circuited mechanical resonance of the bar, f_p. The fact that the frequency f_p corresponds to the mechanical resonance has been confirmed by measurements made on resonators vibrating not only in a direction perpendicular to the applied field but parallel to the applied electric field as well. Because mechanical resonant circuits are represented by parallel-tuned circuits and we are using the mobility analogy, we will use the equivalent circuit of Figure 2.5(c) to represent the piezoelectric transducer. Note that this circuit differs from the magnetostrictive case only in that L_0 is replaced by C_0.

Turning next to the electromechanical equivalent circuit of the piezoelectric transducer shown in Figure 2.5(d), we see that the conversion element is a network designated by $(j\kappa)/A$. This network is composed of a gyrator of resistance value A and an inverter of reactance $j\kappa$. Because a piezoelectric transducer is a so-called nonreciprocal device, the gyrator is needed; but because the gyrator converts parallel circuits into series circuits, an inverter is also needed. The two circuits (gyrator and inverter) in cascade act like a transformer with an imaginary turns ratio $(jA)/\kappa$. If a filter is built with a piezoelectric transducer at one end, and a magnetostrictive transducer at the other, it will be nonreciprocal and could be used in an isolator circuit. Let us now look at the relationships between the electromechanical coupling coefficient, the element values, and the reactance pole and zero frequencies (f_p and f_s).

Piezoelectric ceramics are generally high coupling-coefficient materials; radial and thickness modes have coupling coefficients on the order of 40 to 70 percent. Therefore, let us start with the exact expressions for the coupling coefficient, element value, and frequency relationships namely,

$$\frac{k_{em}^2}{1 - k_{em}^2} = \frac{f_p^2 - f_s^2}{f_s^2} = \frac{C_0}{C_1}. \qquad (2.11)$$

For low values of coupling,

$$k_{em}^2 \simeq \frac{2(f_p - f_s)}{f_s} \simeq \frac{C_0}{C_1}. \qquad (2.12)$$

In Equations (2.10) and (2.12) the frequency difference between f_p and f_s is called the pole-zero spacing Δf. The two equations are basic to transducer and mechanical filter design and are used often in subsequent sections.

MAGNETOSTRICTIVE TRANSDUCERS

Magnetostrictive transducers have played and are still playing an important part in the realization of high-performance filters for single-sideband radios and telephone equipment. Early mechanical filters used both metal-alloy and ferrite transducers. Although filters are still being manufactured with alloy transducers, all new magnetostrictive designs employ the lower-loss ferrite materials. Therefore, the major emphasis in this section is on ferrite magnetostrictive transducers.

The Magnetostrictive Effect

The change in dimensions of ferromagnetic material when exposed to a magnetic field is termed the *magnetostrictive effect*. The term *magnetostrictive effect* also applies to the inverse phenomenon, that is, there is a change in magnetization or the generation of a magnetic field when the material is externally stressed. Magnetostrictive effects can be looked at in various ways: in terms of variations of circuit parameters such as inductance, frequency, and electromechanical coupling; or as a function of material properties like permeability, Young's modulus, and the magnetostrictive strain coefficient or even on a microscopic level in terms of domains and crystals.

Ferromagnetic materials are crystalline in form. Each crystal is composed of a large number (on the order of 10^5) of domains which are aligned along cube edges or cube diagonals, depending on the material. In the demagnetized state, the domains are randomly aligned in the edge or diagonal directions. When a weak external magnetic field is applied, the domains oriented most nearly parallel to the field direction grow in size, at the expense of their neighbors. This so-called domain-wall displacement causes small but significant changes in the magnetic, elastic, and magnetostrictive strain characteristics of the material. When a large magnetic field is applied, the domains rotate into alignment with the magnetic field. It is this rotation process that causes large changes in permeability and Young's modulus and is the major factor in the expansion (or contraction) of the material. In the case of nickel-iron alloys, the material expands when subjected to a magnetic field, whereas ferrite materials contract in the presence of a magnetic field. This means that the magnetostrictive strain, for example in the iron-nickel alloy case, is positive regardless of the direction of the applied magnetic field. This is shown in Figure 2.6 where magnetostrictive strain λ is defined as the change in length per unit length and B is the flux density along the length axis of the bar [3]. Figure 2.6 also shows the effect of magnetic bias on the strain amplitude response to an applied magnetic field.

Transducer magnetic bias serves a similar purpose as application of voltage bias in an amplifier circuit. (1) the output signal becomes a linear function of

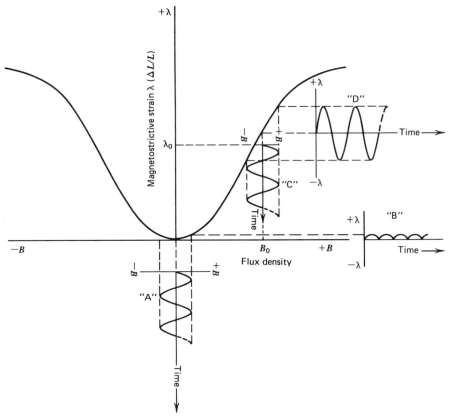

FIGURE 2.6. Magnetostrictive strain as a function of magnetic flux density showing the effect of magnetic bias on the strain amplitude.

the input, and (2) the output signal amplitude is maximized. Superimposed on the B versus λ curve of Figure 2.6 are a series of curves showing the strain time-response to applied sinusoidal flux inputs. These curves are for the unbiased and biased cases. Curve "B" shows the strain output to the input flux signal of curve "A." Note that the amplitude is small and the frequency has "doubled." In other words, the positive and negative swings of the flux amplitude result only in positive variations of the strain. Furthermore, as the input signal amplitude is changed, the strain output tracks in a nonlinear manner. If instead, a fixed bias field B_0 is applied to the transducer through the use of a permanent magnet (or direct current applied to the coil), we then have the response "D" to the applied signal "C." Comparing curve "D" to curve "B" we see that the "D" response is linear, larger in amplitude, and is the same frequency as the applied signal. In a microscopic sense, the alternating field under biased conditions causes the domains to rotate toward and away from alignment with the magnetic field, causing a relative expansion and contraction. It can be seen from Figure 2.6 that the maximum strain for a fixed

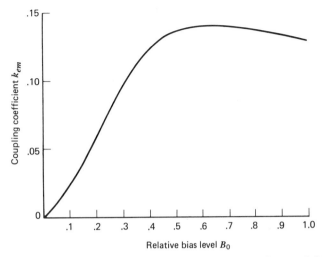

FIGURE 2.7. Electromechanical coupling-coefficient as a function of magnetic bias level.

input signal level occurs where the slope of the magnetostriction curve is greatest. Beyond that point the output signal decreases as the material becomes "saturated" in the sense that the domains become almost fully aligned with the applied field. Parenthetically, it should be understood that the applied signal to the transducer will normally be an exciting field ($H = NI$) which generates the magnetic flux field B, which in turn causes the bar to expand or contract. If λ had been plotted as a function of H, the magnetostrictive strain curve would have shown a hysterisis effect due to the $B - H$ losses.

Since the electromechanical coupling coefficient k_{em} is the ratio of the energy stored in the mechanical circuit to the applied electrical energy, we would expect that k_{em} would vary with magnetic bias. This is indeed the case; k_{em} is maximum where the slope of the $\lambda - B$ curve is greatest. Figure 2.7 shows a k_{em} versus bias-level curve for a typical magnetostrictive ferrite transducer.

Magnetostrictive Transducer Materials

The main emphasis in this section is on materials that have temperature and aging characteristics suitable for filter applications. This means that we restrict our discussion to stable iron-nickel-cobalt ferrite and iron-nickel alloy magnetostrictive rods and wires, as well as stable iron alloy and ceramic permanent magnets used for magnetic bias. Let us first look at magnetostrictive ferrites.

Magnetostrictive Ferrites

Ferrite materials exhibit resistivities on the order of 10^{10} times that of metals. This high resistivity leads to low eddy-current losses in the material and, therefore, moderate-to-high transducer inductance Q's. In addition, Fe-Ni-Co ferrites have mechanical Q values on the order of 2000. The low electrical and acoustic losses make it possible to design and manufacture ferrite-transducer filters with low passband insertion loss. Typically the loss is from 3 dB to 6 dB, as contrasted with alloy-wire filter loss values of greater than 20 dB. In addition to the low insertion loss, ferrite transducers exhibit low temperature coefficients of both resonance frequency and electromechanical coupling.

Figure 2.8 shows the temperature dependence of mechanical resonance frequency, coupling coefficient, and inductance, as a function of the amount (x) of cobalt in $Ni_{1-x}Co_xFe_2O_4$ ferrite [4]. These curves are based on toroidal ring resonators biased to remanence, that is, the material is saturated by means of a large direct current, the current is removed, and the ring remains magnetized at what is called the remanence value. Although this is not the usual transducer configuration, the curves are very similar those of rod-type resonators and also form a basis for comparing various types of ferrite materials and processes. Comparing the frequency shift and coupling-coefficient curves, we see that the optimum characteristics, about the room temperature value, occur for different percentages of cobalt. In the case of frequency shift, the optimum occurs at a value slightly less than 1 percent, whereas the coupling curve is flat at about 0.6 percent, and the inductance curve has a minimum shift at 1.4 percent cobalt. As is discussed in more detail in a subsequent section, it is possible to temperature-compensate or at least dilute the effects of mechanical resonance frequency shifts and inductance changes. But compensating for variations in coupling coefficient is difficult in a practical, manufacturable way, so the 0.6 percent cobalt formulation is often used.

The curves of Figure 2.8 are based on a specific ferrite manufacturing process. Variations in this process, in particular density variations, affect both the shape and the magnitude of the curves. Density and the related porosity parameter are dependent on the way the metal oxides are ball-milled into powders as well as the firing (sintering) temperatures and the furnace atmosphere. Control of these parameters, as well as strict control of impurities, is essential for obtaining ferrite transducers with consistent properties. Table 2.3 shows the characteristics of two types of magnetostrictive ferrite materials. The superscripts H, in relationship to velocity and elastic modulus, signify that the characteristics are measured under open-circuit conditions. This means that when acoustic measurements are made on the ferrite material these measurements are made with the coil leads open-circuited. Making measurements with the electrical terminals shorted results in a stiffening effect and the velocity and elastic modulus increase. Relationships between coupling coefficient, velocity, and the elastic modulus are

$$k_{em}^2 = \left[\frac{v_p^B}{v_p^H}\right]^2 - 1 = \left[\frac{E^B}{E^H}\right] - 1, \tag{2.13}$$

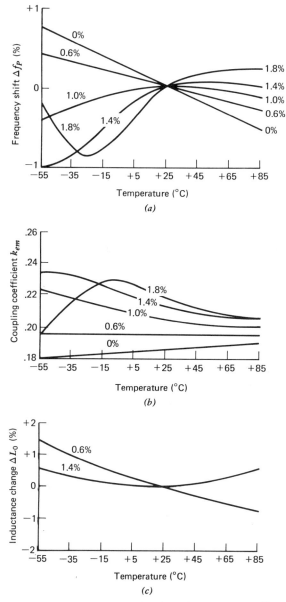

FIGURE 2.8. Variation of circuit parameters as a function of cobalt content (%) and temperature. (*a*) frequency shift (Δf_p), (*b*) coupling coefficient k_{em} and, (*c*) inductance change ΔL_0. [Adapted from C. M. van der Burgt, "Performance of ceramic ferrite resonators as transducers and filter elements," *J. Acoust. Soc. Amer.*, **28**, 1020–1032 (Nov. 1956).]

TABLE 2.3.   Properties of Magnetostrictive Ferrite Materials

Characteristic	Symbol	Units	Ferroxcube 7A1 and Rockwell 6A	Tesla W001
Velocity (open coil, $H = 0$)	v_p^H	(cm/s)	5.45×10^5	$5.71 \times 10^{5\,a}$
Density	ρ	(g/cm^3)	5.25	5.17
Elastic modulus (open coil condition)	E^H	$\dfrac{\text{dynes}}{\text{cm}^2}$	1.5×10^{12}	$1.68 \times 10^{12\,a}$
Mechanical quality	Q	—	> 2500	> 3500^a
Saturation flux density	B_s	(Gauss)	3250	3100
Permeability	μ_p	—	15	15
Electromechanical coupling coefficient	k_{em}	—	.21	.18^a

^{a}Measurements made on a bar sample; all other measurements made on a toroidal sample.

where

$$v_p^{B,H} = \sqrt{\frac{E^{B,H}}{\rho}} . \tag{2.14}$$

Equations (2.13) and (2.14) relate the material constants through the coupling in much the same way the circuit parameters are related by the coupling in Equation (2.10).

Iron-Nickel Alloy Materials

Some early mechanical filter designs made use of iron-nickel metal-alloy transducers in the form of both plates and wires. Two of the advantages of these transducers are that they can be fabricated in much the same way as the resonator materials and they are easily bonded to the other resonators by resistance welding. They also have the advantage of being potentially more temperature stable than magnetostrictive ferrite or piezoelectric ceramic transducers. For instance, the frequency shift of a 44 percent nickel alloy is less than 0.2 percent over a 140°C temperature range. The major disadvantages of these transducers are their high loss due to eddy currents and large variation of coupling coefficient with changes in temperature. Because of eddy-current losses, plate thickness and wire diameters are kept below 0.25 mm when used in the 100 to 500 kHz frequency range.

Returning to the general discussion of nickel-iron alloy transducers, we should look at the curves of Figure 2.9. These curves show the variations of the

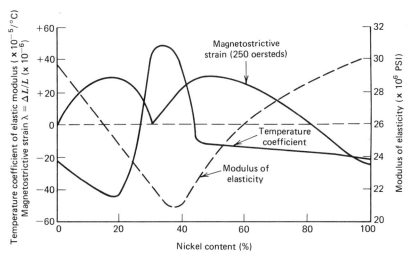

FIGURE 2.9. Variation of Young's modulus, temperature coefficient of Young's modulus, and magnetostrictive strain as a function of nickel content in an iron-nickel alloy.

temperature coefficient of the elastic modulus, the magnetostrictive strain, and the room temperature value of the elastic modulus, each as a function of nickel content. Because of the large dependence of mechanical resonance frequency on Young's modulus [$\sim (E^{1/2})$], the temperature coefficient axis also signifies the temperature coefficient of resonance frequency. Note that at about 27 percent and 44 percent nickel content the temperature coefficient is zero, that is, the slope of the $\Delta E/E$ (or $2\Delta f/f$) versus ΔT curve at room temperature is zero. Although the magnetostrictive strain is finite at both the 27 and 44 percent points, the strain (and therefore the coupling coefficient) at 44 percent is close to the maximum possible value. It has also been found that the 44 percent alloy has a smaller mechanical Q value when properly heat treated and cold worked. The wire can be made from either a powder-sintering process or by induction melting. The powdered wire seems to have the more consistent properties, in terms of both frequency and Q.

A major drawback of the iron-nickel alloy transducer is the steep temperature coefficient curve near the 27 and 44 percent points. Slight changes in the iron-nickel ratio or the presence of impurities can cause large changes in the temperature coefficient of frequency. For this reason, other elements such as chromium or molybdenum have been added to lower the peak of the curve to the extent that it is tangent with the zero temperature-coefficient line. These physically hard, high-Q materials are not used as transducers but are used as the low temperature coefficient (of frequency) resonant elements in almost all mechanical filters. These materials are looked at more closely in the next chapter.

Permanent Magnetic Materials

Permanent magnets provide the bias field needed for proper operation of a magnetostrictive transducer. Because the mechanical resonance frequency, coupling coefficient, and inductance are all quite sensitive to the magnetic bias field, it is important that the permanent magnet material be very stable. Table 2.4 shows characteristics of three magnet materials: Alnico 5, an iron-chromium-cobalt material similar to Alnico 5, and a barium ceramic. It can be seen from the table that these magnets all have excellent stability in terms of aging and shock. When properly stabilized, the irreversible flux density change due to temperature variations is practically zero. These magnets also have high Curie temperatures. The energy product, which is a measure of the magnets strength, is higher in the Alnico and Fe-Cr-Co cases, but this only means that the ceramic magnet must be larger to supply the same field. The ceramic magnets are less sensitive to demagnetizing fields as measured by their large

TABLE 2.4.   Alnico 5, Fe-Cr-Co, and Barium Ceramic Permanent
Magnet Characteristics

Characteristic		Alnico 5 (Cast)	Fe-Cr-Co (Cast Indalloy[a] 5)	Ceramic (Indox[a] 5)
Maximum energy product (Gauss-Oersted)		5.5×10^6	5.4×10^6	3.4×10^6
Coercive force (Oersted)		640	590	2400
Reversible temperature coefficient $(\Delta B/B)/\Delta T\,(\%/°C)$		0.02	0.027	0.19
Irreversible change in flux density of nonstabilized magnets	$-60°C$	-2.5	$-0.4\,(-50°C)$	0
due to temperature variation (%)	$+200°C$	-0.8	-0.5	0
Flux density change due to aging (%/decade)		0.2	Estimated to be similar to Alnico 5	0.1
Flux density change due to mechanical shock (%)		< 0.5	Similar to Alnico 5	0
Curie temperature (°C)		890	760	450
Composition (%)		Al(8) Ni(14) Co(24) Cu(3) Fe(51)	Fe(58) Cr(22) Co(15) V(4) Other (1)	Fe_2O_3 BaO

[a] Trade name of Indiana General Corporation.

coercive force but have the major disadvantage of having a poor reversible temperature coefficient of flux density of 0.19 percent per degree centigrade.

Ferrite Rod Transducers

The most common transducer design is the ferrite rod configuration shown in Figure 2.4. Making use of the transmission-line Equations (2.8) we obtain the stress, strain, velocity, and displacement diagram in Figure 2.10. In this case, stress has replaced force, stress being the force per unit of cross-sectional area. It is important to note that strain is directly related to stress (by the elastic constants) rather than to displacement or velocity. Also, strain is related to flux density by the curve in Figure 2.6. Therefore, regions of maximum strain will contribute most to the magnetic flux field generated in the bar and the resultant induced voltage across the transducer coil. As a result, to obtain maximum electromechanical coupling, the transducer coil should be centered on the velocity nodal points, that is, the points of maximum stress and strain. Figure 2.10 shows two different vibration conditions: the bar vibrating in a

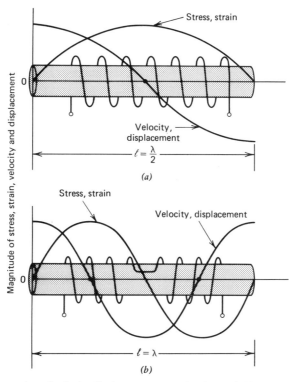

FIGURE 2.10.   Stress-strain and velocity-displacement curves for the (a) half-wavelength longitudinal mode, and (b) the full-wavelength longitudinal mode of vibration.

half-wavelength $\lambda/2$ mode and in a full-wave λ mode. The resonance frequency of these modes can be found from the equation

$$f_n = \frac{nv_p}{2\ell} = \frac{n}{2\ell}\sqrt{\frac{E}{\rho}} \, , \tag{2.15}$$

where n is the number of half-wavelengths, v_p is velocity, E is Young's modulus, ρ is the density, and ℓ is the bar length. In the half-wavelength case, the transducer coil is wound in a single direction and is centered over the nodal point. The full-wavelength transducer is composed of two windings, one being reverse-wound. The reverse winding is necessary because the strains at the two nodal points are opposite in phase with each other, the strains each being referenced to the biased value λ_0 shown in Figure 2.6. In fact, all measurements of parameters such as stress, strain, and displacement are based on the transducer being magnetically biased. The phase reversal therefore produces a condition where the induced strains are in phase with the strains of a full-wavelength bar vibrating in its natural mode.

So far, we have discussed the ferrite rod transducer problem in somewhat general terms, but we have not determined the effects of dimensions, flux densities, and so on, on coupling, resonance frequency, inductance, and Q, that is, the circuit parameters of Table 2.1. Figure 2.11 shows the important elements of a typical magnetostrictive transducer and Table 2.5 shows the relationships of these components to the circuit parameters. Let us first discuss the physical relationships between the component parts of the transducer of Figure 2.11.

A typical mechanical filter transducer design uses a coil wound on a plastic (rexolite or the equivalent) bobbin to drive the ferrite transducer rod. The magnetic bias can be provided in a number of ways such as through the use of a cylindrical magnet surrounding the coil [5], one or more rod or bar magnets outside of the coil as shown in Figure 2.11, or short rod or plate magnets near the end of the ferrite transducer [6]. In addition, the coil and magnet must be supported in a position that allows the ferrite to vibrate freely. This support

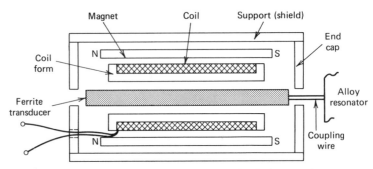

FIGURE 2.11.   Elements of a magnetostrictive ferrite transducer.

TABLE 2.5. Sensitivity Relationships Between Physical and Circuit Parameters in Magnetostrictive Rod-Type Transducers

Physical Parameters		Circuit Parameters[a,b]					Other Parameters
		1	2	3	4	5	
Tranducer rod	Length						
	Diameter						Spurious flexure vibration modes
	Material characteristics						
	Number of turns						Intermodulation distortion
Transducer coil	Wire diameter						
	Distance from ferrite						
	Length						
	Position						
Magnet	Dimensions						
	Flux density						Intermodulation distortion
Support (shield) and end cap	Material characteristics						Spurious modes of vibration and input-to-output coupling
	Dimensions						

[a] ■, First order sensitivity; ◨, Second order sensitivity; □, Very small to zero sensitivity.

[b] 1, Mechanical resonance frequency; 2, Coupling coefficient; 3, Inductance; 4, Mechanical Q; 5, Electrical Q.

often acts as an RF shield, and when a magnetic material is used for the support it can be used as a path for the bias and coil generated signal flux. The end-caps perform the same shielding and flux path functions as the support. The support structure is sometimes used to support the metal resonator assembly, which results in simple alignment of the ferrite transducer and the coil. If it is not necessary for the support to act as a flux path, it can be made from a conductive metal such as aluminum or brass. If it is to support the resonators, all or part of the support must be made of a weldable material. It is important that the metal be thick enough to act as an adequate shield to prevent the magnetic flux field of the input transducer coil from coupling to the output transducer coil.

Next, we look at the relationships of the parts to one another in terms of their effect on the circuit parameters. Starting with the coupling coefficient we have the following guidelines:

To maximize the electromechanical coupling coefficient:

1. The coil should be approximately 0.70 to 0.75 times the length of the ferrite and centered over the vibration node. This assumes a magnet designed for maximum coupling.
2. The magnet should provide a magnetic field that is parallel to the axis of the rod, uniform over its cross section, and have a flux density $B \simeq 0.7\, B_s$ (saturation).
3. The coil turns should be as close to the ferrite rod as is possible.

To minimize the sensitivity of the coupling coefficient to parts variations, the transducer should be designed on a zero-slope basis, that is, the change of coupling versus the varying-parameter curve should be parabolic about the origin. For example:

1. The coil length should be of such a value as to maximize the coupling and should be centered over the nodal point of the ferrite transducer. In this way any changes will reduce the coupling (i.e., the change is parabolic).
2. The magnet should have optimum length, be centered, and provide the bias flux level that maximizes the coupling while being far enough from the ferrite transducer so that small changes in the magnet's position, relative to the ferrite, will not result in a large change in flux density.

The meaning of the preceding paragraphs will become clearer through the use of an example. Figure 2.12 shows how the electromechanical coupling coefficient k_{em} varies with changes in both magnet length and coil length [5]. In this example the magnet is a cylindrically shaped ceramic that surrounds the coil form. The magnet strength is chosen to provide a maximum coupling coeffi-

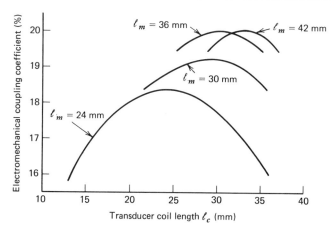

FIGURE 2.12. Dependence of coupling coefficient on both transducer coil length (ℓ_c) and magnet length (ℓ_m). The ferrite length is 46.4 mm. (Courtesy of Tesla, C.S.R.)

cient in each case. The curves clearly show the parabolic change of coupling around the optimum coil length for each length magnet.

Looking at another example, Figure 2.13 shows a series of curves relating to a transducer design at 250 kHz. This transducer makes use of two bar magnets mounted on the coil form. The ratio of the coil length to the ferrite length is 0.7, and magnet length to ferrite length is 0.9. Frequency and Q are linear functions of magnet strength, whereas inductance is inversely proportional to the square of the flux density at the high end of the range. The relative flux density can be measured by using a fixed capacitor to tune with the inductance. In fact, because of the inverse-square relationship, the resonance frequency of the tuned circuit is proportional to the flux density. In a practical

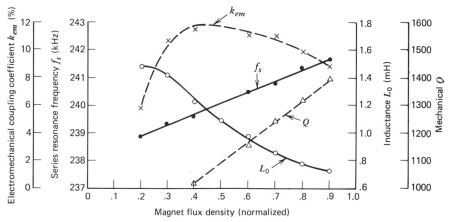

FIGURE 2.13. Variation of circuit parameters with magnet strength. $\ell_c/\ell_f = 0.7$, $\ell_m/\ell_f = 0.9$, Rockwell-5 ferrite, Alnico-5 magnets. (Courtesy of Rockwell International, USA.)

factory situation, the transducer magnets are "degaussed" to the point where the tuned circuit frequency drops to a specified design value (actually to a narrow set of limits). This degaussing can be done with a coil surrounding the entire transducer and after the filter is fully assembled. This eliminates measuring and degaussing individual magnets and reduces the chance of the magnets being degaussed before or during assembly. In addition, this method eliminates some magnet positioning problems which result in flux density changes in the ferrite. This method, whereby we measure and degauss according to the tuned circuit frequency, eliminates all first-order errors in determining the magnetic-bias flux density in the ferrite transducer.

Figure 2.14 shows, for the above example, the variation of the circuit parameters f_s and k_{em} as a function of temperature. Since it is difficult to adjust the value of k_{em} in production, the k_{em} curves show both the total one-sigma variation of a production lot, as well as the one-sigma variation from the room temperature value.

Returning to the subject of physical variables and circuit parameters, rather than a further detailed analysis of the relationship between all of the physical variables of the parts and the circuit parameters, Table 2.5 can be used as a guide to the sensitivity of one, relative to the other. Each first- or second-order entry can be viewed as a design consideration or a manufacturing variable. The table is actually more complex than it seems when we consider that a parameter such as "transducer rod material characteristics" includes density, porosity, elastic modulus, permeability, loss tangents, cracks, voids, and inclusions. But the table does act as an outline for a more detailed analysis of the transducer.

PIEZOELECTRIC CERAMIC TRANSDUCERS

Transducers composed of piezoelectric ceramics have been designed for a broad range of filter applications. Tuning-fork mechanical filters at frequencies as low as 200 Hz, and filters with extensional-mode transducers at 500 kHz, utilize ceramic transducers. Not only do piezoelectric ceramics cover a broad frequency range, but they are very versatile in terms of being used to realize numerous modes of vibration such as extension, thickness, shear, and flexure. Although mechanical filters have been designed with other piezoelectric materials, the ceramics have had, by far, the most popularity. We will therefore concentrate on piezoelectric ceramics and more specifically the lead-zirconate-titanates (PZT's).

The Piezoelectric Effect

A piezoelectric material under a mechanical load develops a charge that is proportional to the applied stress. This phenomenon, which is called the

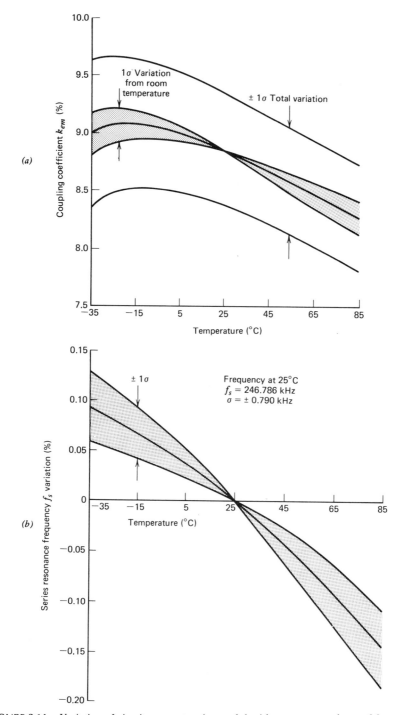

FIGURE 2.14. Variation of circuit parameters k_{em} and f_s with temperature change. Magnet flux density (Figure 2.13) is 0.9.

piezoelectric effect or *piezoelectricity*, was discovered in 1880 by Jacques and Pierre Curie [7]. These materials also exhibit the reverse effect; a mechanical strain is developed when the material is subjected to an electric field. The relationships between the variables such as stress, strain, force, charge, current, and so on, can be described in terms of the circuit parameters, material constants, or on a microscopic level, as is discussed next.

Like magnetostrictive ferrites, piezoelectric ceramics are a crystalline material; the crystals each contain many domains. If the ceramic has not been exposed to an electric field in the manufacturing process, the domains and their electric dipoles will be randomly oriented within each crystal. This means that the net electric dipole for each crystal, and for the ceramic as a whole, is equal to zero. This unpolarized material, when subjected to a mechanical stress, develops only small changes in the total dipole moment, that is, little change in the overall alignment of the electric charge. But, if the material is subjected to high electric field during the cooling process in its manufacture, its domains align in the direction of the field and each crystal, as a whole, exhibits an electric dipole. When the ceramic is stressed, the crystal lattices become distorted and some domains grow at the expense of others. This causes molecular changes in the domains and a resultant variation of the charge distribution. Conversely, if a small electric field is applied, the domains change shape thus causing mechanical strain.

The phenomena described above are very similar to those of magnetostrictive ferrites. In both cases, the electromechanical relationship is parabolic, as in Figure 2.6, where the nonbiased or unpoled material has small strain sensitivity to the applied electrical signal. In the ferrite case, the domain orientation is usually the result of an external permanent magnet bias field whereas in the ceramic case the orientation is due to an internal polarization. This polarizing (poling) is done in an electric field of 10 to 50 kV/cm at an elevated temperature not exceeding the Curie temperature. The poling time may vary from a few seconds to many minutes. Because of the large applied field, the poling is done while immersed in a high dielectric material, such as silicon oil, to prevent voltage breakdown between the upper and lower electrodes. The electrodes are formed by various processes which include vacuum-evaporative plating of gold, silver, or platinum and fired-on paint of silver, gold, palladium, or platinum.

At this point, some words of caution to both the mechanical-filter user and manufacturer. As the ceramic is poled by application of voltage, temperature, and time, it can also be depoled by combinations of the three, as well as by mechanical means [8]. Some guidelines for processing poled ceramics are as follows:

1. Maintain the temperature of the material as far below the Curie point as possible. For continuous operation this means an ambient temperature below 0.5 T_c (in degrees C).

2. If the thermal environment is greater than 0.5 T_c, the time duration of exposure should be minimized.

3. The transducer should not be exposed to large alternating or direct electric fields greater than 5 to 10 kV/cm.

4. The ceramic should not be mechanically stressed beyond 50 to 100 MN/m^2 (about 7,000 to 15,000 lbs/in^2).

Piezoelectric Ceramic Materials

In this section we concentrate on so-called PZT materials. These widely used materials have a chemical composition of $PbTiO_3$-$PbZrO_3$ with small additives of the form, $Pb(XYZ)O_3$ where X, Y, and Z may be elements such as calcium, strontium, barium, niobium, lanthanum, antimony, or manganese. These additives have a large effect on material constants such as acoustic velocity, dielectric constant, quality factor, and coupling as well as on hysteresis, temperature characteristics, and aging. Because of the narrow-bandwidth nature of mechanical filters, we will concentrate only on highly stable PZT materials. Examples of some of these materials are shown in Table 2.6. Each of the materials has its own special characteristic or set of characteristics that makes it unique and useful for a particular filter application. As an example, some of the materials exhibit a short-term temperature hysteresis effect which may not be a problem in a telephone exchange environment, but would have to be considered in an aerospace application.

Although some of the parameters shown in Table 2.6 have been defined in previous sections, some have not. Also, the fact that modes other than extensional are used in ceramic transducers, makes it necessary to discuss the special subscripting. In the following discussion we make use of Figure 2.15, which shows three of the most commonly used modes of vibration.

The three modes of vibration that we consider are length-extension with transverse bias, thickness-extension, and thickness-shear. Other modes are possible, such as the length-extensional with the bias and excitation in the direction of the motion and the radial mode, which is similar to thickness-extension but the particle motion is in a radial direction from the center of the disk. This second extensional mode suffers from a small value of static capacitance because of the distance between the electrodes, whereas the radial mode does not easily couple to other mechanical elements. Other even less commonly used configurations are cylinders and rings vibrating in length, thickness and radial modes. Of course all of these should be looked at when considering a new design. But let us return to the vibration modes of Figure 2.15.

The length-extension mode with transverse bias is commonly called the three-one mode. The 31 is a subscript corresponding to the fact that the polarization and excitation are in the 3-direction and the main component of motion is in the 1-direction. The frequency equation for this mode is

$$f_s = \frac{N_1^E}{\ell} = \frac{v_p^E}{2\ell} = \frac{1}{2\ell}\left(S_{11}^E\rho\right)^{-1/2}, \tag{2.16}$$

TABLE 2.6. Properties of PZT Piezoelectric Ceramic Transducers

Characteristic	Symbol	Units	Material			
			Tokin 314	TDK 61A	Channel Products CPI-1000	Rosenthal P-61
Frequency constant ($v_p/2$)	N_1^E N_{33}^D/N_5^D	Hzcm	1.716×10^5 $-/1.230 \times 10^5$	1.913×10^5 $1721 \times 10^5/820 \times 10^5$	1.780×10^5	1.704×10^5
Coupling coefficient	k_{31} k_{33}/k_{15}	— —	0.220 $-/0.52$	0.259 0.64/0.65	0.212	0.227 $0.430/$—
Mechanical quality factor	Q_m	—	1071	478	530	548
Electrical quality factor	Q_e	—	133	125		
Relative permittivity $\varepsilon_0 = 8.85 \times 10^{-14}$F/cm	$\varepsilon_{33}/\varepsilon_0$	—	1210	1350	1000	769
Temperature coefficient of frequency (slope at 25°C)	TC_{f_p}	ppm/°C	+110	+176	$\simeq 0$	-39
Temperature coefficient of coupling (slope at 25°C)	$TC_{k_{31}}$	ppm/°C	-370	-500	-100	-185
Temperature coefficient of dielectric const. (at 25°C)	TC_K	ppm/°C	+2550	+1600	+2100	+1800
Resonance frequency stability	—	%/decade		0.15	< 0.15	≤ 0.25
Density	ρ	gm/cm^3	7.80	7.6	7.4	7.8
Compliance	S_{11}^E S_{33}^D S_{55}^D	cm/dyn	1.07×10^{-14}	1.35×10^{-14} 0.87×10^{-14} 2.13×10^{-14}	1.05×10^{-14}	
Curie temperature	T_C	°C		320		

VIBRATION MODES	FREQUENCY AND COUPLING EQUATIONS

LENGTH-EXTENSION BAR WITH TRANSVERSE BIAS AND EXCITATION

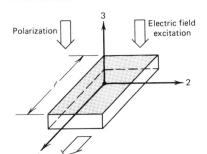

Particle motion and propagation

$$f_s = N_1^E/\ell = \frac{1}{2\ell}\left(S_{11}^E\rho\right)^{-1/2}$$

$$k_{31}^2 \simeq \frac{2\left(f_p - f_s\right)}{f_s}$$

THICKNESS-EXTENSION DISK AND PLATE

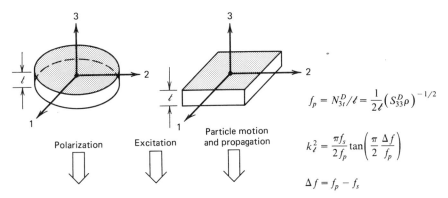

$$f_p = N_{3t}^D/\ell = \frac{1}{2\ell}\left(S_{33}^D\rho\right)^{-1/2}$$

$$k_\ell^2 = \frac{\pi f_s}{2 f_p}\tan\left(\frac{\pi}{2}\frac{\Delta f}{f_p}\right)$$

$$\Delta f = f_p - f_s$$

THICKNESS-SHEAR MODE PLATES

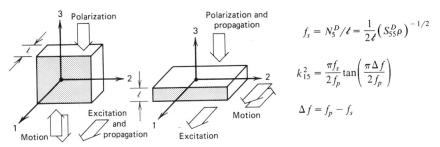

$$f_s = N_5^D/\ell = \frac{1}{2\ell}\left(S_{55}^D\rho\right)^{-1/2}$$

$$k_{15}^2 = \frac{\pi f_s}{2 f_p}\tan\left(\frac{\pi \Delta f}{2 f_p}\right)$$

$$\Delta f = f_p - f_s$$

FIGURE 2.15. Commonly used piezoelectric ceramic modes of vibration.

where f_s is the series resonance frequency (impedance zero), ℓ is the length of the bar, v_p^E is velocity, S_{11}^E is compliance, and ρ is density. The symbol N_1^E represents the frequency constant of the material, which when divided by length gives the frequency. Note that it is equal to one-half the velocity. The superscript E represents a short-circuit condition where the two transducer electrodes are at the same potential resulting in a constant or zero electric field. The superscript D represents an open-circuit condition where the displacement current is constant or equal to zero. The symbol S_{11}^E corresponds to the compliance in the 1-direction under short-circuit conditions. The 11 subscript corresponds to a strain in the 1-direction caused by a stress also in the 1-direction. The symbol S rather than C is used in order to follow the common notation regarding piezoelectric devices.

Referring back to Figure 2.15, we see that the resonance frequencies of thickness-extension and the thickness-shear modes are inversely proportional to the thickness ℓ. The thickness modes have very high electromechanical coupling coefficients, and therefore they require more complex expressions when relating coupling to the series and parallel resonance frequencies. In fact, the equations are identical with exception of the subscripts. At this point it should be mentioned that in the case of long, slim composite transducers, where a thickness-extension disk is sandwiched between two metal rods, the k_{33} coupling coefficient rather than the k_ℓ value is used. Therefore, since this is the most common application of disks in mechanical filter transducers, the k_{33} values are listed in Table 2.6.

The shear modes in Figure 2.15 are described by the subscript 5 which refers to motion around the 2-axis. k_{15} corresponds to excitation in the 1-direction and shear about the 2-axis.

Let us look at some of the other data shown in Table 2.6. Note that the frequency constants are all very close in value so that there is little use in choosing a material on the basis of size (length) reduction. But the differences in coupling coefficient are significant. The maximum bandwidth of a mechanical filter not externally tuned with a coil is proportional to the square of the coupling coefficient. Comparing the square of the highest and lowest values of k_{31} in the table we find the ratio to be 1.5/1, which is significant if a material is to be used for a broad range of designs. Also, notice that the three lowest coupling materials have the highest Q_m and have the lowest temperature coefficients of resonance frequency. These are parameters that are important to the performance of narrow-bandwidth mechanical filters.

Although the permittivities of the different materials vary by almost 2/1, this is usually not important except for special cases, where the terminating conditions are specified by the filter user and all other variables have been tied down to meet other portions of the specification. Looking at the temperature-coefficient of coupling coefficient, we see a considerable variation. But all of these materials are quite good, in that a 500 ppm/°C variation over a 100°C temperature range is equivalent to a ± 2.5 percent variation of coupling. Translating this to an equivalent effect in terms of termination, this would be

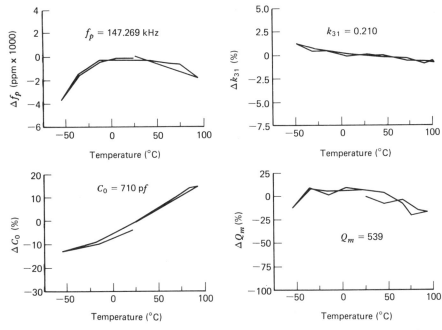

FIGURE 2.16. k_{31}-mode temperature characteristics of a piezoelectric ceramic (PZT) transducer; Channel Products CPI-1000, $\ell = 12$ mm, $w = 2.5$ mm, $\ell = 0.3$ mm.

like a ± 5.0 percent variation of the terminating resistance, which in most applications is acceptable; flat-passband telephone channel filters are a notable exception. The variation of dielectric constant is positive in all cases and is most important when the transducer is externally tuned with inductors. If the external tuning is critical, it is sometimes necessary to use negative temperature-coefficient capacitors to obtain a stable electrical resonance.

Some final comments regarding Table 2.6 are necessary. The materials to be included in the table were chosen because they were designed for use in filters, but also because they exhibited a wide range of characteristics. The materials shown in Table 2.6 are only a few of numerous materials manufactured by the companies shown and other manufacturers of PZT's who make comparable materials. In addition, due to differences in measuring techniques, conditions, and definitions, the data in Table 2.6 may differ somewhat from that listed by the manufacturer.

At this point let us look at the temperature characteristics of one of the materials listed, namely the Channel Products CPI-1000. Figure 2.16 shows the characteristics of a typical sample measured in a "pi" circuit by a computer controlled network analyzer. Some of the variation, for instance at room temperature, is due to measurement accuracy and some due to the fact that there is a time recovery period of one to two hours necessary for the material to return to a stable state. The amount of parameter shift is also a function of the rate of temperature shift from one measurement point to another. Therefore, care must be taken when making temperature measurements, with regard to both thermal shock and recovery time.

Composite Transducers

In most mechanical filter applications, the piezoelectric ceramic is bonded to a stable metal alloy to form a composite transducer. The reasons for forming a composite structure, rather than using the ceramic alone, vary with the application and are summarized below:

1. The composite transducer has the advantage of improved frequency stability with both time and temperature because of the use of highly stable iron-nickel alloy parts. Examples of two-resonator filters are shown in Figure 2.17 [9].

2. The composite structure results in the excitation of desired modes, such as flexure in bar and tuning-fork transducer resonators, and torsion in rod-type transducer resonators.

3. The metal attaches easily to other metal resonators of the filter structure.

Figure 2.18 shows various transducer configurations as a function of the basic piezoelectric-ceramic mode and the composite transducer-resonator mode. In each example, the applied (or induced) electric field is in a direction perpendicular to the metal-ceramic bonded surface. In four of the transducers [Figure 2.18(c) and (e)], the ceramic is in two parts, where each part is poled in opposite directions. In the shear-mode cases, the poling is at right angles to the applied field.

The most common bonding methods used to attach the ceramic to the metal are soldering and the use of epoxy. The solder bonds have the best long-term stability but require a more difficult assembly process. Both the solder and epoxy have negative temperature coefficients of the elastic modulus, which must be taken into account when analyzing the frequency versus temperature behavior of the composite transducer. For composite transducers, used in wide-bandwidth flexural-mode filters, epoxy is sometimes used to compensate the positive temperature coefficient of frequency of the ceramic material (such as TDK 61A). Let us next look, in detail, at the characteristics of the bar flexural-mode transducer.

Bar Flexural-Mode Transducer

The first composite transducer we look at is the flexural-mode resonator of Figure 2.1. This transducer has numerous forms, some of which are shown in Figure 2.19. The resonator of Figure 2.19(a) vibrates at the lowest-frequency flexure mode, that is, the fundamental mode. For the case where the ceramic is

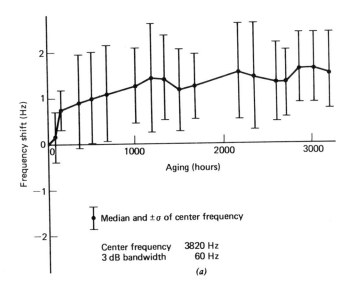

Center frequency   3820 Hz
3 dB bandwidth   60 Hz

(a)

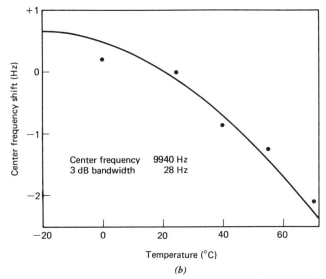

Center frequency   9940 Hz
3 dB bandwidth   28 Hz

(b)

FIGURE 2.17. Typical frequency shift curves, of composite-resonator filters of the type shown in Figure 7.34, due to aging and temperature variations. [© 1974 IEEE. Reprinted with permission, from *Proc. 1974 IEEE Ultrason. Symp.*, Milwaukee, 599–602 (Nov. 1974).]

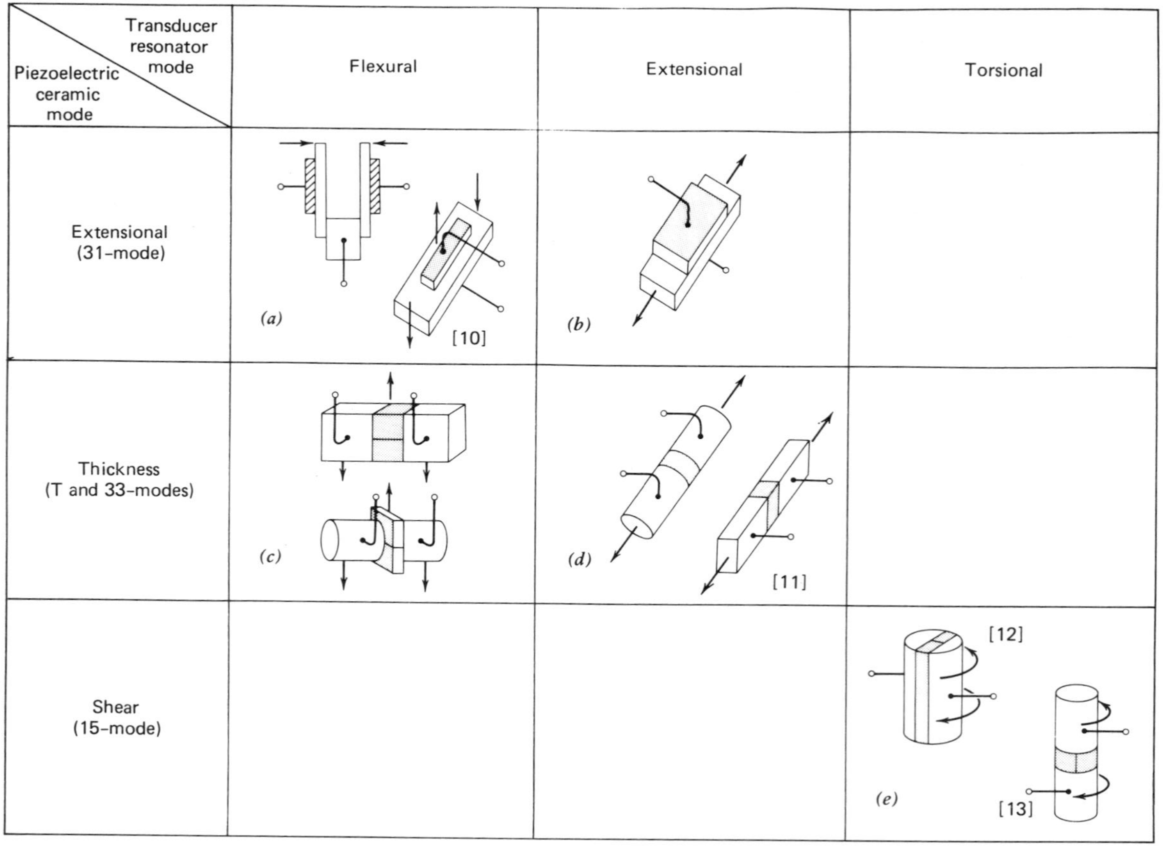

FIGURE 2.18. Composite piezoelectric ceramic and metal transducer resonators. The arrows show the direction of motion at the maximum velocity points during a single half-cycle. Bracketed numbers correspond to references.

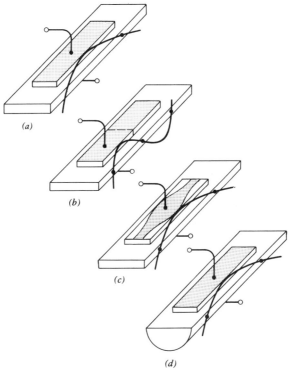

FIGURE 2.19. Bar flexural-mode transducer resonators: (*a*) fundamental mode, (*b*) first overtone, (*c*) electrode shaped for spurious rejection, and (*d*) bar shaped from rod stock.

small compared to the metal bar, the resonance frequency of the composite transducer is

$$f \simeq \frac{\alpha_n^2}{\ell^2} R \sqrt{\frac{E}{\rho}} \qquad \alpha_n = 4.73, 7.85, 10.99, \ldots, \tag{2.17}$$

where n corresponds to the mode, ℓ to the length of the bar and R to the radius of gyration of the cross section of the bar. In the case of a rectangular cross section, $R = \ell/(12)^{1/2}$, where ℓ is the bar thickness. For a circular cross section, $R = d/4$, where d is the rod diameter. A more exact frequency analysis is described in Reference [10].

A first-overtone resonator can be constructed from a single ceramic bar, as shown in Figure 2.19(*b*), where the bar has been poled in opposite directions on either side of the dashed line [14]. This resonance can also be excited with two separated bars of opposite polarity or a single bar located over either region of maximum amplitude. Use of the single bar shown in Figure 2.19(*b*) eliminates, by cancellation, excitation of the fundamental mode, the excitation

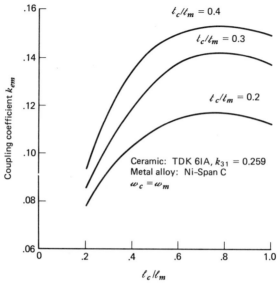

FIGURE 2.20. Coupling coefficient as a function of ceramic length ℓ_c to metal bar length ℓ_m for various values of metal and ceramic thickness (measured data).

due either to electrical signals or to external mechanical vibrations through the support structure. Kusakabe and Konno's idea of shaping the electrode shown in Figure 2.19(c) eliminates selected higher-order modes, again by cancellation of the signals generated by each portion of the ceramic. Figure 2.19(d) differs from (a) only in the shape of the metal part. The segmented-rod construction allows the transducer to be fabricated from uniform rod stock which can be made to close dimensional tolerances. This method of fabrication is compatible with that of other resonators, in the case of torsional- and flexural-mode telephone channel filters.

Looking at the coupling coefficient of the composite flexural-mode transducer, we see from Figure 2.20 how coupling varies as a function of ceramic-to-metal length and ceramic-to-metal thickness. Note that the coupling is maximum at a ceramic-to-metal length ratio between 0.7 and 0.8 and that the coupling increases diminish as the ceramic thickness to metal thickness increases. The coupling coefficient values for other ceramic materials can be roughly scaled by the ratio of the k_{31} values. The curves in Figure 2.20 are based on measurements using the test circuit in Figure 2.22 and include the effects of finite solder thickness and electrode plating thickness.

Langevin Transducers

The Langevin transducer is composed of a thin ceramic plate or disk bonded between two metal bars or rods, as shown in Figure 2.18(d). This type of transducer utilizes the strong k_{33}-coupling mode, which reduces the necessary amount of the low-Q and less stable ceramic needed to obtain a particular coupling value. For the case where the ceramic is centrally located and its thickness ℓ_1 is small compared to the metal length ℓ_2, equations for coupling and resonance frequency ω_p are [11]

$$k_{em} = k_{33} \sqrt{\frac{2}{\pi}} \sqrt{\frac{\rho_2 v_{p2}}{\rho_1 v_{p1}}} \sqrt{\frac{\omega_p \ell_1}{v_{p1}}}, \qquad (2.18)$$

and

$$\tan \frac{\omega_p \ell_1}{2 v_{p1}} = \left(\frac{\rho_1 v_{p1}}{\rho_2 v_{p2}} \right) \left(\frac{1}{\tan \dfrac{\omega_p \ell_2}{v_{p2}}} \right), \qquad (2.19)$$

where subscript 1 refers to the ceramic and the subscript 2 corresponds to the metal. Equation (2.19) is easily solved with a calculator or computer by varying $\omega_p = 2\pi f_p$ until the left and right sides of the equation are equal. Figures 2.21(a) and 2.21(b) are based on exact equations, and show the variation of k_{em} as a function of k_{33}, ceramic thickness, and ceramic location. These curves clearly show the high coupling that can be obtained with a small transducer. The use of a small transducer leads to a low value of temperature coefficient of frequency. For the centered case ($\alpha = 0$),

$$T_c(f_p) = \delta T_{c1}(\text{ceramic}) + (1 - \delta) T_{c2}(\text{metal}), \qquad (2.20)$$

where

$$\delta = \frac{\dfrac{2}{\pi} \times \dfrac{\omega_p \ell_1}{v_{p1}} \times \dfrac{\rho_2 v_{p2}}{\rho_1 v_{p1}}}{1 + \dfrac{1}{\pi} \times \dfrac{\omega_p \ell_1}{v_{p1}} \times \dfrac{\rho_2 v_{p2}}{\rho_1 v_{p1}}}.$$

The overall mechanical resonator Q is related to the Q_1 of the ceramic and the Q_2 of the metal bar by the expression

$$Q = Q_1 \times \frac{1 + \dfrac{\ell_1}{2\ell_2} \times \dfrac{\rho_2}{\rho_1} \left(\dfrac{v_{p2}}{v_{p1}} \right)^2}{\dfrac{Q_1}{Q_2} + \dfrac{\ell_1}{2\ell_2} \times \dfrac{\rho_2}{\rho_1} \left(\dfrac{v_{p2}}{v_{p1}} \right)^2}. \qquad (2.21)$$

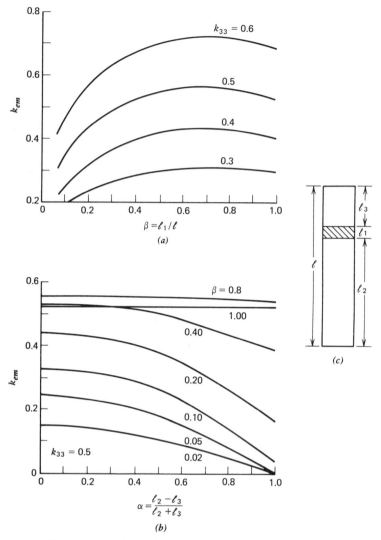

FIGURE 2.21. Coupling coefficients of Langevin longitudinal transducers as a function of: (*a*) ceramic length for $\alpha = 0$, (*b*) ceramic position with $k_{33} = 0.5$, and (*c*) transducer dimensions (calculated values).

Example 2.1.   Let us look at the problem of designing a Langevin transducer for 100 kHz with an electromechanical coupling coefficient of 0.20 and a minimum temperature coefficient of frequency. We will also calculate Q and static capacitance C_0 assuming that the ceramic is the TDK-61A PZT of Table 2.6, and the metal alloy is Ni-Span C of Table 3.5. We will assume that the ceramic is centered.

Ceramic	Metal

$$v_{p1} = 2N_{33}^D = 3.442 \times 10^5 \text{cm/sec} \quad v_{p2} = (E/\rho)^{1/2} = \left(\frac{1.85 \times 10^{12}}{8.3}\right)^{1/2}$$

$$\rho_1 = 7.6 \text{ g/cm}^3 \qquad\qquad\qquad\qquad = 4.721 \times 10^5 \text{ cm/sec}$$

$$k_{33} = 0.64$$

$$Q_1 = 478 \qquad\qquad\qquad\qquad \rho_2 = 8.3 \text{ g/cm}^3$$

$$T_{c1} = +110 \text{ ppm/}^\circ\text{C} \qquad\qquad Q_2 = 20,000$$

$$\varepsilon_{33}/\varepsilon_0 = 1350 \qquad\qquad\qquad\qquad T_{c2} \simeq 0$$

Setting $k_{em} = 0.20$ and solving (2.18) for ℓ_1 we obtain the value of the minimum ceramic thickness and therefore the design with the lowest temperature coefficient of frequency. Thus,

$$\ell_1 = \left(\frac{k_{em}}{k_{33}}\right)^2 \left(\frac{1}{4}\right)\left(\frac{\rho_1 v_{p1}}{\rho_2 v_{p2}}\right)\left(\frac{v_{p1}}{f_p}\right) = (0.0976)\left(\frac{1}{4}\right)(2.298)$$

$$= 0.0561 \text{ cm.}$$

Equation (2.19) can be solved for ℓ_2 because ω_p and ℓ_1 are now known. Therefore, from (2.19),

$$\tan(0.0512) = 0.668 \times \frac{1}{\tan(1.33\ell_2)},$$

or

$$\ell_2 = 1.123 \text{ cm} = \ell_3.$$

The temperature coefficient of frequency from (2.20) is,

$$T_c(f_p) = \delta T_{c1} \text{ (ceramic)} + (1 - \delta)(0)$$

where

$$\delta = \frac{0.637 \times 0.102 \times 1.50}{1 + 0.318 \times 0.102 \times 1.50} = 0.0931,$$

or

$$T_c(f_p) = 0.0931 \times 110 = 10.2 \text{ ppm / }^\circ\text{C.}$$

From (2.21) we can calculate Q;

$$Q = 478 \frac{1 + 0.0513}{0.0239 + 0.0513} = 668.$$

The static capacitance is related to the relative permittivity by

$$C_0 = \left(\frac{\varepsilon}{\varepsilon_0}\right)\varepsilon_0\left(\frac{A}{\ell}\right). \tag{2.22}$$

The capacitance per square cm of cross section of the transducer (where $\ell = \ell_1$) is

$$\frac{C}{A} = 1350 \times 8.85 \times 10^{-14} \times \frac{1}{0.0561} = 2130 \text{ pF} / \text{cm}^2.$$

A Modified Langevin Transducer for Flexural-Mode Resonators

Figure 2.18 shows two flexural-mode resonators that employ thickness-mode transducers. These composite resonators are used where a high electromechanical coupling coefficient is needed to eliminate tuning coils. Examples of the use of these transducers are shown in Figure 7.15 and Reference [15]. Equations for the value of coupling coefficient have not been published, but are based on the Langevin equations of the last section. From correspondence with A. Günther, "the Langevin equations are enhanced by two geometric coefficients: the first one representing the nonuniform stress distribution over the cross section and the second one taking into account the shape of the deflection of the bar. Both factors are experimentally determined."

Torsional-Mode Transducers

Two methods of construction torsional-mode composite transducer resonators are shown in Figure 2.18. Both transducers are capable of realizing inductor-less filters with fractional bandwiths of more than 3 percent. Equations describing the split-disk transducer are shown in the example of Figure 4.35. Applications of this transducer are shown in Figures 7.22 and 7.26. When the ceramic is attached to the end of the metal rod, as in Figure 7.26, the coupling coefficient is reduced, but the stability and mechanical Q are increased. The long split ceramic plate transducer is described in Reference [12].

MEASUREMENT CIRCUITS

There are numerous methods used to determine the circuit parameters f, k_{em}, C_0 (or L_0) and Q. These test methods involve the use of bridges, oscilloscopes or network analyzers and can be performed manually or, in some cases, under computer control.

The Pi-Circuit Method

The pi-circuit method involves using the transducer as a driving-point imped-ance element in the series arm of a pi-network, as shown in Figure 2.22. The

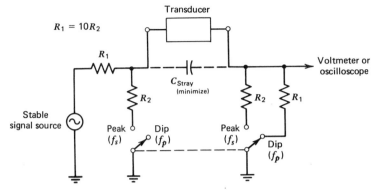

FIGURE 2.22.   Pi-method transducer test circuit.

oscillator and meter can be separate elements or can be part of a network analyzer test set. The transducer can be a piezoelectric ceramic or a magnetostrictive ferrite, the only difference being that the pole and zero frequencies of the input impedance are reversed. Note than $R_1 = 10R_2$, and that a switch is used to emphasize the amplitude of the peak and the dip of the admittance of the transducer. As the transducer's impedance goes through a minimum (a zero), the current through the output resistor is maximum resulting in an output voltage peak (i.e., a peak representing an admittance pole). The frequency f_s of this peak and the frequency f_p of the dip are shown for the piezoelectric case in Figure 2.23. Using Equations (2.10) and (2.12), the coupling coefficients of the piezoelectric and magnetostrictive transducers can be found.

Usually more than adequate results can be obtained using the test circuit of Figure 2.22, if certain precautions are observed. First, the stray capacitance across the transducer should be minimized. One of the strong points of this test circuit is that stray capacitance to ground has almost no effect on the series f_s and parallel f_p resonance frequency measurements and that the stray capacitance across the transducer terminals, which does have an effect, is usually very small. In cases where capacitance C_0 of the piezoelectric transducer is on the order of a few picofarads, the stray capacitance will cause f_p to move down in frequency, causing the coupling coefficient to appear to be a smaller value than it actually is. Stray capacitance causes an apparent increase in coupling coefficient in the magnetostrictive case. The sensitivity of the measurement to stray capacitance can easily be checked by adding a small capacitor across the transducer terminals and measuring the frequency shift of f_p and the resulting change in k_{em}.

A second precaution is to check that the attenuation difference δ_1 or δ_2 between the peak or dip amplitudes and the level above and below the acoustic (peak or dip) response should be on the order of 10 dB or greater, to achieve accurate results. As this level decreases, the frequency difference between f_p and f_s increases, giving erroneous coupling coefficient and resonance frequency values.

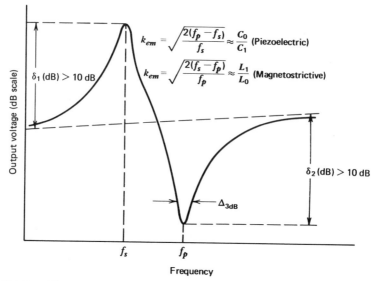

FIGURE 2.23. Piezoelectric transducer admittance measured with the pi-circuit of Figure 2.22.

The mechanical quality factor Q can be calculated in two different ways. One method is to measure the 3 dB points on the curve of Figure 2.23, in the region of f_p. From the 3 dB frequency difference $\Delta_{3\,dB}$ and the frequency f_p, the Q can be calculated from the equation

$$Q = \frac{f_p}{\Delta_{3\,dB}}. \tag{2.23}$$

Although this method is very simple, it can also be very inaccurate if the attenuation spread δ_2 is too small. δ_2 can sometimes be increased by lowering the value of R_1, but only within limits. The measurement accuracy is a function of the coupling coefficient, the greater the coupling, the greater the measurement accuracy.

A second method of determining the value of Q is through use of the equations

$$Q \text{ (piezoelectric)} = \frac{1}{2\pi f_s C_0 k_{em}^2 R} \tag{2.24}$$

and

$$Q \text{ (magnetostrictive)} = \frac{2\pi f_s L_0}{k_{em}^2 R}, \tag{2.25}$$

where R is the value of resistance, which when used in the place of the transducer in Figure 2.22 gives the same attenuation at frequency f_s. In other words, the attenuation A_s, at f_s, with the transducer in the circuit, is measured. The transducer is then replaced by a resistor and through a trial-and-error process (or by another calculation) a resistance value R is found that results in an attenuation A_s. Q is then found from Equation (2.24) or (2.25). The static capacitance C_0 can easily be measured on a capacitance meter, inductance L_0 on a reactance bridge, and electrical Q on various bridges or a Q-meter.

If the stray capacitance across the transducer, in the circuit shown in Figure 2.22, causes a problem in accurately measuring the coupling coefficient, the transducer can be tuned with a shunt coil (piezoelectric case) or a capacitor (magnetostrictive case), and the coupling coefficient can be calculated from

$$k_{em} = \frac{\Delta f}{f_0} \text{ (tuned transducer case)}, \qquad (2.26)$$

where Δf, in this case, is the frequency difference between the two admittance zeros, and f_0 is the frequency of the admittance peak. We should remember that an output voltage peak of the circuit of Figure 2.22 corresponds to a transducer admittance peak.

Example 2.2. We are given a piezoelectric ceramic composite transducer and are asked to measure its circuit parameters: frequency, coupling, and so on. With a capacitance meter and the network of Figure 2.22 we measure the following:

$$f_s = 5120 \text{ Hz} \qquad C_0 = 1050 \text{ pF}$$

$$f_p = 5158 \text{ Hz} \qquad R = 5.1 \text{ k}\Omega$$

From Figure 2.23,

$$k_{em} = \sqrt{\frac{2(5158 - 5120)}{5120}} = 0.122,$$

and from Equation (2.24),

$$Q = \frac{1}{6.283 \times 5120 \times 1050 \times 10^{-12}(0.122)^2 \times 5100} = 391.$$

Much of the effort of using various pieces of test gear to make transducer measurements can be eliminated by use of a calculator and network analyzer. With the use of the pi-circuit, shown in Figure 2.22, between the input and output terminals of a network analyzer, the frequency response of the network can be analyzed. By coupling the digital output of the analyzer to a calculator,

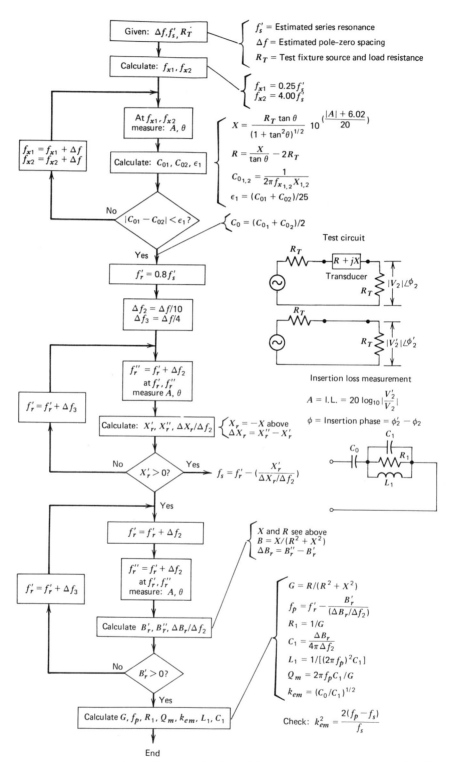

FIGURE 2.24. Test circuit and flow chart for determining transducer circuit parameters based on work of Yano (NEC).

the transducer circuit parameters can be calculated from the amplitude and phase characteristics measured at two or more frequencies. Figure 2.24 shows the test circuit and flow chart for making the measurements and calculations.

REFERENCES

1. H. M. Trent, "On the construction of schematic diagrams for mechanical systems," *J. Acoust. Soc. Amer.*, **30**, 795–800 (Aug. 1958).

2. R. A. Johnson, "Mechanical bandpass filters," in *Modern Filter Theory and Design*, G. C. Temes and S. K. Mitra, Eds. New York: Wiley, 1973.

3. B. A. Wise, *The Design of Nickel Magnetostrictive Transducers*. New York: International Nickel Company, 1955.

4. C. M. van der Burgt, "Performance of ceramic ferrite resonators as transducers and filter elements," *J. Acoust. Soc. Amer.*, **28**, 1020–1032 (Nov. 1956).

5. J. Červený, "Magnetostrictive transducers and their application in EMF," in *Proc. SSCT 71*, Tále, Czechoslovakia, 13/1–12 (Sept. 1971).

6. D. Beaudet, "Un transducteur magnétostrictif miniaturisé," *L'Onde Electrique*, **58** (6–7), 470–474 (1978).

7. B. Jaffe, W. R. Cook, and H. Jaffe, *Piezoelectric Ceramics*. New York: Academic Press, 1971.

8. *Piezoelectric Ceramics*, J. van Randaraat and R. E. Setterington, Eds. Eindhoven: N. V. Philips, 1974.

9. D. P. Havens and P. Ysais, "Characteristics of low frequency mechanical filters," in *Proc. 1974 Ultrasonics Symp.*, 599–602 (Nov. 1974).

10. M. Konno and H. Nakamura, "Equivalent electrical network for the transversely vibrating uniform bar," *J. Acoust. Soc. Amer.*, **38**, 614–622 (Oct. 1965).

11. M. Börner and H. Schüssler, "Miniaturisierung mechanischer Filter," *Telefunken-Z.*, **37**, 228–246 (Fall 1964).

12. I. Takahaski, N. Yoshida, and Y. Ishizaki, "An analysis of a torsional mode transducer for electromechanical filters," in *Proc. 1976 IEEE Ultrason. Symp.* (Sept. 1976)

13. K. Sawamoto, S. Kondo, N. Watanabe, K. Tsukamoto, M. Kiyomoto, and O. Ibaraki, "A torsional-mode pole-type mechanical channel filter," in *Modern Crystal and Mechanical Filters*, D. F. Sheahan and R. A. Johnson, Eds. New York: IEEE Press, 1977.

14. K. Yakuwa, T. Kojima, S. Okuda, K. Shirai, and Y. Kasai, "A 128 kHz mechanical channel filter with finite frequency attenuation poles," *Proc. IEEE*, (1) **67**, 115–119 (Jan. 1979).

15. J. Deckert, "Bauelemente auf magneto-elastischer basis," *Workshop on Magnetic Materials and Applications in Speech and Data Transmiasion*, Bad Nauheim (Apr. 1980).

Chapter Three _____

RESONATORS AND COUPLING ELEMENTS

In the previous chapter, we saw that the primary roles of the mechanical filter transducer were that of energy conversion and termination. In this chapter, we see that the purpose of the resonators and coupling elements is to provide the bandpass filtering. Although the interactions between the resonators and coupling elements can be quite complex, their functions as circuit elements are roughly as follows:

Circuit Element Parameter		Functions
Resonator frequencies	→	Establish the center frequency
Number of resonators	→	Determines the shape factor
Resonator equivalent mass	→	Determines the bandwidth
Coupling-wire compliance	→	Determines the bandwidth
Resonator-to-resonator coupling	→	Determines the passband shape

Figure 3.1 is helpful in explaining the meaning of the above relationships. The filter structure shown in Figure 3.1(a) is composed of seven wire-coupled resonators having mirror-image symmetry about the center line. The resonator-to-resonator coupling variation is achieved by varying the coupling wire lengths, ℓ_{12}, ℓ_{23}, and ℓ_{34}, between the weld points. Because the wire lengths vary, the resonator frequencies f_1, f_2, f_3, and f_4 generally also differ but fall within the 3 dB passband shown in Figure 3.1(b). The filter's position in the frequency spectrum is defined by the center frequency, f_0, which is determined by the resonator frequencies. The number of resonators determines the shape factor, the shape factor being the ratio of the bandwidth at a high attenuation level (for instance, 60 dB) to the passband bandwidth, most often defined at 3 dB.

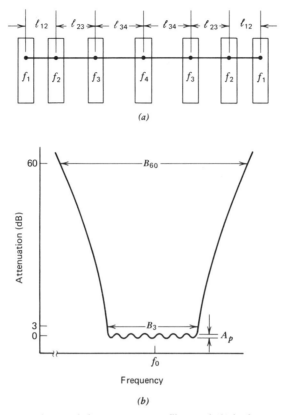

(a)

(b)

FIGURE 3.1.   (a) A wire-coupled, seven-resonator filter, and (b) its frequency response.

The filter bandwidth is a function of the equivalent mass of the resonator and the compliance of the coupling wire. The bandwidth increases as either the equivalent mass or the compliance decrease. The amplitude versus frequency response shape of the passband is related to the variation of coupling between the resonators; greater coupling between the outside resonators, relative to the inside resonators, results in less passband ripple A_p and eventually a rounded passband shape.

In this chapter we look at the characteristics of resonators and coupling wires that most affect their use as circuit elements. These characteristics, such as equivalent mass and compliance, frequency, size, stability, sensitivity, and Q, will be analyzed for various modes of vibration, shapes, and materials by using equations, equivalent circuits, and electrical analogies. In the first sections of this chapter we treat resonators and coupling elements separately, whereas in the last three sections we deal with the subject of coupled resonances. It is a coupled array of resonators that provides bandpass filtering.

RESONATOR MODES OF VIBRATION

Many books and chapters of books have been written on the subject of mechanical resonators. For instance, Rayleigh's book *The Theory of Sound*, written over 100 years ago, is still an excellent resource text on the vibrations of slender bars and rods as well as thin plates and disks [1]. Therefore, we are not developing new theory in this section, but we are looking at the work done over the past 100-plus years that applies to our subject of mechanical filters. Specifically, we are looking at the following types of resonators:

1.   Bodies where one dimension is either very large or very small with respect to the other two dimensions.
2.   Thick bodies where the wave motion is more complex than simple longitudinal, torsional, flexural, or radial modes.
3.   Complex shapes such as dumbbell resonators and tuning forks.

The Classical Method of Calculating Frequencies and Amplitudes

The majority of literature dealing with the vibration of distributed and continuous resonators has followed what we call the classical method. In using this method of analysis we make the following assumptions:

1.   The amplitude of vibration is small and the stress-strain (or force-velocity) characteristics are linear.
2.   There are no internal losses and the vibration occurs in vacuum.
3.   There are no body forces such as gravity or magnetic fields applied to the resonator.

Assuming these three hypotheses, our starting point for the analysis is a set of linear partial differential equations called wave equations, which describe the motion within the resonator. The description of wave motion within each small region of the resonator is in terms of dilatational (longitudinal) and equivoluminal (shear) waves. These equations are usually second or fourth order in terms of the coordinates (x, θ, etc), and second order in time t. The motion of a particle within the small region is called displacement, for example the displacement u of a particle from its initial or rest position at a point defined by a coordinate x of the system.

The next step in the analysis is to eliminate the wave equations' time dependency by substituting, for example, $u = u_1 e^{j\omega t}$ into the term $\partial^2 u / \partial t^2$ and obtaining $-u_1 \omega^2 e^{j\omega t}$. After canceling the $e^{j\omega t}$ terms, which leaves only displacement variables, we can analyze the resulting equation using basic linear differential-equation solution methods.

The general solutions for the displacements normally involve trignometric, hyperbolic, or Bessel functions that are functions of the product of the propagation constant and the coordinate ($A \sin kx + B \cos kx$, for example). If the body is not continuous, it may be necessary to write solutions for each region.

Next, the boundary conditions are defined; the total number is equal to the sum of the orders of the differential equations. When the resonator is free to vibrate, the boundary conditions are derived from the fact that the surfaces are traction-free; in other words, the forces and bending moments, and therefore the stresses and strains at the surfaces, are equal to zero. This allows us to set the derivatives of the displacements, with respect to the coordinates, equal to zero at the surfaces. For example, the strain $du/dx = 0$ at $x = 0$.

By applying the boundary conditions to the displacement equations, we are able to eliminate some of the amplitude constants and write the solution in a form that applies to our specific resonator. Further application of the boundary conditions allows us to write an equation, or a set of equations, from which we can find our basic frequency equation. In matrix form the equations may look like

$$\begin{bmatrix} a_{11} & a_{12} \\ a_{21} & a_{22} \end{bmatrix} \begin{bmatrix} A \\ B \end{bmatrix} = 0, \tag{3.1}$$

where, for example, in the flexural-mode bar case $a_{11} = \cosh k\ell - \cos k\ell$, where ℓ is the length of the bar. A solution to this, generally $n \times n$, set of homogeneous equations is found by setting the determinant $a_{11}a_{22} - a_{21}a_{12}$ equal to zero. After combining terms and applying trigonometric identities we can find the so-called frequency equations, which for a flexural-mode bar is a single equation $\cos k\ell = 1/\cosh k\ell$. The roots of these equations are used to determine the natural frequencies of the resonator.

In order to find the natural frequencies of the resonator, it is necessary to find a relationship between the frequency ω and the propagation constant k. This relationship is found by substituting the specific solution for the displacement into the wave equation and then canceling the displacement terms. The remaining terms are called the dispersion equation, which in the longitudinal-mode resonator is simply $k = \omega(\rho/E)^{1/2}$. Substituting the roots of k into the dispersion equation enables us to find the natural frequencies ω_n.

Having found the natural frequencies of the resonators, we can find the amplitude (displacement) at specific points by substituting the roots $k_n\ell$ of the frequency equation into Equation (3.1), and then solve for the ratio of the constants A and B. Knowing the relative values of the constants, we can then find the amplitude at any point, from the displacement equations.

In the next three sections, we apply the classical method of solution to resonators of increasing complexity. We start with longitudinal-mode and torsional-mode rod resonators.

Longitudinal- and Torsional-Mode Rod Resonators [2]

If one resonator dimension becomes very large with respect to the other two, the vibration modes are simplified and become those of pure longitudinal, torsional, or flexural motion. The wave equation for longitudinal and torsional modes in slender bars or rods is a second-order equation with respect to the length coordinate, whereas the flexural-mode equation is of fourth order and is discussed in the next section.

Longitudinal-Mode Resonators

We begin our analysis by deriving the wave equation for the propagation of longitudinal waves in a thin bar or rod. The force applied to one side of the segment shown in Figure 3.2 is equal to the stress S_x times the cross-sectional area A, whereas on the opposite surface the force is $S_x A + dS_x A$. The force due to the displacement u of the midpoint of the section is equal to its mass times its acceleration. Summing forces we can write

$$(S_x A + dS_x A) - S_x A - \rho A \, dx \frac{\partial^2 u}{\partial t^2} = 0,$$

where ρ is the density of the resonator material.

Combining terms and dividing by $A \, dx$ we obtain

$$\frac{\partial S_x}{\partial x} = \rho \frac{\partial^2 u}{\partial t^2}. \tag{3.2}$$

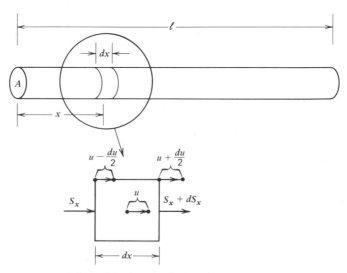

FIGURE 3.2.   Longitudinal-mode bar resonator.

Since the stress S_x is equal to the modulus of elasticity E times the strain du/dx (where du is the difference in the displacement from one side of the section to the other), from (3.2) we obtain the wave equation

$$\frac{E}{\rho}\frac{\partial^2 u}{\partial x^2} = \frac{\partial^2 u}{\partial t^2}. \tag{3.3}$$

From the previous section, our first step in finding the resonance frequencies and amplitudes is to eliminate the time dependency in Equation (3.3). Since u is a function of $e^{j\omega t}$,

$$\frac{E}{\rho}\frac{\partial^2 u}{\partial x^2} = -\omega^2 u. \tag{3.4}$$

From the basic theory of linear differential equations, the solution of (3.4) is

$$u(x) = A \sin kx + B \cos kx, \tag{3.5}$$

where x is the distance from the end of the bar, and k is the propagation constant.

Since we are only considering the case of Figure 3.2, where the ends of the bar are free to vibrate, we can differentiate (3.5) and write

$$\left.\frac{\partial u}{\partial x}\right|_{\substack{x=0 \\ x=\ell}} = Ak \cos kx - Bk \sin kx = 0. \tag{3.6}$$

At $x = 0$, since $\sin kx = 0$,

$$Ak \cos kx - 0 = 0, \text{ or } A = 0.$$

Therefore

$$u(x) = B \cos kx. \tag{3.7}$$

Our next step is to apply the boundary condition at $x = \ell$, which from (3.6) results in

$$\sin k\ell = 0. \tag{3.8}$$

Equation (3.8) is our resultant frequency equation which has roots at

$$k_n \ell = n\pi, n = 1, 2, 3, \ldots . \tag{3.9}$$

Next we establish a relationship between k and the frequency, by substituting the solution (3.7) into the wave equation (3.4) and differentiating the left-hand side. After canceling the cos kx terms we obtain

$$k = \omega \sqrt{\frac{\rho}{E}} \, , \tag{3.10}$$

which is the so-called dispersion equation for this resonator. Note that the phase velocity, defined as ω/k, and the group velocity, defined as $d\omega/dk$, are both constant and equal to $(E/\rho)^{1/2}$. Because the velocity is constant, regardless of frequency, the longitudinal mode is considered nondispersive.

Next we substitute the values of k_n from Equation (3.9) into the dispersion equation (3.10) giving us the equation for the resonance frequencies of the bar,

$$f_n = \frac{\omega_n}{2\pi} = \frac{n}{2\ell} \sqrt{\frac{E}{\rho}} \, . \tag{3.11}$$

The displacement $u(x)$ can be found for each mode n by substituting the value of k_n from (3.9) into Equation (3.7). This gives us

$$u(x) = B \cos \frac{n\pi x}{\ell} \, . \tag{3.12}$$

Equivalent Mass.   The equivalent mass of a resonator is defined as the lumped-mass equivalent of the distributed-parameter rod or bar, at a point in a given direction and in the region of a specified natural frequency. In other words, we replace the transmission-line equivalent circuit described in Chapter 2 by a mass and a spring that resonate at ω_n.

One method of finding the equivalent mass of a longitudinal-mode resonator is to calculate the total kinetic energy in the bar at resonance, and then divide this number by one-half of the squared velocity at a given point and direction. This is the result of the fact that the kinetic energy in the bar is invariant; that is, the kinetic energy is not dependent on the point or direction chosen for calculating the equivalent mass. Therefore we can equate the kinetic energy in the spring-mass system to that in the bar resonator. From Equation (3.7) and the equation $V_x = du(x)/dt$,

$$\tfrac{1}{2} M_{eq\,x} V_x^2 = \tfrac{1}{2} \int_0^\ell \left(V_0 \cos k_n x \right)^2 \rho A \, dx,$$

where $M_{eq\,x}$ is the equivalent mass at a point x as related to the velocity V_x in the x-direction. V_0 is the velocity at $x = 0$ and ρ is the density. Therefore the equivalent mass at the ends of the bar is

$$M_{eq\,0,\ell} = \frac{\tfrac{1}{2} \rho A V_0^2 \int_0^\ell \left(\cos^2 k_n x \right) dx}{\tfrac{1}{2} V_0^2} \, , \tag{3.13}$$

or, after substituting $k_n = n\pi/\ell$ into (3.13) and integrating, we obtain

$$M_{eq\,0,\ell} = \frac{\rho\ell A}{2} = \frac{M_{static}}{2}. \qquad (3.14)$$

Therefore the equivalent mass at the end of a longitudinal-mode resonator, in the direction of the axis of the rod or bar, is simply one-half of its static mass.

Example 3.1. Our problem is to find the fundamental resonance frequency and the equivalent mass of the longitudinal-mode resonator shown in Figure 3.2 where

ℓ(length) = 3.0 cm

d(diameter) = 0.3 cm

ρ(density) = 8.3 g/cm^3

E(Young's modulus) $= 1.9 \times 10^{12}$ dyn/cm^2$\}$ From Table 3.5 (Ni-Span C)

From Equation (3.11), keeping all constants in cgs units, we obtain

$$f_1 = \frac{1}{2\ell}\sqrt{\frac{E}{\rho}} = \frac{1}{2 \times 3.0}\sqrt{\frac{1.9 \times 10^{12}}{8.3}} = 79,742 \text{ Hz}$$

and

$$M_{eq\,0,\ell} = \frac{\rho\ell A}{2} = \frac{8.3 \times 3.0}{2} \times \frac{\pi(0.3)^2}{4} = 0.88 \text{ gms.}$$

Sometimes E is expressed in English units or in terms of the force/unit area. The conversion factors, for the conditions where the acceleration due to gravity is 980.665 cm/sec^2, are:

Multiply E (lbs/in^2) by 6.895×10^4 to obtain E (dyn/cm^2)

Multiply $\frac{E}{g}$ (kg/mm^2) by 9.8×10^7 to obtain E (dyn/cm^2).

Nonslender Rod Solution

If the radius of a rod resonator is greater than one-tenth of a wavelength, a correction factor should be applied to the frequency equation. It has been shown by Rayleigh [1] and Mason [3] that the displacement along the axis of the rod is

$$u_x = Z(r)\sin k_n x,$$

where $Z(r)$ is a function of the radial distance r from the axis. At the resonance frequencies f_n, the propagation constant k_n is equal to $n\pi/\ell$. Considering only the major correction, which is for motion perpendicular to the axis of the rod, that is, lateral inertia, the frequency equation becomes

$$f_n = \frac{n}{2\ell}\sqrt{\frac{E}{\rho}}\left[1 - \left(\frac{n\mu\pi a}{2\ell}\right)^2\right], \qquad (3.15)$$

where μ is Poissons' ratio and a is the radius of the rod. As the radius a approaches zero, the frequency equation becomes that of the slender bar (see Equation 3.11). For very large diameter bars, higher order terms must also be included, as discussed in Mason.

With regard to equivalent-mass calculations, the effect of lateral inertia (i.e., the effect of kinetic energy terms in a direction perpendicular to the axis of the rod) is to cause an increase in the equivalent mass from the value of Equation (3.14). The reason for the increase is the following: of the total kinetic energy in the rod [the numerator of the equivalent-mass Equation (3.13)], a proportionally smaller amount is in the direction x of the axis. Therefore the ratio of the total energy to the x-direction velocity-squared term in the denominator is increased. Conversely, the equivalent mass in a direction perpendicular to the axis becomes smaller as the ratio of diameter to length increases (assuming the static mass remains constant).

Torsional-Mode Resonators

Torsional-mode rod resonators are widely used in radio and telephone voice-bandwidth applications, in the frequency range of 100 to 250 kHz. The resonators are coupled with small diameter wires, which are welded along the major surfaces, and which vibrate in a longitudinal mode. It is therefore important for us to look, not only at the frequency equations, but to look also at the equations for the equivalent mass, of the torsional rod, in the direction of the axis of the coupling wire.

Frequency and Amplitude Equations.   The frequency and amplitude characteristics of a torsional-mode resonator are found by application of the same techniques that were used in the longitudinal-mode derivations. The resultant equations for the thin rod are in terms of the shear modulus G and the angular displacement θ:

$$f_n = \frac{n}{2\ell}\sqrt{\frac{G}{\rho}} \qquad (3.16)$$

and

$$\theta = \theta_1 \cos \frac{n\pi x}{\ell},$$ (3.17)

where θ_1 is the angular displacement at the ends of the rod.

The velocity of propagation $V_p = (G/\rho)^{1/2}$ of the torsional mode is that of an equivoluminal (shear) wave in an infinite medium. Not only is the velocity the same, but as in the infinite medium, there is no change in volume, and the frequency is therefore not affected by the radius-to-length ratios. Therefore the Equations (3.16) and (3.17) also apply to the large-diameter case.

Equivalent Mass.   Figure 3.3 shows a cross section of a torsional-mode rod of radius a. For small displacements, the linear displacement of a point along the radius is $r\theta$. Therefore the linear velocity at any point is

$$V_{x,r} = r\dot{\theta}_1 \cos \frac{n\pi x}{\ell}.$$ (3.18)

To find the equivalent mass at a point in a tangential direction (to the circumferential lines of radius r) we can divide the total kinetic energy $K.E.$ by

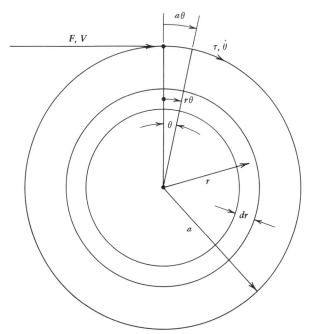

FIGURE 3.3.   Torsional resonator cross-section dimensions for calculating the equivalent mass in the direction of the applied force F and velocity V or the applied torque τ and angular velocity $\dot{\theta}$.

one-half the squared velocity at that point. This involves the double integral

$$K.E. = \int_0^\ell \int_0^a \tfrac{1}{2} [\rho \, dx (2\pi r \, dr)][r^2 \dot\theta_1^2 \cos^2(n\pi x/\ell)],$$

where the two bracketed terms correspond to the mass and velocity squared of a small region of the bar.

Performing the above integration and then dividing by one-half of the square of the velocity (3.18), we obtain the relationship

$$M_{eq}\big|_{r,x} = \frac{K.E.}{\tfrac{1}{2}V_{x,r}^2} = \frac{\rho\pi\ell a^4}{4r^2\cos^2(n\pi x/\ell)}. \tag{3.19}$$

The equivalent mass moment of inertia J_{eq}, in the direction of the applied torque τ in Figure 3.3, is related to M_{eq} at that same point and in the same direction by

$$J_{eq} = r^2 M_{eq}.$$

Therefore,

$$J_{eq}\big|_{r,x} = \frac{\rho\pi\ell a^4}{4\cos^2(n\pi x/\ell)}.$$

Note that the equivalent mass moment of inertia is independent of the radial position r.

Since J_{eq} and M_{eq} are simply related by r^2, we will concentrate only on the equivalent mass. For the case where $r = a$, that is, a point on the major surface of the bar,

$$M_{eq}\big|_{a,x} = \frac{\rho\pi\ell a^2}{4\cos^2(n\pi x/\ell)}, \tag{3.20}$$

and at the ends of the bar,

$$M_{eq}\big|_{r,\ell} = \frac{\rho\pi\ell a^4}{4r^2}, \tag{3.21}$$

or finally when $r = a$ and $x = \ell$,

$$M_{eq}\big|_{a,\ell} = \frac{\rho\pi\ell a^2}{4} = \frac{M_{static}}{4}. \tag{3.22}$$

Equation (3.20) is used in calculating coupling-wire dimensions of longitudinally coupled filters. Equation (3.21) is used to determine the amount of mass removal, from the ends, needed to tune a resonator an amount $\Delta f/f_0$ (see Chapter 6).

At zero frequency the entire bar rotates so the displacement is no longer dependent on the distance x.

By using Equations (3.18) and (3.19), we find that

$$M_{eq}\big|_{a,\,\omega=0} = \frac{M_{static}}{2}.$$

Flexural-Mode Resonators

In this section and the section on nonclassical solutions we look at three types of flexural-mode resonators: the bar or rod, the disk resonator, and the tuning fork. The frequency range of these resonators varies from a few hundred Hz for the tuning fork, to above 500 kHz in the case of the disk resonator. Resonance frequency in flexural-mode resonators is directly proportional to thickness, which means that we can achieve low frequencies by reducing one of the dimensions. This is in contrast with all other resonance modes, where a decrease in a dimension will, at most, result in a second-order reduction of frequency.

Flexural-Mode Bar and Rod Resonators

Let us start our analysis by simply stating, rather than deriving, the wave equation of a thin flexural-mode bar or rod,

$$\frac{\partial^2 u}{\partial t^2} = \left(\frac{EI}{\rho A}\right)\frac{\partial^4 u}{\partial x^4}, \tag{3.23}$$

where the bending moment of inertia I is

$$I = \frac{w\ell^3}{12} \ \text{(rectangular bar)}$$

or

$$I = \frac{\pi a^4}{4} \ \text{(circular rod)}.$$

In the above equations and Figure 3.4, u is the displacement in the y-direction, E is the elastic modulus, ρ is the density of the material, w is the bar width, ℓ is the bar thickness, and a is the radius of the circular rod. Next we eliminate the time dependency by substituting $u = u_1 e^{j\omega t}$ into (3.23). Dropping

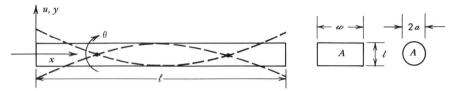

FIGURE 3.4.   Fundamental flexural-mode bar or rod resonator.

the subscript we can write

$$\frac{\partial^4 u}{\partial x^4} = \left(\omega^2 \frac{\rho A}{EI} \right) u, \tag{3.24}$$

which is a fourth-order differential equation having the solution

$$u(x) = \mathscr{A} \cosh kx + \mathscr{B} \sinh kx + \mathscr{C} \cos kx + \mathscr{D} \sin kx. \tag{3.25}$$

For a free bar, the boundary conditions are:

At $x = 0$	At $x = \ell$	
$\dfrac{\partial^2 u}{\partial x^2} = 0$	$\dfrac{\partial^2 u}{\partial x^2} = 0$	$M = 0$ (Bending moment)
$\dfrac{\partial^3 u}{\partial x^3} = 0$	$\dfrac{\partial^3 u}{\partial x^3} = 0$	$\dfrac{\partial M}{\partial x} = 0$ (Shearing force)

Applying the boundary conditions at $x = 0$ to our general solution (3.25) we obtain the relationships between the constants, namely $\mathscr{A} = \mathscr{C}$ and $\mathscr{B} = \mathscr{D}$. For $x = 1$ we can write the remaining equations in matrix form

$$\begin{bmatrix} (\cosh k\ell - \cos k\ell) & (\sinh k\ell - \sin k\ell) \\ (\sinh k\ell + \sin k\ell) & (\cosh k\ell - \cos k\ell) \end{bmatrix} \begin{bmatrix} \mathscr{A} \\ \mathscr{B} \end{bmatrix} = 0. \tag{3.26}$$

Setting the determinant composed of the matrix elements equal to zero, we obtain the frequency equation

$$\cos k\ell = \frac{1}{\cosh k\ell},$$

which has roots at

$$k_1 \ell = 4.730$$

$$k_2 \ell = 7.853 \tag{3.27}$$

$$k_3 \ell = 10.996.$$

Next substituting the solution (3.25) into Equation (3.24) we obtain the dispersion relationship,

$$k^4 = \frac{\rho A}{EI}\omega^2,$$ (3.28)

and the equation for the resonance frequencies,

$$f_n = \frac{(k_n\ell)^2}{2\pi\ell^2}\sqrt{\frac{EI}{\rho A}}.$$ (3.29)

Substitution of the roots (3.27) into (3.29) allows us to calculate the resonance frequencies. The ratios of the higher-order modes to the fundamental mode, f_n/f_1, are shown in Table 3.1.

TABLE 3.1. Characteristics of Flexural-Mode Bar and Rod Resonators

Mode	n	Nodal Points	$k_n\ell$	f_n/f_i
Fundamental (f_1)	1	2	4.730	1.000
1st Harmonic	2	3	7.853	2.757
2nd Harmonic	3	4	10.996	5.404
3rd Harmonic	4	5	14.137	8.932
4th Harmonic	5	6	17.279	13.344

The displacements along the x-axis are found by substituting the roots (3.27) of each mode into the Equation (3.26) and then calculating the amplitude ratio,

$$\frac{\mathscr{A}}{\mathscr{B}} = \frac{\sin k\ell - \sinh k\ell}{\cosh k\ell - \cos k\ell}.$$ (3.30)

From (3.25) and the fact that $\mathscr{A} = \mathscr{C}$ and $\mathscr{B} = \mathscr{D}$,

$$u_x = \mathscr{B}\left[\left(\frac{\mathscr{A}}{\mathscr{B}}\right)(\cosh kx + \cos kx) + (\sinh kx + \sin kx)\right].$$ (3.31)

For the fundamental mode where $k_1\ell = 4.73$, the normalized amplitude $u_1(x)/u_1(0)$ at various points x is shown in Table 3.2. Also shown is the normalized rotational displacement $\theta_1(x)\ell/u_1(0)$ where for small θ,

$$\theta \simeq \sin\theta = \frac{du}{dx}.$$ (3.32)

Equivalent Mass. Knowing the displacement $u(x)$ and $\theta(x)$, we can find the equivalent mass and the mass moment of inertia of the resonator in the y

TABLE 3.2.   Linear and Rotational Displacements u and θ of a Freely Vibrating Bar in Its Fundamental Flexural Mode (x is the Distance from a Free End)

x/ℓ	$u_1(x)/u_1(0)$	$\theta_1(x)/\ell u_1(0)$	x/ℓ	$u_1(x)/u_1(0)$	$\theta_1(x)\ell/u_1(0)$
0.00	1.000	−4.647	0.30	−0.272	−3.171
0.05	0.767	−4.637	0.35	−0.416	−2.485
0.10	0.537	−4.573	0.40	−0.519	−1.733
0.15	0.312	−4.414	0.45	−0.586	−0.494
0.20	0.097	−4.132	0.50	−0.608	0.000
0.25	−0.100	−3.719	0.55	−0.586	+0.494

and θ directions. Since $V(x)$ and $\dot{\theta}(x)$ are simply the time derivatives of $u(x)$ and $\theta(x)$, that is, $V(x) = j\omega u(x)$ and $\dot{\theta}(x) = j\omega\theta(x)$, we can use Equations (3.31) and (3.32) to find

$$M_{eq\,x} = \frac{K.E.}{\frac{1}{2}V_x^2} = \frac{\frac{1}{2}\rho A \int_0^\ell V^2(x)\,dx}{\frac{1}{2}V_x^2} \qquad (3.33)$$

and

$$J_{eq\,\theta} = \frac{K.E.}{\frac{1}{2}\dot{\theta}_x^2}, \qquad (3.34)$$

where $J_{eq\,\theta}$ is the mass moment of inertia.

The kinetic energy $K.E.$ is he same for both the equivalent mass and equivalent mass moment of inertia, assuming that the vibration mode and the displacements are the same. In other words, the total energy in the system remains constant, only the velocities in the y and θ directions vary as we evaluate the equivalent mass and the mass moment of inertia in those directions and at points x along the bar.

The kinetic energy is found by integrating and evaluating the square of Equation (3.31) or by simple numerical integration using Table 3.2. In both cases we then multiply the integral by ω^2. The resultant equations for the fundamental mode and all higher modes of vibration are:

$$M_{eq\,x} = \frac{\rho A\ell}{4\left(\dfrac{u_n(x)}{u_n(0)}\right)^2} \qquad (3.35)$$

$$J_{eq\,x} = \frac{\rho A\ell}{4\left(\dfrac{\theta_n(x)}{u_n(0)}\right)^2} = \frac{\rho A\ell}{4\left(\dfrac{[\theta_1(x)\ell/u_1(0)]}{\ell}\right)^2}, \qquad (3.36)$$

where $u_n(0)$ is the displacement at the ends of the bar. The displacements of Table 3.2 can be used to evaluate the equivalent mass and mass moment of inertia of (3.35) and (3.36). For example,

$$M_{eq\,0} = \frac{\rho A \ell}{4\left(\dfrac{1.0}{1.0}\right)^2} = 0.250\rho A \ell$$

$$M_{eq\,0.5} = \frac{\rho A \ell}{4\left(\dfrac{-0.608}{1.0}\right)^2} = 0.676\rho A \ell,$$

and of importance in torsional coupled low-frequency filters,

$$J_{eq\,0.224} = \frac{\rho A \ell}{4\left(\dfrac{-3.90}{\ell}\right)^2} = \frac{\rho A \ell^3}{60.84}.$$

Thick Bars and Rods.   We have assumed, in the above analysis, that the bar or rod thickness was small in comparison to the length. This assumption allowed us to neglect the effects of shear displacement and rotation of the mass elements of the bar. The latter effect is called rotary (or rotatory) inertia. Including shear and rotary inertia, the differential equation (3.23) becomes

$$\frac{\partial^2 u}{\partial t^2} + \left(\frac{EI}{\rho A}\right)\frac{\partial^4 u}{\partial x^4} + \frac{\rho I}{k'AG}\frac{\partial^4 u}{\partial t^4} - \frac{I}{A}\left(1 + \frac{E}{k'G}\right)\frac{\partial^4 u}{\partial x^2 \partial t^2} = 0, \qquad (3.37)$$

where G is the shear modulus of the bar material and k' is a constant related to the cross section of the bar ($k' = \pi^2/12$ for a regular cross section).

   Näser has solved Equation (3.37) for the resonance frequencies of freely vibrating rectangular bars, in terms of the length to thickness ratio ℓ/ι and the fundamental shear-mode frequency

$$\omega_0 = \frac{\pi}{\iota}\sqrt{\frac{G}{\rho}}, \qquad (3.38)$$

where the shear modulus G is related to Young's modulus by

$$G = \frac{E}{2(1 + \mu)}, \qquad (3.39)$$

and where μ is Poissons' ratio [4].

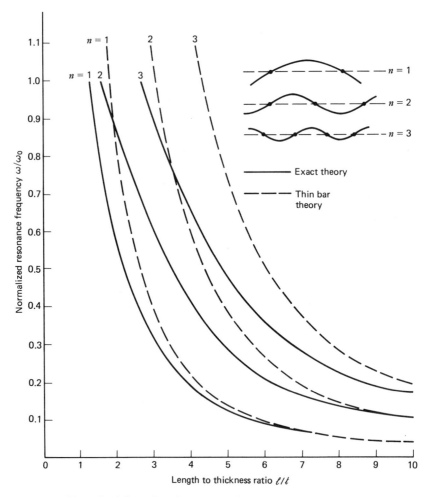

FIGURE 3.5.   Normalized flexural-mode resonance frequency as a function of bar length/thickness ratio. [Adapted, with permission, from *Hochfrequenztech. u. Elektroak.*, **74**, 30–36, (1965).]

Figure 3.5 shows the normalized frequency ω/ω_0 plotted as a function of ℓ/t for the exact theory (where we take shear displacement and rotary inertia into account) and the thin-bar theory of Equation (3.29). Note, that as ℓ/t decreases or the order of the mode increases, the frequency using the exact theory drops with respect to the thin-bar approximation, but at values of ℓ/t above 10/1 for the first two modes they are almost identical. Poisson's ratio is assumed to be that of Ni-Span C which is equal to 0.312. For circular cross-section bars of radius a, the curves of Figure 3.5 can be used, with only a small error, by setting $t = 3^{1/2}a$.

The slender-bar curves are obtained by dividing Equation (3.29) by Equation (3.38). Using the relationship (3.39) and the fact that for a rectangular cross section $E/A = \ell^2/12$,

$$\frac{\omega_n}{\omega_0} = 0.150(k_n\ell)^2\left(\frac{\ell}{\ell}\right)^2.$$

For example, when $n = 3$ and $\ell/\ell = 1/6$,

$$\frac{\omega_3}{\omega_0} = 0.150(10.996)^2(1/6) = 0.5038.$$

The resonance frequency can be found from the curves of Figure 3.5 and the use of Equation (3.38) to calculate ω_0. Even better results have been claimed by Konno et al. [5] because they include vaulting of the cross sections.

Example 3.2. We want to calculate the dimensions of a fundamental flexural-mode resonator, having a rectangular cross section and a resonance frequency f_1 of 128 kHz. We would like as large a resonator as possible, but the package size limits us to a length of 1 cm. The resonator material is Ni-Span C (Table 3.5). Our problem is to find the bar thickness ℓ.

Knowing that $\omega_1 = 2\omega f_1$ and using Equation 3.38 we can write

$$\frac{\omega_1}{\omega_0} = \frac{2\pi f_1}{\dfrac{\pi}{\ell}\sqrt{\dfrac{G}{\rho}}} = \frac{2f_1\ell}{\ell\sqrt{\dfrac{G}{\rho}}}. \tag{3.40}$$

Inserting the values

$$f_1 = 1.28 \times 10^5 \text{ Hz} \quad G = 0.72 \times 10^{12} \text{ dyn/cm}^2$$

$$\ell = 1.0 \text{ cm} \quad \rho = 8.3 \text{ g/cm}^3$$

into 3.40 we obtain

$$\left(\frac{\omega_1}{\omega_0}\right)\left(\frac{\ell}{\ell}\right) = 0.882. \tag{3.41}$$

Equation (3.41) can be solved by finding, through iteration, a pair of values of ω_1/ω_0 and ℓ/ℓ that both fall on the exact ($n = 1$) curve of Figure 3.5 and satisfy (3.41). The ratio of ℓ/ℓ that satisfies these criteria is $\ell/\ell = 3.0$ and therefore $\ell = 1.0/3.0 = 0.33$ cm.

Flexural-Mode Disk Resonators

The analysis of a freely vibrating thin disk is considerably more complex than that of a flexural-mode bar. The added complexity is, for the most part, due to the evaluation of the boundary conditions. In this section we rough out the analysis and show the most important results.

The wave equation in rectangular coordinates involves derivatives in both the x and y directions and with respect to time t,

$$\frac{\partial^4 u}{\partial x^4} + \frac{\partial^4 u}{\partial y^4} + 2\frac{\partial^4 u}{\partial x^2\,\partial y^2} = \frac{1}{c^2}\frac{\partial^2 u}{\partial t^2}, \tag{3.42}$$

where the propagation velocity squared is

$$c^2 = \frac{E\ell^2}{12\rho(1 - \mu^2)}, \tag{3.43}$$

and where ℓ is the disk thickness, and u is the displacement in the z-direction.

Eliminating the time dependency from (3.42) and factoring, we obtain the equation

$$\frac{\partial^2 u}{\partial x^2} + \frac{\partial^2 u}{\partial y^2} \pm k^2 u = 0, \tag{3.44}$$

where the flexural-mode propagation constant k is related to frequency and propagation velocity by

$$k^2 = \frac{\omega}{c}. \tag{3.45}$$

In showing the dispersion relationship (3.45) at this time, we have actually stepped ahead of ourselves by showing the result of substituting (3.49) into (3.46) (written in terms of ω/c).

We can most easily obtain a solution by transforming to polar coordinates. Equation (3.44) becomes

$$r^2\frac{\partial^2 u}{\partial r^2} + r\frac{\partial u}{\partial r} + \frac{\partial^2 u}{\partial \theta^2} \pm k^2 r^2 u = 0, \tag{3.46}$$

where r is the radial distance from the center of the disk and θ the angular distance shown in Figure 3.6.

The solution to (3.46) can be written in terms of the product of two solutions: χ, which is only dependent on (r), and ψ, which is only dependent on θ. Therefore, $u(r, \theta) = \chi(r)\psi(\theta)$. Applying this solution to (3.46) we can write the three ordinary differential equations,

$$\frac{d^2\chi}{dr^2} + \frac{1}{r}\frac{d\chi}{dr} \pm \left(k^2 \mp \frac{n^2}{r^2}\right)\chi = 0 \tag{3.47}$$

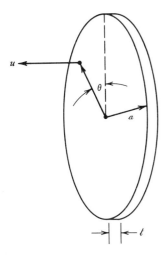

FIGURE 3.6. Flexural-mode disk resonator.

and

$$\frac{d^2\psi}{d\theta^2} + n^2\psi = 0, \qquad (3.48)$$

where n is an integer.

Solutions to (3.47) are in the form of Bessel functions, whereas the solution of (3.48) is a sinusoidal function. Eliminating those functions that go to infinity at $r = 0$, we can write our solution as

$$u(r, \theta) = \chi(r)\psi(\theta) = \left[AJ_n(kr) + BI_n(kr)\right]\cos n\theta, \qquad (3.49)$$

where n is the number of nodal diameters, and the Bessel functions are related by $I_n(kr) = i^{-n}J_n(ikr)$.

The propagation constant k has a different value for each combination of nodal diameters and nodal circles s. Sketches of the four lowest vibration modes are shown in Figure 3.7. The shaded region represents, for example, motion out of the page, whereas the clear region represents motion into the page.

The next step is to apply the boundary conditions to our general solution (3.49). Because of the complexity of these calculations, we simply state the results determined by Rayleigh [1]. For $n = 0$, which is the case of zero nodal lines,

$$\begin{bmatrix} J_1(ka) & I_1(ka) \\ (\mu - 1)J_1(ka) + kaJ_0(ka) & -(\mu - 1)I_1(ka) + kaI_0(ka) \end{bmatrix} \begin{bmatrix} A \\ B \end{bmatrix} = 0.$$

$$(3.50)$$

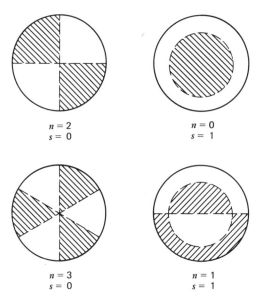

FIGURE 3.7. First four disk-flexural modes. The dashed nodal lines separate the in-phase and out-of-phase regions.

Setting the determinant of the matrix elements equal to zero, we obtain the frequency equation

$$2(\mu - 1) + \left[ka\frac{J_0(ka)}{J_1(ka)} + \frac{I_0(ka)}{I_1(ka)} \right] = 0. \tag{3.51}$$

The roots of the frequency equation are a function of Poisson's ratio μ; this is in contrast with the thin bar and rod solutions where the effects of the lateral dimensions are not included. The roots ka for two values of Poisson's ratio 0.2 and $1/3$ and for the one and two nodal-circle modes ($s = 1$ and $s = 2$) are

$$ka = 2.97(\mu = 0.2, s = 1), 6.18(\mu = 0.2, s = 2),$$

$$3.01(\mu = 1/3, s = 1), 6.21(\mu = 1/3, s = 2).$$

Note that there is only a small difference in the values of ka for widely different values of Poisson's ratio.

The frequency equation is found by substituting the plate flexural-mode velocity of Equation (3.43) into the dispersion Equation (3.45) giving us

$$\omega_s = 2(ka)_s^2 \frac{t}{d^2} \sqrt{\frac{E}{3\rho(1 - \mu^2)}}. \tag{3.52}$$

The resonance frequency for a particular nodal circle mode of vibration, $s = 1, 2, 3, \ldots$, can be calculated by substituting the roots $(ka)_s$ into (3.52).

The amplitudes, that is relative displacements, can be found for the nodal-circle mode case by writing the first of the Equations (3.50) as

$$\frac{A}{B} = -\frac{I_1(ka)}{J_1(ka)}$$

and then substituting this ratio into (3.49), where for the case of $n = 0$ (no nodal lines),

$$u(r) = A J_0(kr) + B I_0(kr).$$

Therefore

$$u(r)_s = B\left[I_0(kr) - \frac{I_1(ka)_s}{J_1(ka)_s} J_0(kr)\right], \qquad (3.53)$$

where B is simply an amplitude constant and

$$kr = \frac{r}{a}(ka)_s. \qquad (3.54)$$

From Equation (3.53), the amplitude distribution from the center of the disk, $r = 0$, to the edge of the disk, $r = a$, is shown in Table 3.3 for the first two nodal-circle modes.

The equivalent mass of the thin disk can be calculated exactly by integrating the square of the displacement of Equation (3.53) or numerically by use of Table 3.3. The resultant equation, for the ratio of the equivalent mass of the

TABLE 3.3. Displacement $u(r)/u(0)$ For One and Two Nodal-Circle Modes of Vibration ($\mu = 14$)

r/a	$s = 1$ $n = 0$	$s = 2$ $n = 0$	r/a	$s = 1$ $n = 0$	$s = 2$ $n = 0$
0.0	+1.000	+1.000	0.6	+0.187	-0.370
0.1	+0.974	+0.907	0.7	-0.045	-0.303
0.2	+0.896	+0.654	0.8	-0.283	-0.109
0.3	+0.771	+0.312	0.9	-0.518	+0.163
0.4	+0.605	-0.026	1.0	-0.748	+0.455
0.5	+0.407	-0.273			

TABLE 3.4. Frequency Constant ka Values as a Function of the Number of Nodal Circles and Nodal Diameters (Poisson's Ratio is 13)

Number of Nodal	Number of Nodal Diameters—n									
Circles—s	0	1	2	3	4	5	6	7	8	9
0			2.29	3.50	4.70	5.66	6.80	7.77	8.93	9.94
1	3.01	4.53	5.94	7.27	8.51	10.42	11.06			
2	6.21	7.74	9.45	10.55						
3	9.34	10.89								

actual static mass, is for $s = 1$,

$$\left(\frac{M_{eq}}{M_{st}}\right)_{s=1} = \frac{0.247}{\left[u(r)/u_0(0)\right]^2}.$$ (3.55)

So far we have ignored the diameter nodes of vibration by setting n equal to zero. The more general solution, where both n and s are finite, is outlined in Rayleigh [1] and is discussed in Timoshenko [6]. Table 3.4 shows ka values for up to three nodal circles and nine nodal diameters. These values can be used in conjunction with Equation (3.52) to find the thin-disk natural resonance frequencies.

Thick Disk Equations   As in the case of the flexural-mode bar, the thin-disk equations ignore the effects of rotary inertia and shear. Dealing with axisymmetric (circle) modes of vibration, Mindlin and Deresiewicz [7] included the rotary inertia and shear terms in their solutions for the natural resonance frequencies; Sharma [8] added a solution for the equivalent mass. Onoe and Yano [9] solved the non-axisymmetric (diameter nodal lines included) mode problem by including torsional waves in their solution. Onoe and Yano's work has been extended to include displacement and equivalent mass solutions [10].

A set of normalized frequency curves, similar to those of Figure 3.5, is shown in Figure 3.8 [11]. In this case we are plotting the normalized frequency ω/ω_0 as a function of the disk diameter-to-thickness ratio d/t. Poisson's ratio μ is that of Ni-Span C. The solid lines correspond to the case where rotary inertia and shear deformation are taken into account and, as with the flexural-mode bar, the resonance frequencies are lower than those of the classical solution.

The thick disk illustrates the complex motion of a resonator, as its dimensions tend to be equal. Figure 3.9 shows a sketch of disk motion and particle displacement, as a function of radial position r and axial position z. The origin of the coordinates is at the center of the disk. This figure roughly applies to all flexure modes where nodal circles are present. Note that the radial and axial

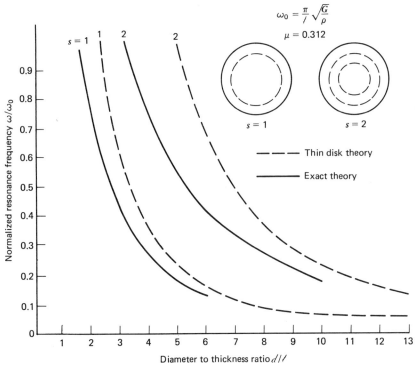

FIGURE 3.8. Circle-mode resonance frequency curves based on Equation 3.52 (thin-disk theory) and equations that take into account rotary inertia and shear (exact theory). [Adapted, with permission, from *Hochfrequenztech. u. Elektroak.*, **71**, 123–132 (Oct. 1962).]

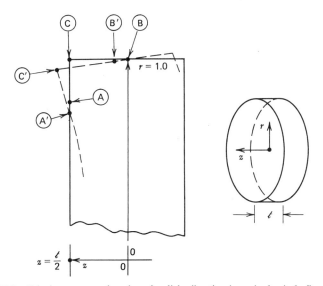

FIGURE 3.9. Displacement at the edge of a disk vibrating in a single-circle flexural mode.

89

components of the displacement vary with both the radial and axial position of the point being considered. For instance, a particle at the static position A is displaced only in the radial direction to point A'. The particle at the center of the disk edge at B is displaced in the z-direction to B' and the corner C moves in both the r and z directions to point C'. It is important to note, that although there is axial motion at point A, the displacement at the point is only in the radial direction.

From the displacements, such as those shown in Figure 3.9, we can plot the equivalent mass of a disk in both the radial and axial directions as a function of position on the disk surface. Figures 3.10 and 3.11 show the equivalent mass at points on the two surfaces of a disk vibrating in a single circle ($s = 1$, $n = 0$) mode. Figures 3.12 and 3.13 show the equivalent mass on the major surface of the disk ($z = \ell/2$) as a function of r, for the $s = 2$, $n = 0$ and the $s = 1$, $n = 2$ modes, respectively. In these curves, Poisson's ratio is 0.3 and the diameter-to-thickness ratio is $4/1$. Remembering that the equivalent mass is a constant (the total kinetic energy) divided by the velocity (du/dt) squared, we can find the

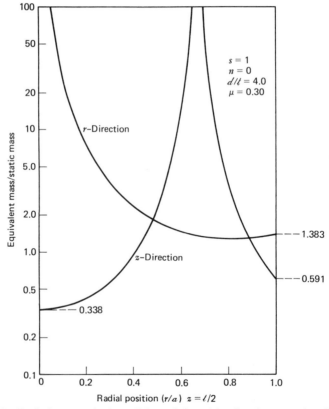

FIGURE 3.10. Equivalent mass in the radial r and the axial z directions, as a function of radial position on the disk face. The disk vibration is a single-circle mode.

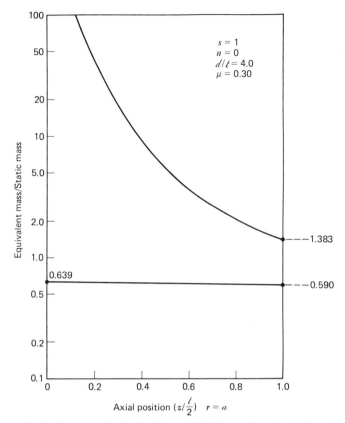

FIGURE 3.11. Equivalent mass of a single-circle mode disk, as a function of the axial position on the disk circumference, in the radial *r* and axial *z* directions.

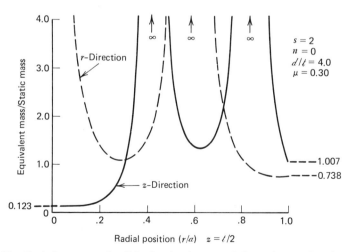

FIGURE 3.12. Equivalent mass of a disk vibrating in a two-circle mode, as a function of radial position on the disk face, in the radial *r* and axial *z* directions.

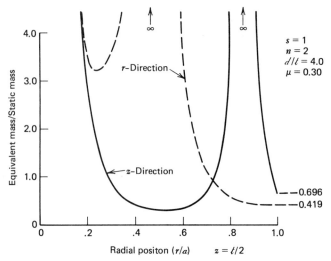

FIGURE 3.13. Equivalent mass of a disk vibrating in a two-diameter, one-circle mode, as a function of radial position on the disk face, in the radial r and axial z directions.

relative displacements u_r and u_z from the equivalent mass curves of Figures 3.10 to 3.13.

Example 3.3. Thick disks are used in the design of filters for single-sideband radios. These filters are typically in the 450 to 500 kHz frequency range. In this example we will calculate the dimensions of a 455 kHz disk vibrating in a two nodal-circle mode. The diameter-to-thickness ratio is 4/1 and the disk material is Ni-Span C.

From Figure 3.8, the normalized frequency based on the 4/1 diameter-to-thickness ratio is approximately 0.80. From Equation (3.38),

$$\frac{\omega}{\omega_0} = \frac{2\pi f}{\frac{\pi}{\ell}\sqrt{\frac{G}{\rho}}} = 0.80,$$

or

$$\ell = \frac{0.80}{2f}\sqrt{\frac{G}{\rho}} = \frac{0.80}{2\times 455\times 10^3}\sqrt{\frac{0.72\times 10^{12}}{8.3}} = 0.26 \text{ cm},$$

and the diameter is

$$d = \left(\frac{d}{\ell}\right)\ell = 4\times 0.26 = 1.04 \text{ cm}.$$

In order to achieve a bandwidth in the range of 2 to 3.5 kHz the point on the disk to which the coupling wire or wires is attached must be chosen so that the wire diameter is sufficiently large to support the structure. Looking at Figure 3.12, we see that a radial position of 0.63 is both a low-sensitivity point (to variations in wire position) and a high-equivalent-mass point. Also the radial displacement at that point is very small, which reduces the effects of radial contour modes of vibration. In the case of edge coupling ($r/a = 1.0$), both axial and radial contributions must be considered.

Contour Modes

We use the term *contour mode* to describe the vibration of a plate where the displacements are in directions parallel to the major surfaces. We will assume that the plate is thin and has either a rectangular or circular contour. In general, the motion can include both extension and shear. Because there are so many different mode shapes, we will not be able to describe all of them in this chapter, but we should remember that other modes exist. The various modes have names such as: length, edge, radial, shear, flexure, or even more specifically, concentric shear, length-width flexure, face shear, and so on. These modes have not only been used as the fundamental resonance of many filter resonators, but they exist as spurious, that is, unwanted, modes.

In this section we will look at the rectangular plate vibrating in the length extensional mode and the circular disk vibrating in radial and non-axisymmetric contour modes.

Length-Extensional Contour Modes

Rectangular plate filters have been popular because their planar construction leads to ease of manufacturing and small physical size. In this section we eliminate derivations and simply state the resonator frequency and displacement equations. The frequency equation is

$$\frac{\cot\left(\alpha \frac{\ell}{2}\right)}{\cot\left(\beta \frac{\ell}{2}\right)} = \frac{-2\alpha\beta k^2(1-\mu)}{(\beta^2 - k^2)(\alpha^2 + \mu k^2)}, \tag{3.56}$$

where

$$\alpha^2 = \frac{(1-\mu)\theta^2}{2} - k^2 \qquad \theta^2 = \frac{2\rho\omega^2(1+\mu)}{E}$$

$$\beta^2 = \theta^2 - k^2 \qquad\qquad k = \frac{\pi}{\ell}$$

and ℓ is the width of the plate [12],[13],[14].

Equation (3.56) is transcendental and must be solved by graphical or iterative methods. Figure 3.14 shows normalized frequency curves based on Equation (3.56). The curves were obtained by setting $\ell = 1.0$ and varying θ for each value of k. The solution is the value of θ that makes the RHS and LHS of (3.56) equal. From the curves of Figure 3.14 we can find the plate length ℓ from a given plate width ℓ and frequency ω.

Figure 3.15 shows the plate coordinates and the motion of the plate at its two lowest natural frequencies, which we will call the first and second modes. The displacement equations corresponding to the two modes are:

$$u = A\left(\cos \alpha y + \frac{B}{A}\cos \beta y\right)\cos kx$$

$$v = A\left(-\frac{\alpha}{k}\sin \alpha y + \frac{Bk}{A\beta}\sin \beta y\right)\sin kx,$$

(3.57)

where A and B are amplitude constants related by

$$\frac{B}{A} = \frac{-2\alpha\beta \sin(\alpha\ell/2)}{(\beta^2 - k^2)\sin(\beta\ell/2)}.$$

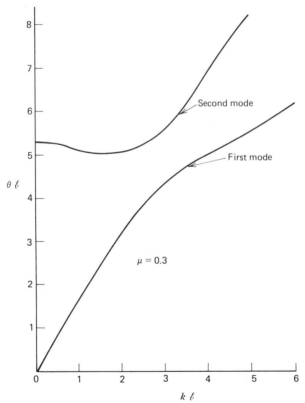

FIGURE 3.14.   Normalized frequency curves for the two lowest frequency modes of flat-plate contour-mode resonators.

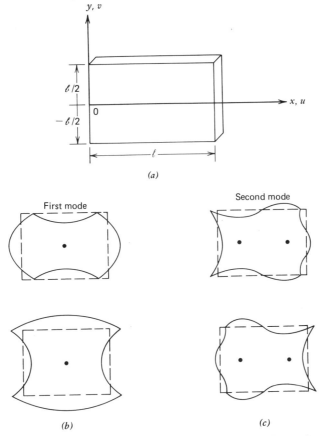

FIGURE 3.15. Thin plate contour modes: (a) coordinate axis, (b) lowest frequency mode showing displacements at half-cycle periods, and (c) second lowest mode (nodal points are denoted by dots).

Example 3.4. As the starting point in a filter design, we need to calculate the approximate dimensions of a first-mode 455 kHz thin-plate resonator having a width-to-length ratio of 0.75. The material is Ni-Span C where $E = 1.9 \times 10^{12}$ dyn/cm², $\mu = 0.312 \simeq 0.3$ and $\rho = 8.3$ g/cm³. In order to use Figure 3.14 we must first calculate the product $k\ell$ where

$$k\ell = \frac{\pi \ell}{\ell} = \pi \times 0.75 = 2.356.$$

From the lower curve of Figure 3.14, $\theta \ell = 3.65$. Next,

$$\theta = \omega \sqrt{\frac{2\rho(1 + \mu)}{E}} = 2\pi \times 455 \times 10^3 \sqrt{\frac{2 \times 8.3 \times 1.3}{1.9 \times 10^{-12}}} = 9.63,$$

$$\ell = \frac{3.65}{\theta} = \frac{3.65}{9.63} = 0.38 \text{ cm},$$

and

$$\ell = \frac{\ell}{0.75} = 0.51 \text{ cm.}$$

Circular Disk Contour Modes

Lower in resonance frequency than the fundamental radial resonance of a thin disk are the two non-axisymmetric modes shown in Figure 3.16(c) and (d). These modes are denoted by the subscripts (n, m) where n corresponds to the circumferential order (related to nodal diameters) and m corresponds to the radial harmonic [15]. Pure symmetric radial modes, such as that shown in Figure 3.16(b), are denoted by (R, m).

The frequency equation for the contour-mode disk is

$$f_{n,m} = \frac{\alpha_{n,m}}{a} \sqrt{\frac{E}{\rho}}, \tag{3.58}$$

where a is the radius of the disk and $\alpha_{n,m}$ is a frequency constant related to the mode (n, m) and Poisson's ratio μ. For $\mu = 0.30$, $\alpha_{2,1} = 0.272$, $\alpha_{1,1} = 0.317$, $\alpha_{R,1} = 0.342$, $\alpha_{3,1} = 0.418$, and $\alpha_{2,2} = 0.493$. For other values of Poisson's ratio, the following approximation can be used: a 10 percent increase in μ results in a 2 percent increase in $\alpha_{R,1}$ and a 1 percent decrease in the constants $\alpha_{n,m}$.

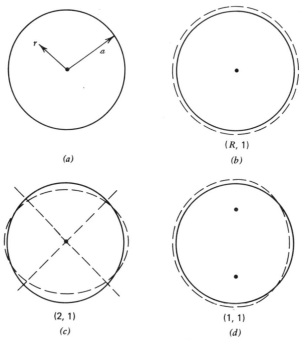

                                           (R, 1)

(a)   (b)

(2, 1)   (1, 1)

(c)   (d)

FIGURE 3.16. Thin-disk contour modes: (a) coordinates, displacement pattern of (b) symmetric radial mode, (c) lowest frequency contour mode, and (d) second antisymmetric mode. Nodal points are denoted by dots.

As an example of the use of Equation (3.58), we calculate the radius of a thin Ni-Span C disk vibrating in the fundamental radial mode R, 1 at 455 kHz;

$$a = \frac{\alpha_{R,1}}{f_{R,1}} \sqrt{\frac{E}{\rho}} \simeq \frac{0.342}{0.455 \times 10^6} \sqrt{\frac{1.9 \times 10^{12}}{8.3}} = 0.36 \text{ cm}.$$

Because Poisson's ratio for Ni-Span C is 0.312 there was no need to apply a correction to $\alpha_{R,1}$, because Young's modulus and the density are only known to two digits.

Non-Classical Solutions

So far in this chapter we have looked only at solutions for resonators having simple shapes such as bars, rods, disks, and rectangular plates. Except in the most simple cases, we had to use more than one set of waves to approximate the boundary conditions. As an example, Onoe and Yano's solution to the flexural-mode thick disk case involved using two flexural waves and one torsional wave [9]. Even this very complex solution for a simple resonator shape was only an approximation.

Partition Solutions

In contrast to the previous methods are those involving a dissection of the resonator into rectangular or triangular elements. Each element is described by a matrix which may include both electrical and mechanical parameters. The individual elements are then combined in a matrix in such a way that there is compatibility at the boundaries. Using a computer, an iterative solution for the natural frequencies that satisfy the matrix equations is then found.

These methods of solution, involving a partitioning of the resonator, include the finite-element method, the finite-difference method, and the series-expansion method. The finite-element method has been used in structure design for analyzing beams and shells. In the case of mechanical filter resonators, this method has been used to analyze plates [16], flexural-mode bars [17], and tuning forks having complex shapes [18].

Tuning-Fork Resonators

Tuning forks are basically folded flexural-mode bars. The folded geometry is used to realize a very low frequency without an excessive resonator length. Various types of tuning forks are shown in Figure 3.17.

The rectangular shapes of Figure 3.17(a) and (b) can be analyzed by means of the extensional-mode and bending-mode matrix Equations (3.65) and (3.70) that are described later in this chapter. Details of this analysis are shown in Nagai and Konno [19]. Analysis of the simple U-shaped fork has been done by

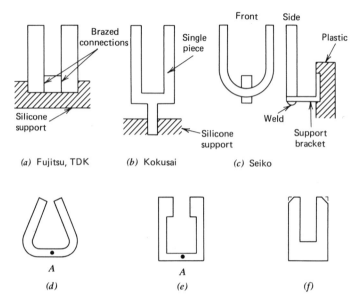

FIGURE 3.17. Tuning-fork resonators. (*a*) and (*b*) are conventional rectangular designs; (*c*) is an inexpensive U-shaped fork; (*d*), (*e*), and (*f*) are high-Q designs of Konno et al.

the finite-element simulation described in Reference [20]. The tuning fork shapes shown in Figure 3.17(*d*) and (*e*) are designed to locate a nodal point at the support point *A*, in order to realize a high resonator *Q*. An analysis of Figure 3.17(*d*) is described in Reference [21]. In order to obtain very high resonator *Q*'s, a corner of the resonator is removed, as shown in Figure 3.17(*f*) [18].

Complex Resonators

By the term *complex resonators*, we are referring to resonators of irregular shapes or to composite resonators of different materials, such as metal and piezoelectric ceramic. In determining the resonance frequencies and amplitudes of these resonators we make use of basic transmission matrix methods where each section of the matrix corresponds to a region of different shape or material. In this section we look only at a cascade of extensional or torsional rods and defer the discussion of flexural-mode elements to our discussion of flexural-mode coupling-wire equivalent circuits.

Complex Extensional and Torsional Resonators

In Chapter 2 we discussed the mechanical transmission line and its matrix representation [Equation (2.8)]. Figure 3.18(*a*) shows a cascade of lines each having a transmission matrix [*T*]. If Figure 3.18(*a*) represents an extensional-mode resonator, its resonance frequencies can be found by calculating the poles of the input mobility (the ratio of V_1 to F_1) under the open-circuit

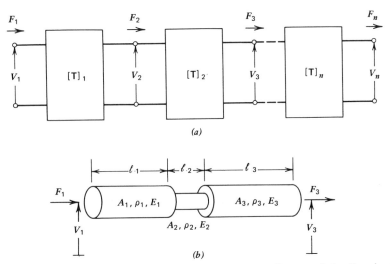

FIGURE 3.18. (*a*) Cascade of transmission matrix sections, (*b*) transmission-line dumbbell resonator.

condition $F_n = 0$. In other words, the natural modes of vibration correspond to frequencies where both the input and output ends of the bar or rod are vibrating freely ($F = 0$). The equivalent mass M_{eq} of each mode can be calculated from the slope of the ratio of F_1 to V_1 at its natural resonance frequency ω_j. In general,

$$M_{eq} = \frac{1}{2}\left(\frac{d(F_1/V_1)}{d\omega}\right)_{\omega=\omega_j}. \tag{3.59}$$

The networks of Figure 3.18 could have been shown as torsional-mode resonators by replacing F by τ, V by $\dot{\theta}$, and E by G. Examples of extensional and torsional mode dumbbell resonators are shown in Figures 5.3(*b*) and 7.15.

Example 3.5. In this example we will outline a method of calculating the resonance frequency of the dumbbell resonator shown in Figure 3.18(*b*). From Equations (2.8), the transmission matrix of the first element is

$$[\mathbf{T}] = \begin{bmatrix} \cos\dfrac{\omega\ell_1}{v_{p1}} & \dfrac{j}{A_1\sqrt{\rho_1 E_1}}\sin\dfrac{\omega\ell_1}{v_{p1}} \\ jA_1\sqrt{\rho_1 E_1}\sin\dfrac{\omega\ell_1}{v_{p1}} & \cos\dfrac{\omega\ell_1}{v_{p1}} \end{bmatrix} = \begin{bmatrix} \mathscr{A}_1 & \mathscr{B}_1 \\ \mathscr{C}_1 & \mathscr{D}_1 \end{bmatrix},$$

where

$$v_{p1} = \sqrt{E_1/\rho_1}\,.$$

Resonator Mode	Structure	Frequency	Displacement	Equivalent Mass
Extensional		$f_n = \dfrac{n}{2\ell}\sqrt{\dfrac{E}{\rho}} = \dfrac{n}{2\ell}v_{pL}$	$u_n = u_0\cos\left(\dfrac{n\pi x}{\ell}\right)$	$M_{eqx} = \dfrac{\rho A\ell}{2\cos^2\left(\dfrac{n\pi x}{\ell}\right)}$
Torsional		$f_n = \dfrac{n}{2\ell}\sqrt{\dfrac{G}{\rho}} = \dfrac{n}{2\ell}v_{pS}$	$u_n = r\theta_0\cos\left(\dfrac{n\pi x}{\ell}\right)$.	$M_{eqx} = \dfrac{\rho A\ell}{4\cos^2\left(\dfrac{n\pi x}{\ell}\right)}$
Torsional		$f_n = \dfrac{n}{2\ell}\sqrt{\dfrac{G}{\rho}}$	$\theta_n = \theta_0\cos\left(\dfrac{n\pi x}{\ell}\right)$	$J_{eqx} = \left(\dfrac{d}{2}\right)^2\dfrac{\rho A\ell}{4\cos^2\left(\dfrac{n\pi x}{\ell}\right)}$

Flexural	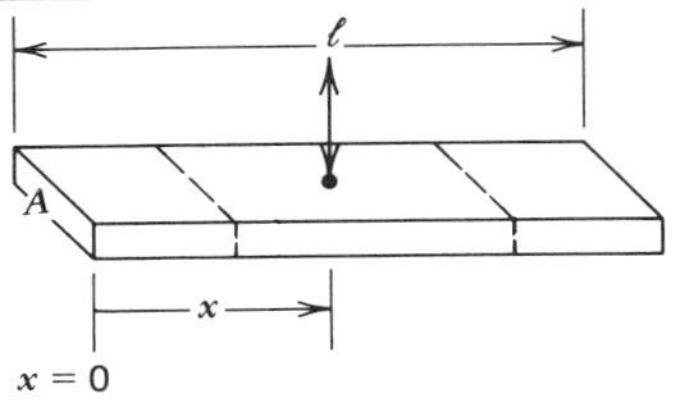	$f_n = \dfrac{(k_n\ell)^2}{2\pi\ell^2}\sqrt{\dfrac{EI}{\rho A}}$ $I = \dfrac{A\ell^2}{12}$ (Bar), $\dfrac{Aa^2}{4}$ (Rod) See Table 3.1	See Equation (3.31) or Table 3.1 for $u_n(x)$	$M_{eqx} = \dfrac{\rho A\ell}{4\left(\dfrac{u_n(x)}{u_n(0)}\right)^2}$ See Table 3.2
Flexural	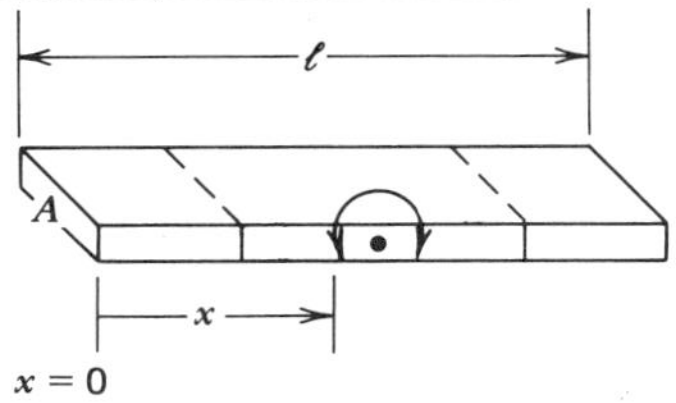	$f_n = \dfrac{(k_n\ell)^2}{2\pi\ell^2}\sqrt{\dfrac{EI}{\rho A}}$ See Table 3.4	$\theta_n(x) = \dfrac{du_n(x)}{dx}$ See Table 3.2	$J_{eqx} = \dfrac{\rho A\ell}{4\left[\left(\dfrac{\theta_n(x)\ell}{u_n(0)}\right)/\ell\right]^2}$ See Table 3.2
Flexural	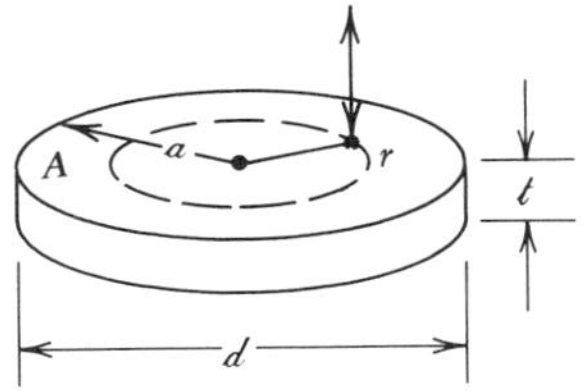	$f_s = \dfrac{(ka)_s t}{\pi d^2}\sqrt{\dfrac{E}{3\rho(1-\mu^2)}}$ See Table 3.4	$u = u_0\left[I_0(kr) - \dfrac{J_0(kr)I_1(ka)_s}{J_1(ka)_s}\right]$ See Tables 3.3 and 3.4	$M_{eq} = \dfrac{0.247\rho a t}{\left(\dfrac{u(r)}{u(0)}\right)^2}$ See Table 3.3

FIGURE 3.19. Resonator frequency, displacement, and equivalent mass equations.

Rather than attempt a closed-form solution to this problem, we can use a digital computer to evaluate each of the matrix elements in the three transmission matrices for increasing values of frequency ω. For each ω, matrix multiplication is performed to obtain the $\mathscr{A}$, $\mathscr{B}$, $\mathscr{C}$, and $\mathscr{D}$ matrix elements of the entire resonator. The driving point mobility V_1/F_1 under open-circuit conditions ($F_3 = 0$) is simply

$$Z_1 = \frac{V_1}{F_1} = \frac{\mathscr{A}}{\mathscr{C}}.$$

The lowest natural resonance is found by increasing the frequency ω by steps $\Delta\omega$ until the mobility becomes infinite or $1/Z_1$ is equal to zero. At the resonance frequency, the equivalent mass is easily calculated from the slope of the inverse mobility vs. ω curve by use of Equation (3.59).

A Summary of Frequency, Displacement, and Equivalent-Mass Equations

Figure 3.19 summarizes the frequency, displacement, and equivalent mass equations derived for some of the simple resonators discussed in the previous sections. The equations are valid for extensional and torsional slender rods and bars and for thin flexural-mode bars and disks. In each case, the displacements u_0 and θ_0 correspond to the displacement at $x = 0$. The heavy arrows show the directions in which the equivalent mass is measured.

RESONATOR EQUIVALENT CIRCUITS

The engineer interested in analyzing a mechanical filter has a choice between using particular integrals of differential equations that describe the entire device or using lumped and distributed elements arranged in the form of an electrical or mechanical circuit. If we limited ourselves to the analysis of a single resonator, there would be little need for equivalent circuits, but in the analysis of coupled systems, the strict use of differential equations is not only very difficult, but this method often obscures the solution. The equivalent-circuit method, although not exact, lends itself to better visualization of the system. It also allows us to construct the electrical equivalent circuit of the filter and then analyze the circuit, using general-purpose computer analysis programs. In this way, numerous configurations can be looked at in a short time.

In this section we look at a wide variety of equivalent circuits starting with a two-terminal spring-mass resonator.

A Generalized Spring-Mass Resonator Equivalent Circuit

A starting point for deriving a generalized equivalent circuit is the simple, lumped spring-mass resonator of Figure 3.20(a). The driving-point mobility Z_1

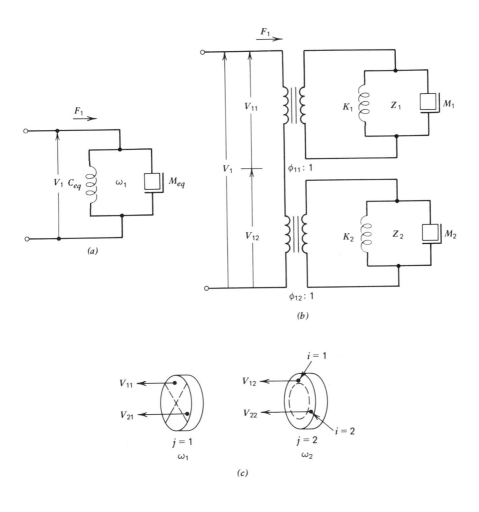

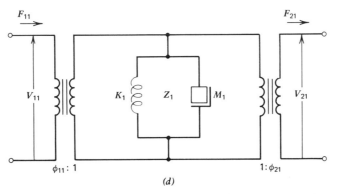

FIGURE 3.20. Mechanical equivalent circuits of mechanical filter resonators. (a) A two-terminal, single-mode resonator; (b) two-terminal, two-mode description; (c) a flexural-mode disk example; and (d) a single-mode, four-terminal equivalent circuit.

is

$$Z_1 = \frac{V_1}{F_1} = \frac{\omega/j}{M_{eq}(\omega^2 - \omega_1^2)}. \tag{3.60}$$

If we choose to represent the driving-point mobility of a distributed-element resonator, such as a disk or bar in terms of lumped elements we can write [22], [23]

$$Z_i = \frac{V_i}{F_i} = \frac{\omega}{j} \sum_{j=1}^{\infty} \frac{1}{M_{ij}(\omega^2 - \omega_j^2)}, \tag{3.61}$$

where the subscript i is the point (and direction) of the measurement, j is the mode of vibration when used as a subscript [otherwise $j = (-1)^{1/2}$], and M_{ij} is the equivalent mass associated with the point i and the mode j. By making use of velocity transformers, having turns ratios ϕ_{ij}, and recognizing Equation (3.61) as the partial-fraction expansion of a reactance function, we can construct the schematic diagram shown in Figure 3.20(b).

The circuit shown in Figure 3.20(b) describes the characteristics of a two-mode resonator at a point (or port, if we choose to use electrical terminology) $i = 1$. The resonator could be a rod vibrating in torsion and flexure or the disk of Figure 3.20(c), which is shown vibrating in a two-diameter flexure mode and a single-circle flexure mode. From Figure 3.20(b) the relationship between velocity and force can be written as

$$V_1 = V_{11} + V_{12} = (\phi_{11}^2 Z_1 + \phi_{12}^2 Z_2)F_1, \tag{3.62}$$

where Z_1 and Z_2 are the mobilities of the tuned circuits and ϕ_{11} and ϕ_{12} are velocity ratios corresponding to the turns ratios of electrical-circuit transformers. Relationships between resonance frequency, equivalent mass, velocity ratio, and mobility can be found by comparing Equation (3.62) to the first two terms of Equation (3.61).

Let us now look at two points on a resonator vibrating at frequency ω_1. From Figure 3.20(d) we can write

$$
\begin{bmatrix} V_{11} \\ F_{11} \end{bmatrix} =
\begin{bmatrix} \phi_{11} & 0 \\ 0 & \dfrac{1}{\phi_{11}} \end{bmatrix}
\begin{bmatrix} 1 & 0 \\ \dfrac{1}{Z_1} & 1 \end{bmatrix}
\begin{bmatrix} \dfrac{1}{\phi_{21}} & 0 \\ 0 & \phi_{21} \end{bmatrix}
\begin{bmatrix} V_{21} \\ F_{21} \end{bmatrix}
$$

$$
=
\begin{bmatrix} \dfrac{\phi_{11}}{\phi_{21}} & 0 \\ \dfrac{1}{\phi_{11}\phi_{21}Z_1} & \dfrac{\phi_{21}}{\phi_{11}} \end{bmatrix}
\begin{bmatrix} V_{21} \\ F_{21} \end{bmatrix}. \tag{3.63}
$$

Combining the results of Equations (3.62) and (3.63) we can write the more general equation [23],

$$[V_i] = [\mathbf{Z}][F_i]_{i=1, M} \ z_{kl} = \sum_{j=0}^{N} \phi_{kj}\phi_{lj}Z_j, \tag{3.64}$$

where $[V_i]$ and $[F_i]$ are column matrices of order M and z_{kl} is an element in the $M \times M$ $\mathbf{Z}$ matrix $[\mathbf{Z}]$. An equivalent circuit based on Equation (3.64) is shown in Figure 3.21. The network of Figure 3.21, because it is so general, suffers from the problem of being somewhat unmanageable when used for computer analysis of a mechanical filter network. For this reason, we will look at some simpler resonator models.

Simplified Resonator Equivalent Circuits

Figure 3.22(a) shows a two-port, three-mode approximation of Figure 3.21 [24]. This model is an electrical equivalent circuit that approximates the effects of in-phase (symmetric) modes with the tuned circuit elements C_S and L_S and out-of-phase (antisymmetric) modes with circuit elements C_A and L_A. By in-phase we mean that the displacements, at the two ports, are in the same direction. The primary mode, which is in the passband region, is represented by C_0 and L_0. Since this network is still somewhat unwieldy for a computer analysis, we can replace it with the network of Figure 3.22(b) or the narrow-bandwidth equivalent circuit of Figure 3.22(c). The circuits of Figure 3.22(a) and (b) approximate both the impedance magnitude and impedance slope of the in-phase and out-of-phase modes at frequency ω_0, whereas the five-element circuit of Figure 3.22(c) approximates only the magnitude at ω_0.

The general equations for transforming the network of Figure 3.21 to Figure 3.22(b) are quite lengthy, but can be found in Reference [24]. A simple but important case is that of a single spurious mode-of-vibration near the filter passband frequency. Figure 3.23(a) shows the basic two-port resonator and Figure 3.23(b) shows its eight-element equivalent. The network of Figure 3.23(b) can be derived by making a series of transformations on the basic two-port. The transformations involve (1) modifying turns ratios by adding additional transformers, and (2) application of Norton transformations [see Figure 4.11(f)]. In this way, the turns ratios associated with each port are set equal and then eliminated.

The circuits of Figures 3.22 and 3.23 involve coupling wires attached at two points, in other words, two-port networks. If we are only dealing with a single wire, the equivalent circuit of Figure 3.22(a) can be reduced to the four-element circuit of Figure 3.24 by leaving the right-hand port open; this eliminates the transformers and allows us to combine the series-tuned circuits.

A second method of obtaining the simplified network of Figure 3.24 is to use Figure 3.21 to analyze a one-port acoustic resonator. From the poles and zeros of the analysis we can synthesize the electrical equivalent circuit. In other words, we approximate the impedance of the entire acoustic system near the

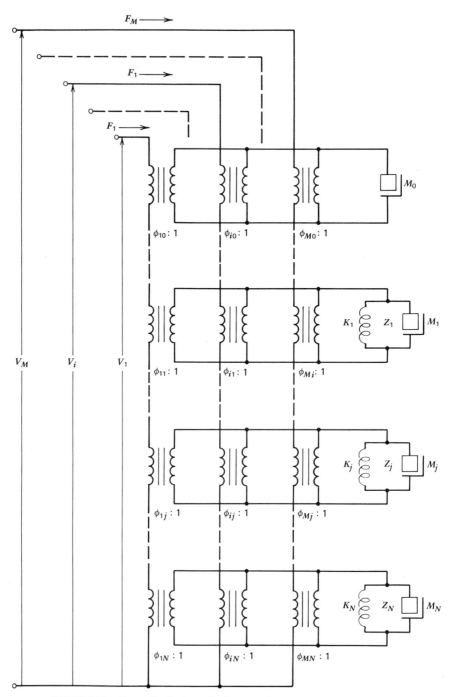

FIGURE 3.21. Generalized schematic diagram of an M-port, N-mode resonator.

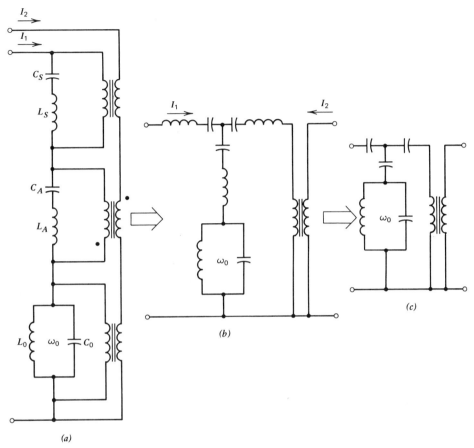

FIGURE 3.22. Simplified two-port electrical equivalent circuits derived from Figure 3.21.

desired resonance with a simple four-element electrical circuit. Let us illustrate this method with an example.

Example 3.6. Spurious modes of vibration generally have more effect on filters with large fractional bandwidths (B/f_0) than on small-bandwidth filters. Therefore, in analyzing telephone voice-channel filters designed at 50 kHz, we must take spurious modes of vibration into account [25]. In this example, we choose a bar resonator with a circular cross section, which vibrates in its fundamental mode at 50.9 kHz, and is wire coupled as shown in Figure 3.25.

The spurious modes that are taken into account are the rotation at zero frequency, the second torsional mode, and the third flexural mode. The first torsional mode and the second flexural mode are not considered because the coupling wire is located at the nodal points of these modes. The resonator bar is Thermelast 5409, which is used for both resonators and coupling wires (see Table 3.6). Its length ℓ is 1.673 cm and the radius a is 0.175 cm. Using the material constants from Table 3.6, we can calculate the flexural- and

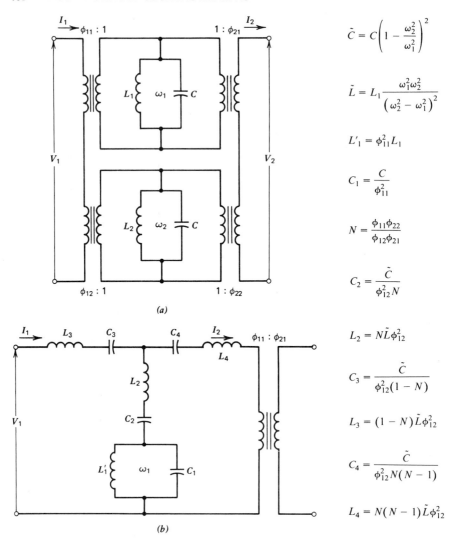

$$\tilde{C} = C\left(1 - \frac{\omega_2^2}{\omega_1^2}\right)^2$$

$$\tilde{L} = L_1\frac{\omega_1^2\omega_2^2}{\left(\omega_2^2 - \omega_1^2\right)^2}$$

$$L'_1 = \phi_{11}^2 L_1$$

$$C_1 = \frac{C}{\phi_{11}^2}$$

$$N = \frac{\phi_{11}\phi_{22}}{\phi_{12}\phi_{21}}$$

$$C_2 = \frac{\tilde{C}}{\phi_{12}^2 N}$$

$$L_2 = N\tilde{L}\phi_{12}^2$$

$$C_3 = \frac{\tilde{C}}{\phi_{12}^2(1 - N)}$$

$$L_3 = (1 - N)\tilde{L}\phi_{12}^2$$

$$C_4 = \frac{\tilde{C}}{\phi_{12}^2 N(N - 1)}$$

$$L_4 = N(N - 1)\tilde{L}\phi_{12}^2$$

FIGURE 3.23. Two-port, two-mode electrical equivalent circuits.

torsional-mode resonance frequencies from the equations of Figure 3.19. For example, for the fundamental and third flexural modes,

$$f_{f1} = \frac{(k_1\ell)^2}{2\pi\ell^2}\sqrt{\frac{EI}{\rho A}} = \frac{(k_1\ell)^2}{2\pi\ell^2}(v_{pL})\sqrt{\frac{a^2}{4}}$$

$$= \frac{4.73^2}{2\pi(1.673)^2}(4.595 \times 10^5)\left(\frac{0.175}{2}\right) = 50.9 \text{ kHz}$$

$$f_{f3} = \left(\frac{k_3\ell}{k_1\ell}\right)^2 f_{f1} = 280 \text{ kHz},$$

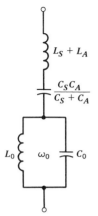

FIGURE 3.24. One-port electrical equivalent circuit taking into account symmetric and antisymmetric spurious modes of vibration.

and for the second torsional mode,

$$f_{t2} = \frac{n}{2\ell}\sqrt{\frac{G}{\rho}} = \frac{n}{2\ell}v_{ps} = \frac{2}{2(1.673)}(2.833 \times 10^5) = 169.3 \text{ kHz.}$$

The equivalent mass values also can be calculated from the equations of Figure 3.19. For instance, the equivalent mass of the first flexural mode is

$$M_{eqf1} = \frac{\rho A \ell}{4\left(\dfrac{u_1(\ell/2)}{u_1(0)}\right)^2} = \frac{M_{st}}{4\left(\dfrac{-0.608}{1.0}\right)^2} = 0.676 M_{st},$$

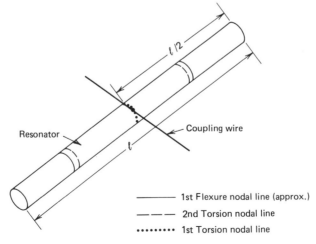

———————— 1st Flexure nodal line (approx.)

— — — — 2nd Torsion nodal line

•••••••• 1st Torsion nodal line

FIGURE 3.25. Flexural-mode resonator with extensional-mode coupling. Motion of the bar-wire weld junction is in the direction of the axis of the coupling wire.

and the second torsional mode is

$$M_{eq\,t2} = \left.\frac{\rho A\ell}{4\cos^2\left(\dfrac{n\pi x}{\ell}\right)}\right|_{x=\frac{\ell}{2}} = \frac{\rho A\ell}{4} = 0.25 M_{st}.$$

We will assume that $u_n(\ell/2)/u_n(0)$ is the same for all higher-order bending modes (this is an approximation, the amplitude ratio of the third flexure mode being 1.17 times that of the first mode). From our discussion of torsional resonators, the equivalent mass at zero frequency is $M_{st}/2$.

From the foregoing calculations, we can draw the lumped-element mechanical equivalent circuit and an electrical analogy as shown in Figure 3.26(a) and (b). By using the actual equivalent-mass values, we were able to eliminate the mechanical transformers shown in Figure 3.21.

The next step, in obtaining the four-element equivalent circuit shown in Figure 3.26(c), is a frequency vs. impedance analysis of the circuit of 3.26(b). Using a computer network-analysis program, the impedance zeros around the pole at 50.9 kHz were found to be at 37.5 kHz and 115.6 kHz. Using these

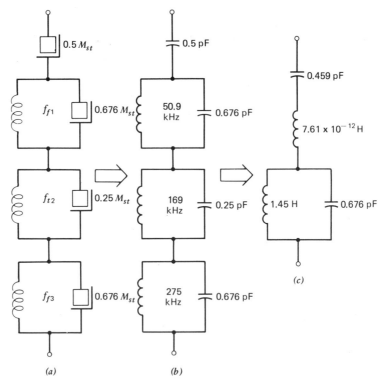

FIGURE 3.26. Equivalent circuits of the flexural-mode resonator of Figure 3.25: (a) mechanical equivalent circuit, (b) electrical analogy, and (c) four-element approximation of the electrical analogy.

frequencies, a Foster driving-point synthesis of the network was performed and the element values shown in Figure 3.26(c) were obtained. An analysis of this network showed very close correlation with the network of Figure 3.26(b) in the frequency region between the two impedance zeros.

RESONATOR MATERIALS

In Chapter 2 we briefly touched on the subject of using special iron-nickel alloys as resonator materials. These materials are useful because of the following three characteristics: a low temperature coefficient of resonance frequency, a high mechanical quality factor Q, and magnetostrictive properties. The extraordinary characteristic of the iron-nickel alloys, having nickel contents between 27 percent and 44 percent, is that the temperature coefficient of the modulus of elasticity is positive. In other words, the material becomes stiffer as the temperature increases. In the region slightly above 27 percent and slightly below 44 percent nickel, the small positive variation of Young's modulus, with changing temperature, cancels the positive temperature coefficient of thermal expansion. An example of how this works is to note that in the equation for the resonance frequency f of a longitudinal-mode bar resonator,

$$f = \frac{1}{2\ell}\sqrt{\frac{E}{\rho}} \, ,$$

that a positive variation of Young's modulus E cancels a positive change in length ℓ.

A problem with the basic iron-nickel alloys is that their temperature coefficients, near the zero crossing of the temperature-coefficient versus nickel-content curve, are extremely sensitive to variations in chemical content. For this reason, other elements are added in order to lower the temperature coefficient curve of Figure 2.9 to the point where the rounded peak is tangent to the zero axis. Under these conditions, small changes in composition have only a small effect on the maintenance of a zero temperature coefficient.

The prototype of these alloys is called *Elinvar*, an invention of C. E. Guillaume in the early 1920s. The word *Elinvar* is derived from "*el*astic *invar*iable." The original Elinvar contained 36 percent nickel, 12 percent chromium, and the rest iron, but most modern materials contain at least one other additive for finely adjusting the temperature coefficient of frequency.

In the case of Ni-Span C material, titanium is added to the iron, nickel, and chromium to produce a material with a controllable temperature coefficient. The alloy is melted while maintaining the composition shown in Table 3.5. The desired temperature-frequency characteristics are then obtained by using cold work (drawing) and heat treatment [26]. The final cold work, which follows an annealing step, is typically a 30 to 50 percent reduction in cross-sectional area and produces internal strains which push the temperature coefficient in the

Table 3.5. Mechanical Filter Resonator Materials

Characteristics	Symbol	Units	International Nickel Ni-Span C (Durinval-C) (Thermelast 4002)	Metalimphy Durinval	Metalimphy Elinvar	Vacuumschmelze Thermelast 5405	Vacuumschmelze Thermelast 4290 (5429)	Tokin TE-3	Sumitomo Sumi-Span EL-3
Longitudinal velocity	v_L	cm/s	4.9×10^5	$4.8-5.0 \times 10^5$	$4.6-4.8 \times 10^5$		$2.72-2.86 \times 10^5$	4.75×10^5	4.9×10^5
Shear velocity	v_s	cm/s	2.9×10^5					2.85×10^5	
Longitudinal mechanical Q	Q	—	20,000	> 12,000	> 11,000	16,000	15,000	10,000	> 15,000
Shear mechanical Q	Q	—	> 25,000			10,000	18,000	30,000	20,000 to 26,000
Young's modulus	E	dyn/cm²	$1.85-2.0 \times 10^{12}$	$1.7-2.0 \times 10^{12}$	1.85×10^{12}	1.9×10^{12}	1.9×10^{12}	1.81×10^{12}	1.93×10^{12}
Shear modulus	G	dyn/cm²	0.72×10^{12}			0.65×10^{12}	$0.61-0.68 \times 10^{12}$	0.67×10^{12}	
Poisson's ratio	μ	—	0.312			0.3	0.3		
Density	ρ	gm/cm³	8.0^a	7.92	8.15	8.3	8.3	8.1	
Temperature coefficient-frequency	TC_f	ppm	$-25°C$ to $+75°C$ 80	$+20°C$ to $+80°C$ ±1ppm/°C		$-20°C$ to $+80°C$ -84	$0°C$ to $+60°C$ 40	$0°C$ to $+50°C$ ±3ppm/°C	
Cold work condition		%	30	50		50	50		
Heat treatment temperature	T_{HT}	°C	500-650 (2 hours)	500		650	550	510-520	
Curie temperature	T_C	°C	165	140	180	175	220	130	
Chemical composition		%	Fe = 49 Ni = 42 Cr = 5.4 Ti = 2.5 Al = 0.5 Mn = 0.4	Ni = 42 Ti = 2 Cr + Mo = 5 Al = 1	Ni = 36 Cr = 9 W = 3 Fe = BAL	Fe = 53.2 Ni = 37 Cr = 8 Ti = 1 Be = 0.8	Fe = 50.5 Ni = 40 Be = 0.5 Mo = 9	Ni = 42.4 Cr = 5.5 Ti = 2.6 Fe = BAL	

^{a}An incorrect but commonly used value of 8.3 is used in the example problems.

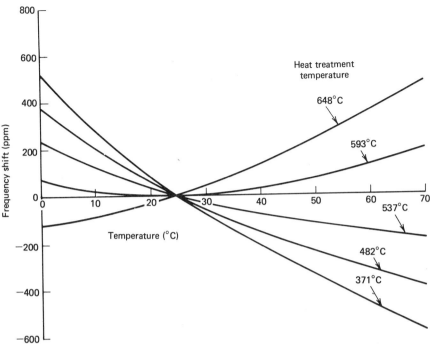

FIGURE 3.27. Frequency shift vs. operating temperature, as a function of the heat-treatment temperature (Ni-Span C).

negative direction. Following a stress-relief heat treatment near 375°C, heating the material to temperatures above 500°C causes precipitation of an intermetallic compound of titanium and nickel, which withdraws nickel from the matrix, causing the temperature coefficient to move in a positive direction, as shown in Figure 3.27. The results of the cold-work and heat-treatment process are shown in Figure 3.28 for Ni-Span C and Thermelast 5409 materials.

Thermelast 5409 is an iron-nickel alloy similar to Ni-Span C but one that makes use of molybdenum, rather than chromium, to reduce the sensitivity of the temperature coefficient of frequency to changes in nickel content [27]. Performing the function of titanium in Ni-Span C, beryllium is used in Thermelast 5409 to adjust the temperature coefficient of frequency by heat treatment. Also, a very important role of the titanium and beryllium is the improvement of both the frequency aging and the quality (Q) of Elinvar-type alloys. Aging rates are typically 0.1 ppm/week or, approximately, 25 Hz at 500 kHz over a period of 10 years. 5409 is also used as a coupling-wire material. Therefore its characteristics are shown in Table 3.6.

High mechanical Q can be achieved through a combination of the proper chemical composition, cold work, and heat treatment. Q's of torsional-mode resonators can be as high as 35,000, whereas the maximum Q values for longitudinal resonators is about 20,000. Although there is a wide variety of chemical compositions listed in Table 3.5, there is very little difference in most

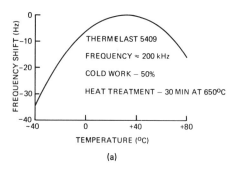

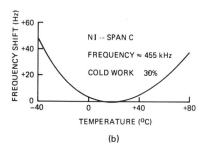

FIGURE 3.28. Temperature characteristics of resonators using constant-modulus materials. (*a*) Thermelast 5409 torsional-rod resonator, (*b*) Ni-Span C 902 flexural-mode disk.

of the physical characteristics, Q being an exception. Therefore, the choice of a resonator material is often based on criteria like weldability, machinability, consistency of material properties, and cost. Let's address the question of material consistency.

With Ni-Span C as an example, the problem of consistency involves variations both in an entire heat (batch) of material and variations from heat-to-heat. The variations within a heat are usually caused by the cold-work process and may involve cracks or voids in the material, or a variation of the percentage of cold work between the ends and the center of the rod. Variations from heat-to-heat most commonly involve differences in chemical composition, which, as with variations in cold work, cause changes in the required heat treatment. Variations of the amount of cold work, or of the heat-treatment temperature, cause changes in Young's modulus, which in turn results in varying resonator dimensions. If the variations in material properties are only on a batch-to-batch basis, then each batch of materials can be treated separately and adjustments can be made. These adjustments include heat-treatment temperature, resonator dimensions, coupling-wire size, and welding voltages. Variations within a batch, caused by the vendor or subsequent machining or heat treatment, are harder to deal with. The parts may have to be measured individually to determine the values of parameters such as the temperature coefficient or equivalent mass. From the foregoing discussion, it should be clear that consistency is an important aspect of material selection.

COUPLING ELEMENTS

The most common coupling method is the attachment of one resonator to another by means of one or more small-diameter wires. But this is not the only coupling means. Filters have been designed with mass coupling, shunt resonator couplers, and various cross sections that approximate simple wire coupling. Being limited in the amount of material we can discuss, we will look only at the coupling-wire element vibrating in extensional, torsional, and flexural modes.

Extensional-Mode and Torsional-Mode Coupling

The characteristics of a small-diameter coupling wire are similar to the characteristics of an electrical transmission line. As shown in Chapter 2, a wire vibrating in an extensional mode can be described in $\mathscr{A} \mathscr{B} \mathscr{C} \mathscr{D}$ matrix form by the acoustic equations

$$\begin{bmatrix} V_1 \\ F_1 \end{bmatrix} = \begin{bmatrix} \cos \alpha\ell & jZ_0 \sin \alpha\ell \\ \dfrac{j \sin \alpha\ell}{Z_0} & \cos \alpha\ell \end{bmatrix} \begin{bmatrix} V_2 \\ F_2 \end{bmatrix}, \qquad (3.65)$$

where ℓ is the length of the wire, Z_0 is the characteristic mobility $[1/A(\rho E)^{1/2}]$, α is the propagation constant ω/v_e, v_e is the velocity $(E/\rho)^{1/2}$ of an extensional wave, and V_1, V_2, F_1, and F_2 are the velocities and forces at the ends of the coupling wire.

Torsional- and extensional-mode lines are described by equations of the same type, but where $V \to \dot{\theta}$, $F \to \tau$, $A \to I_p$, and $E \to G$. I_p is the polar moment of inertia and G is the shear modulus. For a circular cross section, $I_p = \pi d^4/32$. Therefore, the matrix description of a torsional-mode coupling wire is

$$\begin{bmatrix} \dot{\theta}_1 \\ \tau_1 \end{bmatrix} = \begin{bmatrix} \cos \alpha\ell & jZ_0 \sin \alpha\ell \\ j\dfrac{\sin \alpha\ell}{Z_0} & \cos \alpha\ell \end{bmatrix} \begin{bmatrix} \dot{\theta}_2 \\ \tau_2 \end{bmatrix}, \qquad (3.66)$$

where $Z_0 = 32/[\pi d^4 (\rho G)^{1/2}]$; d is the diameter of the wire, α is the propagation constant $\omega/(\rho/G)^{1/2}$, and $\dot{\theta}$ and τ are the angular velocity and torque.

A general transmission-line model, corresponding to Equations (3.65) and (3.66) is shown in Figure 3.29(a). The variation of Y_a and Y_b, as a function of $\omega\ell/v$, is plotted in Figure 3.29(b).

The y parameters of the series and shunt arms are a function of the $\mathscr{D}$ and $\mathscr{B}$ matrix elements of the Equations (3.65) and (3.66). Because

$$y_{11} = \frac{\mathscr{D}}{\mathscr{B}} = \frac{1}{jZ_0} \cot \alpha\ell$$

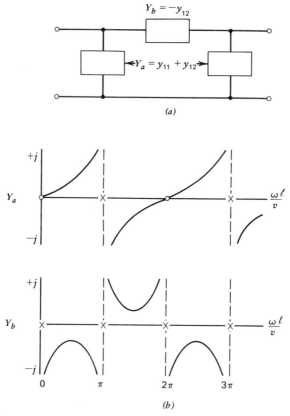

FIGURE 3.29. (a) General coupling-wire model for extensional and torsional modes and (b) variation of mechanical impedances Y_a and Y_b with the parameter $\omega\ell/v$.

and

$$y_{12} = -\frac{1}{\mathscr{B}} = j\frac{1}{Z_0 \sin \alpha\ell},$$

we can use a half-angle formula to obtain the equation

$$y_{11} + y_{12} = \frac{j}{Z_0} \tan \frac{\alpha\ell}{2}. \tag{3.67}$$

For extensional-mode coupling, when the coupling wire is less than one-eighth wavelength, the shunt arm can be replaced by a mass M_a equal to one-half of the static mass of the wire, and the series arm becomes a spring with stiffness AE/ℓ. In the case of short wire torsional-mode coupling, the shunt arm is an inertia $J_a = \pi d^4 \rho\ell/64$ (where d is the wire diameter) and the series arm is a spring with stiffness $K_b = \pi d^4 G/(32\ell)$. The extensional and torsional mode cases are shown schematically in Figure 3.30(a).

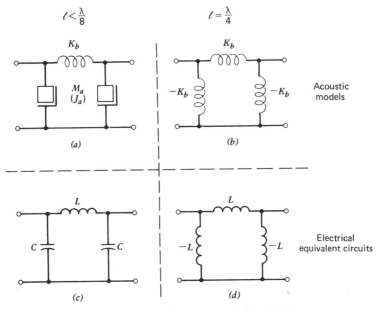

$\ell < \frac{\lambda}{8}$ $\ell = \frac{\lambda}{4}$

K_b K_b

M_a
(J_a) $-K_b$ $-K_b$ Acoustic
                                                           models

(a)   (b)

L L

C C $-L$ $-L$ Electrical
                                             equivalent circuits

(c)   (d)

FIGURE 3.30. Coupling-wire equivalent circuits.

Note in Figure 3.29(b) that the slope of the series arm Y_b is zero at $\omega\ell/v = \pi/2$. At this point the coupling is not sensitive to changes in wire length, which is a benefit in the manufacturing process. At $\omega\ell/v = \pi/2$, the coupling wire is one-quarter wavelength long and the $\mathscr{A}$ and $\mathscr{D}$ elements of the matrix Equations (3.65) and (3.66) are

$$\cos \alpha\ell = \cos\left(\frac{\omega\ell}{v}\right) = \cos\left(\frac{\pi}{2}\right) = 0$$

and

$$\sin \alpha\ell = \sin\left(\frac{\omega\ell}{v}\right) = \sin\left(\frac{\pi}{2}\right) = 1,$$

and the matrices become

$$\begin{bmatrix} \mathscr{A} & \mathscr{B} \\ \mathscr{C} & \mathscr{D} \end{bmatrix} = \begin{bmatrix} 0 & jZ_0 \\ \dfrac{j}{Z_0} & 0 \end{bmatrix}$$

$$\ell = \frac{\lambda}{4}. \tag{3.68}$$

For the extensional-mode quarter-wavelength case, since $Z_0 = 1/A(\rho E)^{1/2}$ and $v = (E/\rho)^{1/2} = \lambda f$, the characteristic mobility can be written as $Z_0 =$

$2\ell\omega/\pi AE$ and Equation (3.68) becomes

$$
\begin{bmatrix} 0 & jZ_0 \\ \dfrac{j}{Z_0} & 0 \end{bmatrix}
=
\begin{bmatrix} 0 & j\omega\left(\dfrac{2\ell}{\pi AE}\right) \\ \dfrac{j}{\omega\left(\dfrac{2\ell}{\pi AE}\right)} & 0 \end{bmatrix}
=
\begin{bmatrix} 0 & j\omega\left(\dfrac{1}{K}\right) \\ \dfrac{j}{\omega\left(\dfrac{1}{K}\right)} & 0 \end{bmatrix},
$$

$$(3.69)$$

where the stiffness $K = \pi AE/2\ell$. The corresponding torsional stiffness is $K = A^2 G/4\ell$. Equation (3.69) describes the equivalent circuit shown in Figure 3.30(b).

The electrical equivalent circuit elements shown in Figure 3.30(c) and (d) are used in conjunction with the electrical circuit designs in Chapter 4.

Flexural-Mode Coupling

The equations describing flexural-mode coupling are very complicated because they include the bending moment M and angular velocity $\dot\theta$ variables as well as forces and linear velocities. Treating the wire as a thin bar, we can write equations similar to those describing extensional and torsional transmission lines. From Reference [27], and Figure 3.31(a), we can write

$$
\begin{bmatrix} F_1 \\ M_1 \\ V_1 \\ \dot\theta_1 \end{bmatrix}
= \frac{1}{2}
\begin{bmatrix}
H_9 & -H_8(\alpha/\ell) & -H_7(K\alpha^3/j\omega\ell^3) & -H_{10}(K\alpha^2/j\omega\ell^2) \\
H_7(\ell/\alpha) & H_9 & H_{10}(K\alpha^2/j\omega\ell^2) & -H_8(K\alpha/j\omega\ell) \\
H_8(j\omega\ell^3/K\alpha^3) & H_{10}(j\omega\ell^2/K\alpha^2) & H_9 & -H_7(\ell/\alpha) \\
-H_{10}(j\omega\ell^2/K\alpha^2) & H_7(j\omega\ell/K\alpha) & H_8(\alpha/\ell) & H_9
\end{bmatrix}
\begin{bmatrix} F_2 \\ M_2 \\ V_2 \\ \dot\theta_2 \end{bmatrix},
$$

$$(3.70)$$

where:

$H_1 = S \cdot s$	$H_6 = S \cdot c + C \cdot s$	$S = \sinh\alpha$
$H_2 = C \cdot c$	$H_7 = s + S$	$s = \sin\alpha$
$H_3 = C \cdot c - 1$	$H_8 = s - S$	$C = \cosh\alpha$
$H_4 = C \cdot c + 1$	$H_9 = c + C$	$c = \cos\alpha$
$H_5 = S \cdot c - C \cdot s$	$H_{10} = c - C$	

$$
\alpha^4 = (\rho A/K)\omega^2\ell^4, \qquad \frac{K\alpha^2}{(j\omega\ell^2)} = \frac{(\rho AK)^{1/2}}{j}, \qquad K = EI,
$$

$$
\frac{\alpha^4}{\ell^2} = k^4 (\text{see } 3.28).
$$

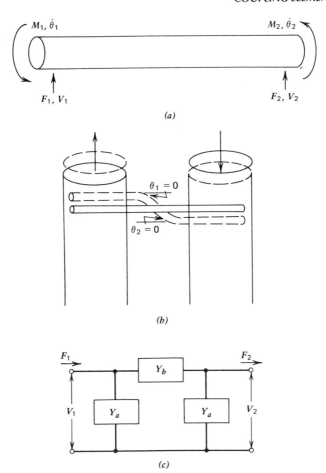

FIGURE 3.31. (a) Flexural-mode rod or wire, (b) flexural-mode coupling between two extensional-mode resonators, and (c) two-port mechanical equivalent circuit.

Equation (3.70) is a general equation which can be used to find the frequency equations (3.29) of the flexural-mode resonator, coupling-wire stiffness and equivalent mass, and in Chapter 5, equations for support wires. In this section, we consider only the special case shown in Figure 3.31(b), where the resonators vibrate in an extensional mode and the rotations of the wire at the connection points are equal to zero, that is, $\theta_1 = \theta_2 = 0$. Setting $\dot{\theta}_1 = \dot{\theta}_2 = 0$ in Equation (3.70) we obtain, after some manipulation,

$$
\begin{bmatrix} V_1 \\ F_1 \end{bmatrix} = \begin{bmatrix} H_6/H_7 & -j(\omega\ell^3/\alpha^3 K)H_3/H_7 \\ (1/-j)(\alpha^3 K/\omega\ell^3)(2H_1/H_7) & H_6/H_7 \end{bmatrix} \begin{bmatrix} V_2 \\ F_2 \end{bmatrix}.
$$

$$(3.71)$$

The $\mathscr{A}\mathscr{B}\mathscr{C}\mathscr{D}$ matrix elements of Equation (3.71) can be transformed to y parameters and then to the Y_a and Y_b elements of Figure 3.31(c), where

$$Y_a = j\left(\frac{K\alpha^3}{\omega\ell^3}\right)\left(\frac{H_6 - H_7}{H_3}\right)$$

$$Y_b = +j\left(\frac{K\alpha^3}{\omega\ell^3}\right)\left(\frac{H_7}{H_3}\right). \tag{3.72}$$

The coupling wire acts as a quarter-wavelength line when $H_6 = 0$. In this case, we see from Equation (3.72) that $Y_a = -Y_b$ and from Equation (3.71) that

$$\begin{bmatrix} V_1 \\ F_1 \end{bmatrix} = \begin{bmatrix} 0 & j(\omega\ell^3/\alpha^3 K)(H_7/2H_1) \\ j(\alpha^3 K/\omega\ell^3)(2H_1/H_7) & 0 \end{bmatrix} \begin{bmatrix} V_2 \\ F_2 \end{bmatrix}.$$

$$\tag{3.73}$$

A Summary of Coupling-Wire Equations

Corresponding to Figure 3.19, which is a summary of resonator equations, is Figure 3.32, which summarizes the coupling-wire equations derived in the previous sections. It is important that the flexural-mode equations be used only for the case where there is no rotation of the ends ($\dot{\theta} = 0$). Note that the extensional-mode equations and the torsional-mode equations are related by the transformations $E \leftrightarrow G$ and $A \leftrightarrow I_p = \pi d^4/32$. Examples of the use of these equations are found in the next chapter in the discussion of Figure 4.21, which is a matrix of resonators and coupling elements.

COUPLING-WIRE MATERIALS

As stated on the first page of this chapter, the compliances (reciprocal of stiffness) of the coupling wires determines the filter bandwidth. If a single wire is used, and its stiffness is in error by 5 percent, the bandwidth of the filter will be 5 percent too wide or too narrow. Therefore, our choice of coupling-wire material is very important.

Looking at Figure 3.32, we see that wire stiffness K_b is proportional to Young's modulus in the extensional and flexural cases, and proportional to the shear modulus in the case of torsional coupling. Also, the stiffness is proportional to the diameter raised to the second or to the fourth powers. Therefore, in looking for a good coupling-wire material, we should look for consistency and stability of both the elastic moduli and wire diameter. With regard to consistency, the wire's properties must be uniform along its length in order to

	SERIES ARM (b-arm)			SHUNT ARMS (a-arm)		
COUPLING-WIRE MODE	General Equation	Short Wire $<\lambda/8$	Quarter-Wave $(\lambda/4)$	General Equation	Short Wire $<\lambda/8$	Quarter-Wave $(\lambda/4)$
Extensional $Z_0 = \dfrac{1}{A\sqrt{\rho E}}$ $\alpha = \omega\sqrt{\dfrac{\rho}{E}}$	$Y_b = \dfrac{-j}{Z_0 \sin\alpha\ell}$	$Y_b = \dfrac{AE}{j\omega\ell}$ $K_b = \dfrac{AE}{\ell}$	$Y_b = \dfrac{\pi AE}{j2\omega\ell}$ $K_b = \dfrac{\pi AE}{2\ell}$	$Y_a = \dfrac{j\tan(\alpha\ell/2)}{Z_0}$	$Y_a = \dfrac{j\omega\rho A\ell}{2}$ $M_a = \dfrac{\rho A\ell}{2}$	$Y_a = \dfrac{\pi AE}{-j2\omega\ell}$ $K_a = -\dfrac{\pi AE}{2\ell}$
Torsional $Z_0 = \dfrac{32}{\pi d^4\sqrt{\rho G}}$ $\alpha = \omega\sqrt{\dfrac{\rho}{G}}$	$Y_b = \dfrac{-j}{Z_0 \sin\alpha\ell}$	$Y_b = \dfrac{\pi d^4 G}{j32\omega\ell}$ $K_b = \dfrac{\pi d^4 G}{32\ell}$	$Y_b = \dfrac{A^2 G}{j4\omega\ell}$ $K_b = \dfrac{A^2 G}{4\ell}$	$Y_a = \dfrac{j\tan(\alpha\ell/2)}{Z_0}$	$Y_a = \dfrac{j\pi\omega\rho d^4}{64}$ $J_a = \dfrac{\pi d^4\rho\ell}{64}$	$Y_a = \dfrac{A^2 G}{-j4\omega\ell}$ $K_a = -\dfrac{A^2 G}{4\ell}$
Flexural $(\dot\theta_0 = 0)$ $I = \dfrac{\pi d^4}{64}$ $\alpha^4 = \dfrac{\rho A\omega^2\ell^4}{EI}$	$Y_b = \dfrac{jEI\alpha^3 H_7}{\omega\ell^3 H_3}$	$Y_b = \dfrac{12EI}{j\omega\ell^3}$ $K_b = \dfrac{12EI}{\ell^3}$	$Y_b = -\dfrac{K\alpha^3 H_7}{j\omega\ell^3 H_3}$ $K_b = -\dfrac{K\alpha^3 H_7}{\ell^3 H_3}$	$Y_a = \dfrac{jEI\alpha^3(H_6 - H_7)}{\omega\ell^3 H_3}$	$Y_a = j\omega\left(\dfrac{\rho A\ell}{2}\right)$ $M_a = \dfrac{\rho A\ell}{2}$	$Y_a = \dfrac{K\alpha^3 H_7}{j\omega\ell^3 H_3}$ $K_a = \dfrac{K\alpha^3 H_7}{\ell^3 H_3}$

E = Young's modulus; G = Shear modulus; ρ = Density; ω = Frequency; H_3, H_7, H_6 [see Equation (3.70)]; A = Area

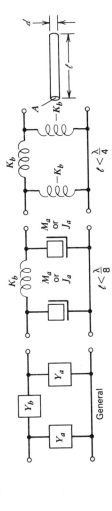

FIGURE 3.32. Equations for series and shunt arm values of two-port coupling-wire mechanical equivalent circuits.

TABLE 3.6.   Mechanical Filter Coupling-Wire Materials

			Materials		
Characteristics	Symbol	Units	Tokin TE-2	Thermelast 5409	Magnetics Inc. 56/44
Velocity	v_L	cm/s	4.8×10^5	$4.59\text{--}4.80 \times 10^5$	4.3×10^5
	v_s	cm/s		2.83×10^5	
Young's modulus	E	dyn/cm^2	1.89×10^{12}	1.79×10^{12}	1.54×10^{12}
Shear modulus	G	dyn/cm^2		0.653×10^{12}	0.60×10^{12}
Poisson's ratio	μ	—			.290
Density	ρ	gm/cm^3	8.1	8.16	8.2
Mechanical Q	Q	—	2.5×10^4	$> 1.5 \times 10^4$	
Heat treatment	—	—	695°C (1 hour)		
Curie temperature	T_c	°C	140	155	
Chemical composition	—	—		Fe(50.5), Ni(40), Be(0.5), Mo(9)	Fe(56) Ni(44)

maintain the correct inter-resonator coupling. Regarding stability, the temperature coefficient of the elastic modulus must be maintained constant; the variation of the diameter with temperature is only a second-order error.

A primary factor in the choice of a coupling-wire material is its weldability; will the welds between the wires and the resonators be consistent? Other mechanical factors are strength and the ability to be formed into headed wires or simply bent into various shapes for coupling or resonator-support purposes. If the coupling-wire stiffness can be adjusted by heat treatment, this is an additional bonus. This is possible with the precipitation-hardenable materials, such as Tokin TE-2 and Thermelast 5409, shown in Table 3.6.

COUPLED RESONATORS

In this section we look at the heart of mechanical bandpass filtering: the acoustic coupling of mechanical resonators. As a starting point, we go back to basic concepts and work with spring-mass mechanical elements. These concepts are our model for much of the remaining material in this book and are be helpful for understanding bandpass filtering, in general. Please take some extra time to learn this material well.

Spring-Mass Systems

Our starting point in dealing with spring-mass systems are the components themselves. We assume that the spring is massless and is linear, that is, it obeys Hook's law, $F = Kx$. The mass elements are assumed to be rigid and con-

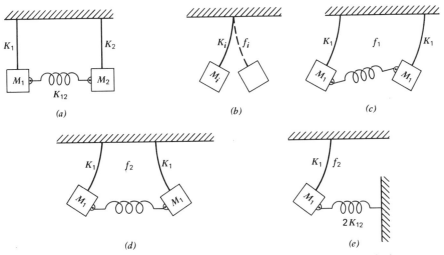

FIGURE 3.33. Coupled spring-mass resonators. (*a*) Resonators at rest, (*b*) single-resonator natural mode, (*c*) in-phase mode f_1, (*d*) out-of-phase mode f_2, and (*e*) single-resonator equivalent to the f_2 mode.

centrated at a single point. Our final assumption is that the vibration amplitude is small and therefore the masses move along a straight line.

Figure 3.33(*a*) shows a coupled spring-mass system composed of two resonators having stiffness and mass K_1, M_1, and K_2, M_2, and the coupling element of stiffness K_{12}. The leaf springs K_1 and K_2 are rigidly attached to the support in a cantilever fashion, whereas the coil spring K_{12} is attached to the masses by a hinge. The motion of the resonator *i* is shown in exaggerated form in Figure 3.33(*b*). Since we are assuming small displacements, we can neglect the rotation of the mass and other second-order effects. Referring to Equation (2.1), the resonance frequency of a simple spring and mass is

$$f_i = \frac{1}{2\pi} \sqrt{\frac{K_i}{M_i}} \, . \tag{3.74}$$

If we next consider two resonators which are tuned to the same frequency f_1, are located side-by-side, and vibrate as shown in Figure 3.33(*c*), we can see that the spring K_{12} can be removed without affecting the vibration of the system. In other words, if we displace the two masses an equal distance in the same direction and let go, there will be no forces applied to spring K_{12} and the resonators will vibrate at frequency f_1 and only at frequency f_1. If the masses are displaced an equal distance in opposite directions, as shown in Figure 3.33(*d*), the spring K_{12} will be stretched, but there will be no displacement at the exact center of the spring. When we release the masses, the spring is alternately compressed and expanded during each cycle, but the center of the spring remains stationary, that is, it becomes a node. In addition, the frequency of the coupled resonators remains constant at a value f_2, as if the system were composed of the single mass and two springs K_1 and $2K_{12}$ shown in Figure 3.33(e). The coupling spring, in Figure 3.33(e), is one-half the length of the original and therefore has twice the stiffness $2K_{12}$. The second resonance frequency is a function of the sum of the two spring stiffnesses and therefore

$$f_2 = \frac{1}{2\pi}\sqrt{\frac{K_1 + 2K_{12}}{M_1}}. \qquad (3.75)$$

The two-resonator circuit has two natural resonances, f_1 and f_2. Each resonance can be excited by applying the proper initial conditions. Any other initial conditions will cause the two natural resonances to be excited simultaneously, but no other resonances will be present.

It is important to note that at frequency f_1 the displacements of the masses are in phase, whereas at frequency f_2, the displacements are 180 degrees out of phase. As additional resonators and coupling elements are added, we find that at the lowest natural resonance of the system, all resonators are vibrating in phase and at the highest natural resonance, all adjacent resonators are vibrating 180 degrees out of phase. Also, we find that there is an additional natural resonance for each resonator added. Resonances between the highest and lowest natural frequencies have displacement patterns where a resonator may be in phase, out of phase, or stationary with respect to its neighbor. Examples of three- and four-resonator systems are shown in Figure 3.34. Note that the natural frequencies are related to the number of phase changes. For instance, the lowest mode has no differences in phase between adjacent resonators, whereas, at the highest frequency mode, all of the resonators are out of phase.

In the next chapter we show how resonators and coupling wires can be combined to realize a specified filter bandwidth. Extensive use is made of the resonator equivalent mass equations from Figure 3.19 and the series and shunt-arm coupling elements of Figure 3.32.

Resonator and Coupled-Resonator Measurements

Usually there is neither the time nor the tools available to calculate all of the relevant frequencies and displacements (amplitudes) of resonators and coupled resonators that are chosen for a particular filter design. Therefore, we must build parts and make measurements in order to obtain exact natural frequency numerical values. Displacement measurements can also be made, in order to identify the various desired modes and spurious modes of vibration. Both measurements employ the methods of resonator excitation described in Chapter 6. Methods of detecting the resonance frequencies are also described in Chapter 6.

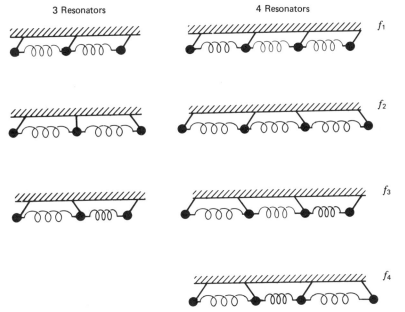

3 Resonators   4 Resonators

f_1

f_2

f_3

f_4

FIGURE 3.34. Phase relationships between resonators in three- and four-resonator coupled systems.

Identification of the resonance modes is accomplished by studying their displacement amplitudes. Using the excitation methods of Chapter 6 we can measure the resulting displacements by means of fine powders or holographs [28], or by simply probing the resonator with a pointed low-Q tool and observing the varying amplitude of the detected response.

Figure 3.35 shows photographs of the vibration patterns of a two-identical-resonator wire-coupled pair. The lycopodium powder, which is sprinkled evenly across the disk, is forced at the resonance frequency to move to the nodal points by the flexural-mode displacements of the disk. Figure 3.35(a), (c), and (d) correspond to the natural modes of both the individual two disks and the coupled pair of disks. The patterns of the three nodal-line responses in Figure 3.35(c) and (d) are poorly defined in the center of the disk because of the smaller displacement amplitudes in that region. The coupled response shown in Figure 3.35(b) [f_2 of Figure 3.33(d)] is a linear combination of the modes shown in Figure 3.35(a) and (c). The mode in Figure 3.35(d) has no effect on the coupled mode because the nodal points are aligned with the coupling wires. Therefore the mechanical impedance (equivalent mass) of the mode in Figure 3.35(d) is infinite. A more detailed discussion of coupled disk modes is found in Reference [23].

Fine powder is used to identify nodal point locations of composite or complex resonators, so the support wires can be attached to the exact nodes in order to maintain a high resonator Q.

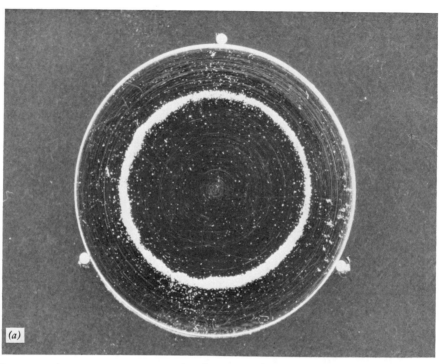

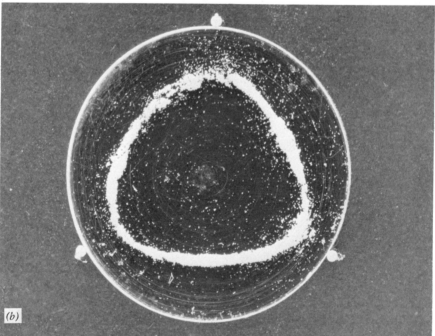

FIGURE 3.35. Photographs of lycopodium powder nodal patterns of a two-identical-disk set. The coupling-wire orientation is symmetrical. The modes are (a) the single-circle mode, $f_1 = 60.06$ kHz; (b) a combination single-circle and low-frequency three-diameter mode, $f_2 = 63.38$ kHz;

126

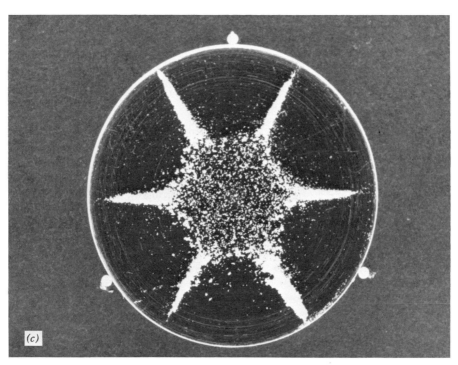

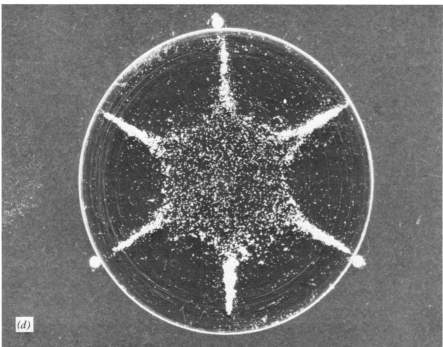

FIGURE 3.35 (continued). (c) the low-frequency three-diameter mode, $f_3 = 77.05$ kHz; and (d) the high-frequency three-diameter mode, $f_4 = 78.05$ kHz.

REFERENCES

1. J. W. S. Rayleigh, *Theory of Sound*. New York: Dover, 1945.

2. P. M. Prache and H. Ernyei, "Vibration des barreaux homogènes en régime harmonique," *Cables et Transm.*, **28**(4), 328–366 (Oct. 1974).

3. W. P. Mason, *Electromechanical Transducers and Wave Filters*. New York: Van Nostrand, 1942.

4. J. Näser, "Exakte Berechnung der Biegeresonanzen rechteckiger und zylindrischer Stäbe," *Hochfrequenztech. u. Elektroak.*, **74**, 30–36 (1965).

5. S. Sugawara, M. Konno, and T. Watanabe, "The equivalent mechanical network of a bar vibrating in flexure derived from Timoshenko's beam theory," *Trans. IECE*, **61-A**(9), 903–909 (Sept. 1978).

6. S. P. Timoshenko, *Vibration Problems in Engineering*. New York: Van Nostrand, 1955.

7. H. Deresiewicz and R. D. Mindlin, "Axially symmetric vibrations of a circular disk," *J. Appl. Mech.*, **75**, 86–88 (1953).

8. R. L. Sharma, "Equivalent circuit of a resonant, finite, isotropic, elastic circular disk," *J. Acoust. Soc. Amer.*, **28**, 1153–1158 (Nov. 1956).

9. M. Onoe and T. Yano, "Analysis of flexural vibrations of a circular disk," *IEEE Trans. Sonics Ultrason.*, **SU-15**, 182–185 (July 1968).

10. J. Klovstad, E. M. Frymoyer, and R. A. Johnson, "Calculation of frequency, equivalent mass, and amplitude of thick vibrating disks," Computer Program SA159, Rockwell International, Newport Beach, California.

11. J. Näser, "Kurze Darlegung der Theorie elektromechanischer Filter mit Plattenresonatoren," *Hochfrequenztech. u. Elektroak.*, **71**, 123–132 (Oct. 1962).

12. S. P. Lapin, "Electromechanical filters," *Radio and Television News, Radio-Electronic Eng. Sec.*, **50**, 9–11 ff (Dec. 1953).

13. H. J. McSkimin, "Theoretical analysis of modes of vibration for isotropic rectangular plates having all surfaces free," *Bell System Tech. J.*, **23**, 151–177 (Apr. 1944).

14. I. Lucas, "Plattenförmige elektromechanische Filter mit Dämpfungspolen," *Arch. Elek. Übertr.*, **17**, 230–236 (May 1963).

15. M. Onoe, "Contour vibrations of isotropic circular plates," *J. Acoust. Soc. Amer.*, **28**(6), 1158–1162 (Nov. 1956).

16. Y. Kagawa and T. Yamabuchi, "Finite element simulation of two-dimensional electromechanical resonators," *IEEE Trans. Sonics Ultrason.*, **SU-21**, 275–283 (Oct. 1974).

17. Y. Kagawa, "Analysis and design of electromechanical filters by finite element technique," *J. Acoust. Soc. Amer.*, **49**(5, part 1), 1348–1356 (May 1971).

18. Y. Tomikawa, S. Sugawara, and M. Konno, "Finite element analysis of displacement at base portion of a quartz crystal tuning fork," *IEEE Trans. Sonics Ultrason.*, **SU-26**(3), 359–361 (May 1979).

19. K. Nagai and M. Konno, *Electro-Mechanical Resonators and Applications*. Tokyo: Corona, 1974.

20. H. Nakamura, Y. Tomikawa, and M. Konno, "Finite element simulation of U-type tuning fork," *Trans. IECE*, **59-A**(1), 21–27 (1976).

21. M. Konno, S. Oyama, and Y. Tomikawa, "Equivalent electrical networks for transversely vibrating bars and their applications," *IEEE Trans. Sonics Ultrason.*, **SU-26**(3), 191–201 (May 1979).

22. E. J. Skudrzyk, "Vibrations of a system with a finite or an infinite number of resonances," *J. Acoust. Soc. Amer.*, **30**, 1140–2152 (Dec. 1958).

23. R. A. Johnson, "Electrical circuit models of disk-wire mechanical filters," *IEEE Trans. Sonics Ultrason.*, **SU-15**, 41–50 (Jan. 1968).

24. A. Günther and K. Traub, "Precise equivalent circuits of mechanical filters," *IEEE Trans. Sonics Ultrason.*, **SU-27**(5), 236–244 (Sept. 1980).

25. F. Künemund, "Channel filters with longitudinally coupled flexural mode resonators," in *Modern Crystal and Mechanical Filters*, D. F. Sheahan and R. A. Johnson, Eds. New York: IEEE Press, 1977.

26. M. Börner, "Magnetische Werkstoffe in elektromechanischen Resonatoren und Filtern," *IEEE Trans. Magn.*, **MAG-2**, 613–620 (Sept. 1966).

27. M. Konno and H. Nakamura, "Equivalent electrical network for the transversely vibrating uniform bar," *J. Acoust. Soc. Amer.*, **38**, 614–622 (Oct. 1965).

28. S. Sugawara and M. Konno, "Spurious responses and their suppression in the mechanical filter with flexure mode resonators," *Trans. IECE*, **61-A**(11), 1122–1127 (1978).

Chapter Four

CIRCUIT
DESIGN METHODS

This chapter is written for a broad audience, which includes filter users, component specialists, filter designers, and people in manufacturing. One of the major thrusts in the chapter is to take the mystery out of mechanical filters by showing similarities to electrical filters as well as by showing how they differ. By understanding the similarities and the differences we can make use of general filter concepts, tables, curves, and equations not developed specifically for mechanical filters. If you are an equipment designer you can use universal curves to make estimates of mechanical filter performance before interacting with the mechanical filter designer. If you are a component specialist who understands LC filters, then you already know a great deal about mechanical filters. So, let's start demystifying the mechanical filter by looking at its similarities to electrical filters.

A zero is a zero! What this means is that an attenuation zero (in other words, a transmission pole or natural frequency) is produced by a resonant circuit whether the circuit is electrical or mechanical. Both circuits have poles and zeros and can be described by mathematical functions that approximate the filter's actual frequency response. If the topology of the two circuits is the same, the mathematical form of the transfer function will be the same. Therefore, whether we are oriented toward electronics or, conversely, toward mechanics, we are only a change in variables away from the less understood field.

Similarities also include the first-order approximation, that both electrical and mechanical circuits can be described by a network of lumped elements or distributed transmission-line elements. Furthermore, these elements can be described in terms of Q, resonance frequency, turns ratios, characteristic impedances, and so on. Electrical and mechanical circuits can be treated alike mathematically, but when realistic element values are used, we find large differences in the response characteristics. Therefore, let us look at some of the unique characteristics of mechanical filter elements.

The two most important differences between electrical and mechanical elements are the high Q's and extraordinary stabilities of mechanical resonators. In both cases, there is roughly a two-orders-of-magnitude difference. With regard to Q, the mechanical filter can usually be treated as an infinite-Q device, which allows the user and designer to estimate selectivity by using already published lossless curves. Because of the excellent stability, a good first-order estimate is that the filter simply does not drift in frequency with either time or temperature.

A second difference between mechanical and electrical filters is that the mechanical filter has rigid constraints on its topology. Most mechanical filters are ladder networks composed of wire-coupled resonators, which sometimes utilize wire over-coupling (bridging), whereas LC topologies are completely general. In addition, it is desirable from a manufacturing standpoint that the mechanical resonators be the same size and that the filter be as physically symmetrical as possible.

Parasitic effects force the mechanical filter engineer to use special design techniques. Examples of these parasitics are unwanted resonator modes of vibration, support-structure modes, input-to-output coupling, and acoustic air resonances. These effects, along with the difficulty in isolating the resonator, coupling wire, and transducer in a welded structure (the welded wire becomes part of the resonator, etc.), often make it more feasible to simply build a filter, make adjustments, measure the modified frequencies and couplings, and repeat this process until a satisfactory design is obtained [1].

Finally, because of unique mechanical filter manufacturing processes and varying stabilities of the electrical, electromechanical, and mechanical elements, special requirements must be put on the design. This is best illustrated in the case of intermediate-band filters discussed in the next section.

GENERAL SYNTHESIS CONCEPTS

Network synthesis is usually described as the mathematical approximation of an ideal response and then the network realization of the mathematical function. We most often follow these steps with network transformations and sometimes add an optimization step. But before starting the design of the electrical model and subsequently the mechanical realization, it is first necessary to make measurements or calculations of dimensions, equivalent masses, and electromechanical coupling of the proposed filter elements. As is shown in Figure 4.1, the synthesis, when time allows, is then followed by a sensitivity analysis.

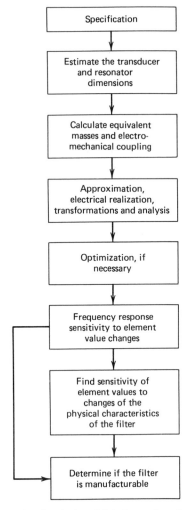

FIGURE 4.1.   Procedure for the electrical-model design and mechanical realization verification.

The Transfer Function

The approximation part of the network synthesis involves finding a mathematical equation that describes the filter response, which can be used to obtain the element values of the electrical equivalent circuit. This equation, which is the ratio of two polynomials, is usually written in terms of the complex frequency variable $s = \sigma + j\omega$. As an example, the equation describing the network of Figure 4.2 can be written as

$$H(s) = \frac{E(s)}{P(s)} = C \frac{s^{20} + e_{19}s^{19} + e_{18}s^{18} + \cdots + e_0}{s^5\left(s^8 + p_6s^6 + p_4s^4 + p_2s^2 + p_0\right)}. \tag{4.1}$$

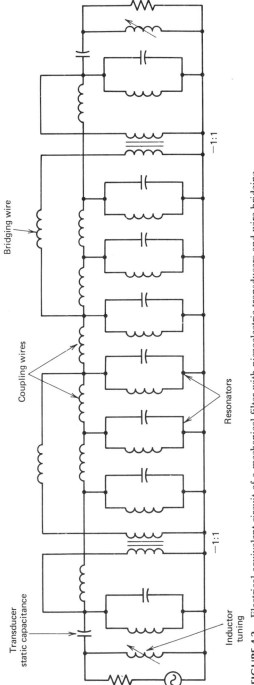

Bridging wire

Coupling wires

Transducer
static capacitance

Inductor
tuning

Resonators

−1:1

−1:1

FIGURE 4.2. Electrical equivalent circuit of a mechanical filter with piezoelectric transducers and wire bridging.

The roots of the numerator of Equation (4.1) are the natural frequencies of the filter and are located in the left half of the s-plane. The denominator has five roots at zero frequency and eight along the $j\omega$-axis. The difference between the number of numerator roots (20) and denominator roots ($5 + 8 = 13$) is the number of poles (7) of $H(s)$ at infinite frequency.

The attenuation function $H(s)$ is the most often used transfer function in the design of passive filters, whereas its reciprocal, the gain function, is most commonly used when designing amplifiers or active filter networks. Therefore, the natural frequencies of a mechanical filter will be expressed as zeros of the attenuation function $H(s)$ rather than as poles of the gain function $1/H(s)$.

The attenuation function can also be written as

$$H(s)H(-s) = 1 + K(s)K(-s) = 1 + \frac{F(s)F(-s)}{P(s)P(-s)}, \qquad (4.2)$$

where $K(s)$ is the characteristic function $F(s)/P(s)$ [2]. The roots of $K(s)$ are complex conjugates and are usually located inside the passband on the $j\omega$-axis, or in the right- and left-half planes. Figure 4.3(a) shows the upper-half plane roots of both $H(s)$ and $K(s)$ for the filter of Figure 4.2. Figure 4.3(b) shows the filter response as a function of frequency. Note that some of the roots of $F(s)$ are off the $j\omega$-axis, and if we include the lower-half plane roots, they have quadrantal symmetry (i.e., they are symmetrical about the origin). These roots have been called zero-quads and are associated with so-called intermediate-band filter designs; more specifically, they relate to the input and output electrical tuned circuits [3].

Intermediate-Bandwidth Designs

The intermediate-bandwidth or *intermediate-band* design is based on the use of coils or capacitors to tune the transducer capacitance or inductance. This tuning adds little to the filter selectivity, but instead allows the filter to be terminated with a lower electromechanical coupling-coefficient transducer than is necessary in the case where no tuning is used. This improves the frequency stability of the transducer resonator. Conversely, care is also taken to maintain a high enough coupling coefficient so that the sensitivity of the filter response to changes in the electrical tuning is not too great. Therefore, the intermediate-band filter is designed to balance the stability of the electrical and mechanical terminating elements. Let's look at these concepts in greater detail.

What we will do is analyze the input circuit of Figure 4.2 and see how it relates to the pole-zero diagram of Figure 4.3. Figure 4.4 is a more detailed version of 4.2 and includes separate parallel-tuned circuits to represent the ceramic and the metal portions of a composite electromechanical transducer (examples of composite transducers are shown in Figure 2.18). R_t is the user-supplied terminating resistance (i.e., the output resistance of the circuit

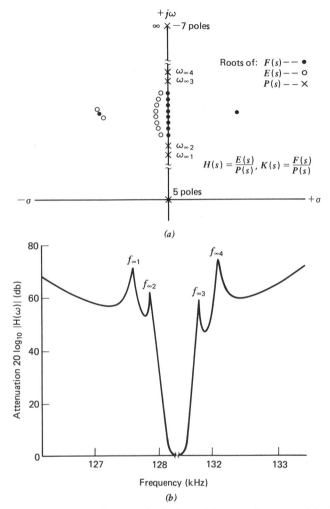

FIGURE 4.3.   Characteristics of the network of Figure 4.2. (*a*) *s*-plane natural frequencies and attenuation poles. (*b*) Amplitude versus frequency response.

that drives the filter). L_0 is the resonating inductance (which has been placed in series with C_0 and R_t for ease of understanding the basic principles), C_0 is the static capacitance of the transducer, and L_c and L_m, and C_c and C_m correspond to the respective compliances and masses of the ceramic and metal portions of the transducer. Showing the mechanical elements as two simple parallel-tuned circuits helps in the understanding of the basic intermediate-band concepts, but the network is only a rough model of the actual device.

Let's start our analysis with a hypothetical example of a filter similar to that of Figure 4.3 that has all of its natural frequencies located close to the passband. In this example, we will increase the value of C_0 of Figure 4.4,

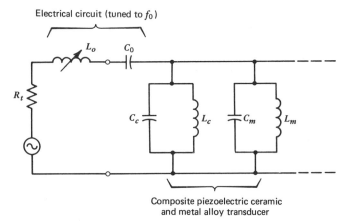

FIGURE 4.4.   Intermediate-band mechanical-filter terminating circuit.

keeping all other element values the same except L_0, which decreases in order to maintain a constant electrical circuit tuning frequency of $f_0 = 1/2\pi(L_0 C_0)^{1/2}$. Increasing C_0 without increasing C_c is equivalent to increasing the electromechanical coupling coefficient of the transducer material $(k^2_{em_t} = C_0/C_c)$ and the coupling coefficient of the composite transducer $[k^2_{em_{tm}} = C_0/(C_c + C_m)]$. As C_0 becomes large, one of the natural frequencies [roots of $E(s)$ in Figure (4.3)] moves in the $-\sigma$ direction and out of the cluster near the passband. The limiting case, though not physically possible, is when C_0 becomes infinite, L_0 goes to zero, and the natural frequency goes to infinity because the tuned circuit disappears. Conversely, if the electromechanical coupling decreases to the point where the reactances of C_0 and L_0 become very large, then the natural frequencies associated with this resonator will move horizontally in close to the $j\omega$-axis. Therefore, in summary, as the electromechanical coupling increases, the natural frequencies associated with the electrical termination move away from the $j\omega$-axis of the passband, and conversely, as the coupling decreases, the natural frequencies move closer to the passband. This means that if we increase the coupling by increasing the size of the relatively unstable ceramic transducer material, we will decrease the sensitivity of the circuit to electrical element value changes by moving the roots of $E(s)$ associated with the electrical tuned circuits away from the passband. The challenge to the designer is finding the composite transducer design that results in a minimization of passband ripple variations due to both electrical and mechanical resonator instabilities.

Narrow-Bandwidth Designs

The narrow-bandwidth or *narrowband* designs are characterized by the absence of tuning coils or capacitors. The tuning coils can be eliminated if the

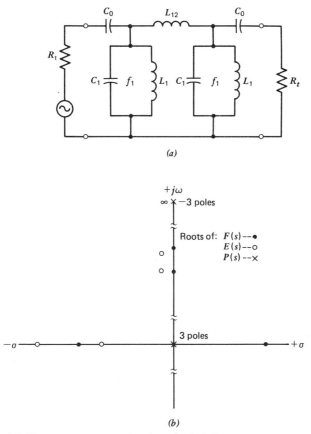

FIGURE 4.5. (a) Two-resonator narrowband mechanical-filter equivalent circuit. (b) s-plane pole-zero diagram.

composite transducer has a high enough electromechanical coupling-coefficient to satisfy the demands of the specified filter bandwidth. Let's look at a typical circuit and then its s-plane pole-zero locations.

Although high selectivity filters similar to that shown in Figure 4.2 have been designed without coils, most narrowband applications involve single resonators or two-resonator filters like that shown in Figure 4.5(a). The untuned capacitance C_0 leads to characteristic-function zeros [roots of $F(s)$] and natural frequencies [roots of $E(s)$] on the σ-axis. These are shown in Figure 4.5(b). Rather than choose values of $F(s)$ on the σ-axis and then synthesize the network, usually the roots appear as the result of narrow-band-width parallel R-C to series R-C transformations that follow the synthesis (see Figure 4.37). The natural frequencies near the passband are not affected by small changes in C_0, as in the tuned case, but this is at the expense of needing a high coupling-coefficient transducer resonator, with its greater instability of the mechanical resonance frequency f_1 (the coupling coefficient requirements for the various filter types are shown in Figure 4.34).

Wide-Bandwidth Designs

On the other end of the bandwidth spectrum is the *wideband* design where the electrical tuning circuits become the end resonators in a "conventional" filter design. An example of what is meant by a conventional filter is an eight-pole (eight-resonator) 0.1 dB equal-ripple design. Designed as a wideband filter, the realization has two electrical end resonators and six mechanical resonators. The eight natural frequencies form an ellipse similar to that in Figure 4.3, but there are no other zeros on the σ-axis, as in the narrowband case, or further out in the complex plane, as with the intermediate-band design. Because the electrical elements play an equal part with the mechanical circuit in providing selectivity, the filter is very sensitive to electrical element-value changes. Fortunately, the high sensitivity is somewhat offset by the fact, that in moderate-ripple filters the response is not quite as sensitive to end-resonator changes as to interior resonator changes. In addition, the sensitivity is inversely proportional to the filter bandwidth, so in wideband designs, electrical tuned-circuit sensitivity becomes less of a problem. In spite of these positive balances, the two-orders-of-magnitude difference between electrical and mechanical resonator stabilities usually precludes the use of the wideband design for stable, low-ripple communication filters.

Attenuation Poles

The filter of Figure 4.2 uses bridging elements to realize the finite $j\omega$-axis attenuation poles shown in Figure 4.3(*b*). If the phase inverters were not used with the bridging inductors (wires of the mechanical circuit), the attenuation poles would have moved from the $j\omega$-axis to the complex plane. This would result in the possibility of delay equalization, but as a practical matter, this is seldom done.

Alternate circuits that can be used to realize attenuation poles on the $j\omega$-axis include simple electrical capacitor input-to-output bridging, parallel-ladder configurations, and shunt resonators in ladder networks. The last configuration is based on the principle that attenuation poles are produced by ladder network shunt-arm impedance zeros (i.e., shorts at the pole frequencies). Although this is a common method of realizing attenuation poles in *LC* filters, attempts at designing manufacturable mechanical filters using this idea have usually failed. We therefore concentrate on bridging and parallel-ladder (twin-tee) circuits. In the next section we look at methods of realizing the element values corresponding to various network configurations.

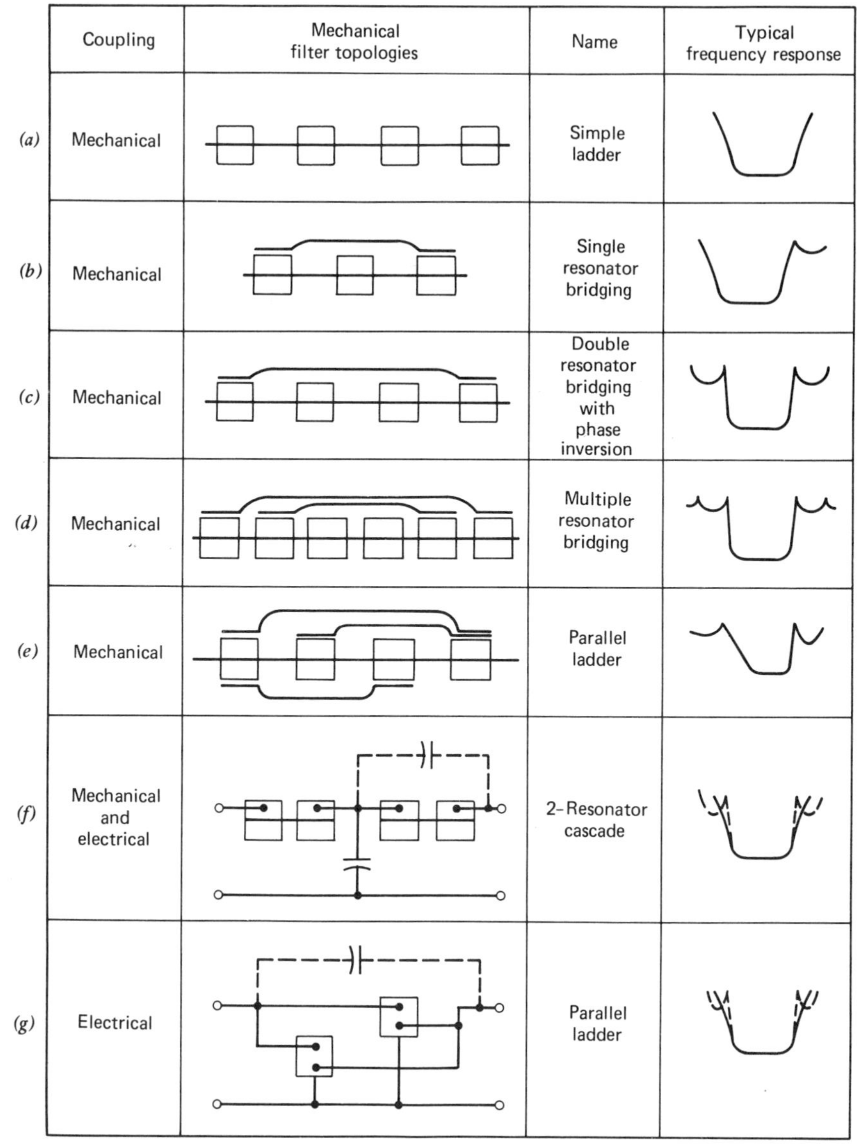

	Coupling	Mechanical filter topologies	Name	Typical frequency response
(a)	Mechanical		Simple ladder	
(b)	Mechanical		Single resonator bridging	
(c)	Mechanical		Double resonator bridging with phase inversion	
(d)	Mechanical		Multiple resonator bridging	
(e)	Mechanical		Parallel ladder	
(f)	Mechanical and electrical		2-Resonator cascade	
(g)	Electrical		Parallel ladder	

FIGURE 4.6. Mechanical filter topologies and their typical frequency response curves. [Adapted, with permission, from *IEEE Trans. Circuits and Syst.*, **CAS-22** (2), 69–89 (Feb. 1975). © 1975 IEEE.]

139

Electrical Equivalent Circuit Realizations

Although methods can be found to realize most electrical network configurations, certain of these are simply not practical—or at least the practical ones have not yet been invented. We therefore confine ourselves to the types of circuits shown in Figure 4.6. These networks can be mechanical or electrical; the blocks correspond to mechanical resonators or parallel LC circuits, and the connecting lines correspond to coupling wires or inductors. In all cases, we will use the mobility analogy described in Chapter 2.

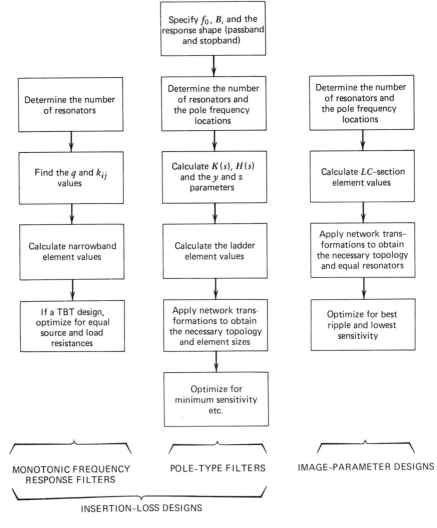

FIGURE 4.7.   Mechanical-filter equivalent-circuit design flow chart.

Figure 4.7 describes methods of realizing the various network topologies. The two "insertion-loss" methods involve exact synthesis calculations for obtaining a prescribed frequency response. The image-parameter method is not exact, but it is simple and has a great deal of flexibility with regard to realizing a variety of network configurations.

Ladder Networks Having Monotonic Response Shapes

In this section we look at ladder networks that produce attenuation poles only at zero and infinite frequencies. These filters have a monotonic response curve in the stopband region, that is, the slope does not change sign. With regard to the passband, we consider both equal-ripple (Chebyshev) and monotonic (from the center frequency) amplitude response curves.

Equal-Ripple Passbands. Chebyshev polynomials, which describe an equal-ripple amplitude response, have the important characteristic of providing the maximum stopband attenuation for a prescribed value of passband response variation. An aspect of designs based on these polynomials that is very important to mechanical filter designers is that the resultant circuits are physically symmetrical. This means that the source and load resistances are equal, the first and last resonator frequencies and couplings are equal, and so on, throughout the filter. In addition, under the lossless condition of infinite-Q resonators the insertion loss of these symmetrical filters is equal to zero. These characteristics also apply to the maximally flat amplitude response (Butterworth) polynomials, which are limiting cases of the Chebyshev polynomials.

Let us now look at a procedure for designing physically symmetrical, equal-ripple filters. Figure 4.8(a) shows an electrical equivalent circuit of an n-resonator filter; the parallel-tuned circuits correspond to mechanical resonators, and the series-arm inductors correspond to coupling wires. The element values of the equivalent circuit can easily be calculated from tables of normalized Q's (q_1, q_n) and coupling values ($k_{j,j+1}$). Tables of the q and k values are found in References [4] and [5]. We note from the tables that in the lossless case the terminating q's are equal, and there is mirror-image symmetry of the k values about the center of the filter.

A starting point in the element-value calculation is to choose $L_{12} = 1.0$, or some other arbitrary number, and proceed according to Equations 4.3 to 4.7.

$$L_{j,j+1} = \frac{k_{12}L_{12}}{k_{j,j+1}} \qquad\qquad j = 1,2,\ldots,n-1 \quad (4.3)$$

$$L_j = L_1 = L_{12}k_{12}\frac{B}{f_0} \qquad\qquad j = 1,2,\ldots,n \quad (4.4)$$

$$L'_j = \frac{1}{1/L_{j-1,j} + 1/L_j + 1/L_{j,j+1}} \qquad\qquad j = 1,2,\ldots,n \quad (4.5)$$

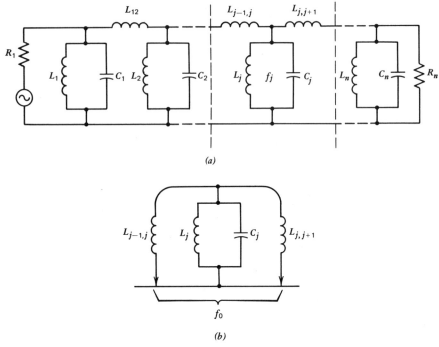

FIGURE 4.8. (*a*) Electrical equivalent circuit of a monotonic-stopband mechanical filter, and (*b*) the resultant circuit after shorting adjacent resonators.

$$C_j = \frac{1}{(2\pi f_0)^2 L_j'} \qquad j = 1, 2, \ldots, n \tag{4.6}$$

$$R_1 = R_n = \frac{q_1}{2\pi B C_1} \tag{4.7}$$

In Equation (4.4) the filter bandwidth B is the 3 dB bandwidth in Hz, and f_0 is the center frequency, also in Hz. Equations (4.5) and (4.6) are related to the fact that each resonator that is formed by shorting to ground the nodes adjacent to the jth node must be resonant at f_0. Therefore, in Figure 4.8(*b*) we show that the capacitor C_j is adjusted to tune at f_0 with the parallel combination of the coupling inductors and the resonator inductance. The first and last resonators are shunted by a single coupling inductor during tuning.

General Passband Shapes.   In this section we look at circuit designs based on mathematical equations other than those having equal-ripple or maximally flat amplitude in the passband. Examples of these functions are equal-ripple phase, maximally-flat delay, Gaussian, and transitional Butterworth-Thompson (TBT). A problem in designing these filters is that they are generally not realizable as symmetrical networks. The two-resonator filter is an exception, but at the expense of additional insertion loss. Symmetrical designs with three or more resonators can only approximate the foregoing functions. Because physical symmetry is important in the manufacturing process, we must look at a procedure for obtaining symmetrical designs that are approximations to the desired ideal general passband shape. The procedure is as follows. First estimate k and q values from curves or tables of k and q versus passband ripple. Then analyze a filter based on these values and compare its response to the ideal amplitude or delay response shapes. (The ideal amplitude or delay response shapes can be found from the nonsymmetrical network realization or the mathematical function.) Then by iteration adjust the k and q values to obtain the best approximation to the ideal response shapes.

This procedure is not only simple in concept, but it is easy to implement because symmetry restricts us to varying only one-half of the couplings. Therefore, in a seven-resonator filter we need to vary only three couplings and the termination.

The two-resonator filter, although an exception to the rule that physical symmetry and nonequal-ripple passbands are not compatible, is so commonly used that it deserves separate attention. We base our analysis on the TBT designs, which include the maximally flat amplitude and maximally flat delay responses [6].

Figure 4.9(a) shows the upper-half s-plane natural frequencies of a two-resonator TBT filter. The amplitude and delay responses are a function of the equal angles θ. When $\theta = 30$ degrees, the amplitude response is rounded and the delay is flat, as is shown in Figure 4.9(b). In the same figure, the maximally flat amplitude response is shown with its corresponding double-peaked delay response. Detailed amplitude, delay, and pulse response curves are shown in Reference [6] for angles of 27 to 48 degrees. Because q and k values are not given in the preceding reference, they are shown in Table 4.1; included is the insertion loss at the center frequency f_0. Element values for the various response shapes can be calculated by using Equations (4.3) to (4.7).

Ladder Networks Having Attenuation Poles

The design of mechanical filters with attenuation poles that are at other than zero and infinite frequency is considerably more difficult than designing monotonic-stopband-response filters. With regard to the electrical equivalent circuit, tabular values of the k's and q's are not generally available for the most common networks realizable with mechanical elements. Therefore, the filters must be designed from basic equations involving $K(s)$, $H(s)$, and the y and z parameters, as shown in Figure 4.7. Because mechanical filters are narrow-bandwidth devices, special frequency transformations are necessary for preserving accuracy if the filter is designed as a bandpass filter [7]. An alternative is to start with a conventional lowpass filter design and transform to a bandpass filter. This method will work if we want the bandpass filter to have frequency symmetry, that is, if we want the attenuation poles on the $j\omega$-axis to be spaced symmetrically about the center frequency. If a dissymmet-

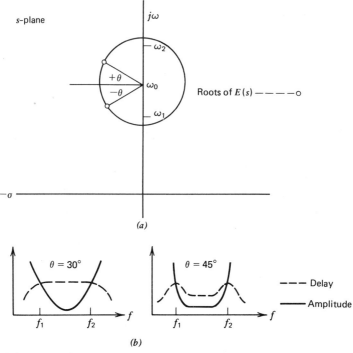

FIGURE 4.9. (*a*) *s*-plane roots, and (*b*) frequency response curves for monotonic stopband filters.

rical frequency response is desired, a lowpass synthesis with constant-reactance elements can be used. Regardless of the method we choose, the final electrical network must correspond to a realizable and manufacturable mechanical circuit of the types shown in Figure 4.6(*b*)–(*d*).

By the use of examples, we will explore circuits that can be used to realize each of the bridged networks shown in Figure 4.6. Let us start with a filter that contains acoustic bridging of a single resonator.

TABLE 4.1.   Normalized Q and Coupling Values As Well As Center Frequency Loss for Two-Resonator Filters Having Various Root Locations and Bandpass Response Shapes

Filter Type or TBT θ Values	Gaussian	25°	30° (Bessel)	35°	40°	45° (Butter-worth)	50°	3 dB Ripple (Chebyshev)	Narrow-band Image Parameter
q_1, q_2	0.770	0.815	0.908	1.03	1.20	1.41	1.70	3.13	1.0
k_{12}	0.532	0.572	0.637	0.678	0.701	0.707	0.703	0.779	1.0
Insertion loss at f_0 (dB)	3.0	2.3	1.2	0.54	0.13	0	0.13	3.0	0

Example 4.1.   Single Resonator Bridging.   Figure 4.10 shows the electrical equivalent circuit of an eight-resonator mechanical filter that in two sections employs bridging across a single resonator. This design is a conventional-symmetric type with a quad of characteristic-function zeros. The zero-quad, in this particular design, is used to reduce the sensitivity of the filter to variations in the ferrite-transducer end resonators, which are wire-coupled to the first and last metal resonators. An additional quad of zeros could have been added, to take into account the electrical circuit termination. In this case we simply chose to insert a series-resonance *LC* circuit to represent the transducer inductance and resonating capacitance. The synthesis was carried out using the computer program FILSIN [8]. This program makes use of the *z*-transformation [7] for generating, first the *y* and *z* parameters shown in Figure 4.7 and then the ladder element values. Based on measurements made on mechanical parts (resonators and wires), specific ratios of element values were chosen before the synthesis was performed. For instance, the interior (metal) resonators have equal capacitance (mass) as do the end resonators (the ferrite transducer bars). The ratio of the transducer equivalent mass to the metal-resonator equivalent mass is 0.14.

Since FILSIN is a general purpose synthesis program, it is necessary to transform the two five-element attenuation-pole-producing sections into their bridged equivalent circuit. This transformation is in two parts and is shown in Figure 4.11(*a*). Figure 4.11 also includes transformations that will be used in subsequent sections. The frequency response of the filter has the same general shape as that shown in Figure 7.25.

When mechanical phase inversion is possible, attenuation poles can be realized on the low-frequency side of the filter passband by making use of the transformations of Figure 4.11(*b*) or the circuits in Reference [9]. If attenuation poles are needed on both sides of the passband, we can use the double-resonator bridging section with phase inversion that was shown in Figure 4.6(*c*).

Example 4.2.   Double Resonator Bridging.   In this example, we will look at the design of an eleven-resonator filter employing bridging across two resonators, in each of two pole-producing sections. Figure 4.12 shows the initial electrical design, and after a series of transformations, the final electrical equivalent circuit. Although this design is at 240 kHz, it is typical of highly selective filters designed for radio applications at 200, 450, 455, and 500 kHz. Because the filter has an odd number of resonators, it is realized as a conventional-symmetric filter, whereas an equal-ripple design having an even number of resonators would have an antimetric realization. The antimetric filter has the input-output relationships:

$$\frac{z_{11}}{y_{22}} = \frac{z_{22}}{y_{11}} = R_1 R_2, \qquad (4.8)$$

where the *z*'s and *y*'s are the open- and short-circuit input and output impedance parameters and R_1 and R_2 are the source and load resistances. In

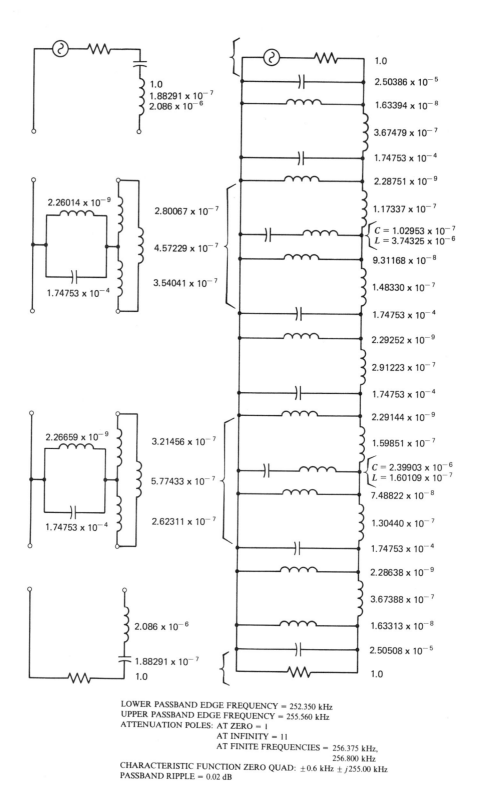

FIGURE 4.10. Electrical equivalent circuit of an eight-resonator mechanical filter with two finite-frequency attenuation poles in the upper stopband.

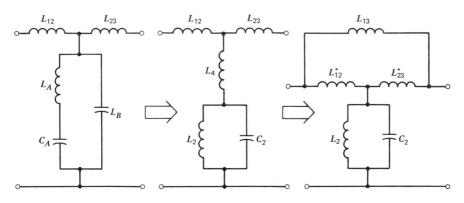

1. $L_4 = L_A L_B / (L_A + L_B)$
2. $L_2 = L_B^2 / (L_A + L_B)$
3. $C_2 = C_A [1 + (L_A / L_B)]^2$
4. $A2 = L_{12} L_{23} + L_{12} L_4 + L_{23} L_4$

5. $L_{13} = A / L_4$
6. $L_{12}^* = A / L_{23}$
7. $L_{23}^* = A / L_{12}$

(a)

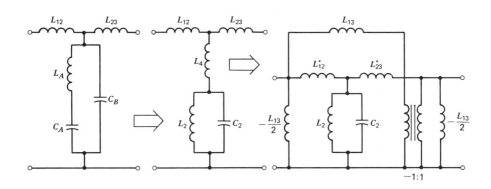

1. $C_4 = C_B + C_A$
2. $L_2 = L_A / [1 + (C_B / C_A)]^2$
3. $C_2 = C_B (1 + C_B / C_A)$
4. $L_4 = -1 / (\omega_0^2 C_4)$

5. $A = L_{12} L_{23} + L_{12} L_4 + L_{23} L_4$
6. $L_{13} = -A / L_4$
7. $L_{12}^* = A / L_{23}$
8. $L_{23}^* = A / L_{12}$

(b)

FIGURE 4.11. (a) Ladder-to-bridged-ladder transformations (upper stopband attenuation pole). (b) Ladder-to-bridged-ladder transformations (lower stopband attenuation pole).

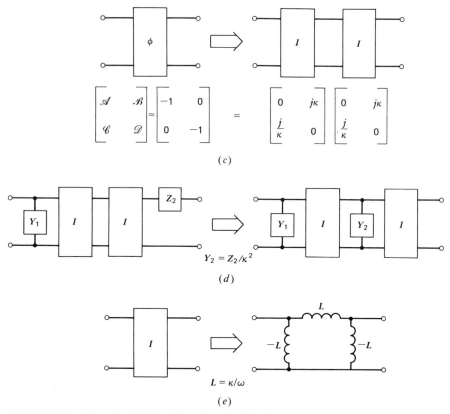

FIGURE 4.11 (continued). (c) Transformation of a phase inverter to a cascade of two impedance inverters. (d) Impedance-to-admittance transformation. (e) Lumped-element equivalent circuit representation of an impedance inverter.

the symmetric case, $z_{11} = z_{22}$, $y_{11} = y_{22}$, and $R_1 = R_2$. Although it may not be obvious, the addition of two impedance inverters, like those shown in Figure 4.11(c), one of which is used to invert one-half of the antimetric network, results in the conversion of the antimetric network into a symmetric network [see the transformation of Figure 4.11(d) and the conversion of the inverter to a lumped-element equivalent circuit in Figure 4.11(e)]. In this section we are using the term *symmetric* to mean that $z_{11} = z_{22}$, and so on, rather than the more restrictive sense that the network must also have physical symmetry about the center. If the attenuation poles are realized in pairs, where the outermost pole in the upper stopband and the outermost pole in the lower stopband are in one-half of the filter and the innermost poles are in the other half of the filter, then the final network will be "close" to being physically symmetrical, as can be seen by studying the element values of Figure 4.12(a). An additional restriction on the final network of Figure 4.12 is that the attenuation poles in each half of the filter have geometric symmetry about the

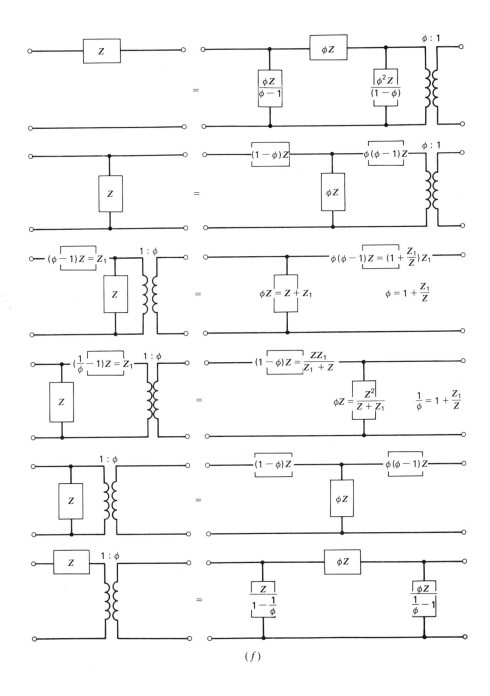

FIGURE 4.11 (continued). (f) Norton transformations.

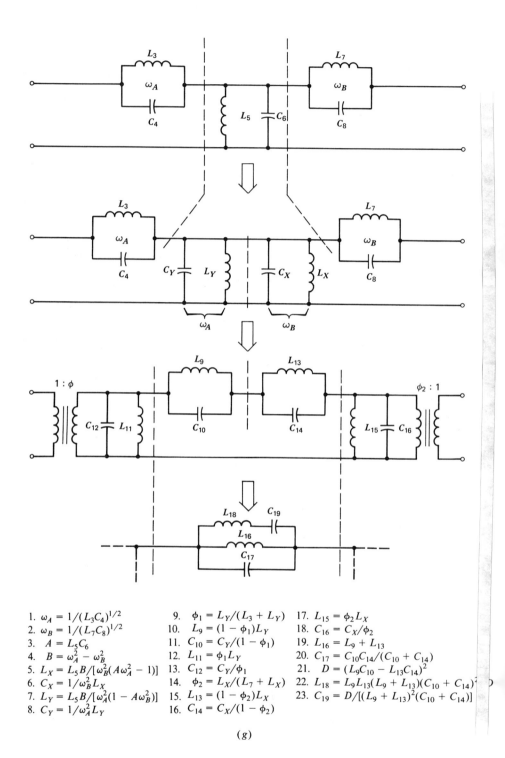

1. $\omega_A = 1/(L_3 C_4)^{1/2}$
2. $\omega_B = 1/(L_7 C_8)^{1/2}$
3. $A = L_5 C_6$
4. $B = \omega_A^2 - \omega_B^2$
5. $L_X = L_5 B/[\omega_B^2(A\omega_A^2 - 1)]$
6. $C_X = 1/\omega_B^2 L_X$
7. $L_Y = L_5 B/[\omega_A^2(1 - A\omega_B^2)]$
8. $C_Y = 1/\omega_A^2 L_Y$

9. $\phi_1 = L_Y/(L_3 + L_Y)$
10. $L_9 = (1 - \phi_1)L_Y$
11. $C_{10} = C_Y/(1 - \phi_1)$
12. $L_{11} = \phi_1 L_Y$
13. $C_{12} = C_Y/\phi_1$
14. $\phi_2 = L_X/(L_7 + L_X)$
15. $L_{13} = (1 - \phi_2)L_X$
16. $C_{14} = C_X/(1 - \phi_2)$

17. $L_{15} = \phi_2 L_X$
18. $C_{16} = C_X/\phi_2$
19. $L_{16} = L_9 + L_{13}$
20. $C_{17} = C_{10}C_{14}/(C_{10} + C_{14})$
21. $D = (L_9 C_{10} - L_{13} C_{14})^2$
22. $L_{18} = L_9 L_{13}(L_9 + L_{13})(C_{10} + C_{14})^2/D$
23. $C_{19} = D/[(L_9 + L_{13})^2(C_{10} + C_{14})]$

(g)

FIGURE 4.11 (continued). (g) Bandpass ladder transformations of a filter having upper and lower stopband attenuation poles.

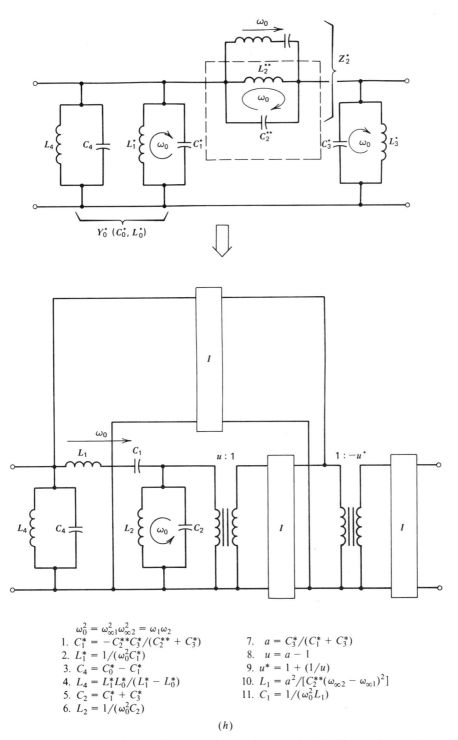

$$\omega_0^2 = \omega_{\infty 1}^2 \omega_{\infty 2}^2 = \omega_1 \omega_2$$

1. $C_1^* = -C_2^{**} C_3^* / (C_2^{**} + C_3^*)$
2. $L_1^* = 1/(\omega_0^2 C_1^*)$
3. $C_4 = C_0^* - C_1^*$
4. $L_4 = L_1^* L_0^* / (L_1^* - L_0^*)$
5. $C_2 = C_1^* + C_3^*$
6. $L_2 = 1/(\omega_0^2 C_2)$

7. $a = C_3^* / (C_1^* + C_3^*)$
8. $u = a - 1$
9. $u^* = 1 + (1/u)$
10. $L_1 = a^2 / [C_2^{**}(\omega_{\infty 2} - \omega_{\infty 1})^2]$
11. $C_1 = 1/(\omega_0^2 L_1)$

(h)

FIGURE 4.11 (continued). (h) Transformation of (g) to a bridged-ladder topology based on work of B. Kohlhammer and H. Schüssler.

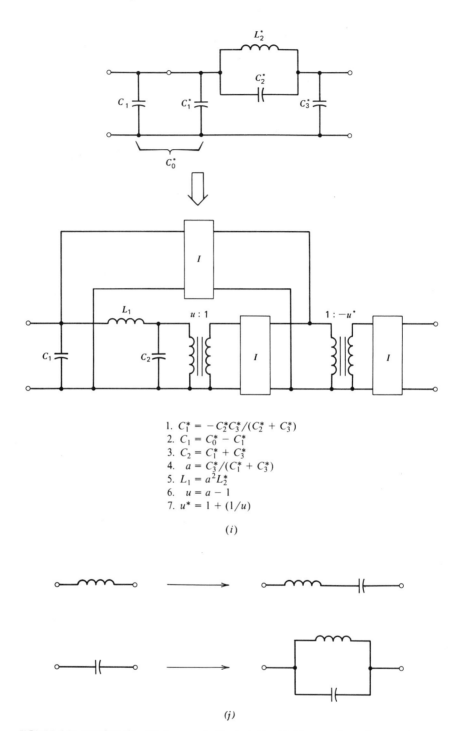

1. $C_1^* = -C_2^* C_3^* / (C_2^* + C_3^*)$
2. $C_1 = C_0^* - C_1^*$
3. $C_2 = C_1^* + C_3^*$
4. $a = C_3^* / (C_1^* + C_3^*)$
5. $L_1 = a^2 L_2^*$
6. $u = a - 1$
7. $u^* = 1 + (1/u)$

(i)

(j)

FIGURE 4.11 (continued). (i) Lowpass ladder-to-bridged-ladder transformation. (j) Lowpass-to-bandpass element transformation.

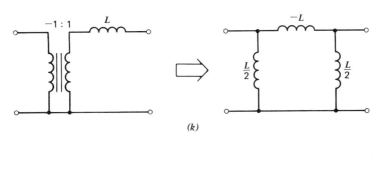

(k)

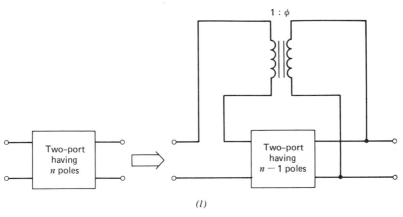

(l)

FIGURE 4.11 (continued). (k) A special case of the Norton transformation. (l) Cauer transformation.

center frequency, that is, the square root of the product of the pole frequencies is equal to the center frequency.

The condition of having an upper-stopband/lower-stopband pair of poles in each half of the filter allows us to use the Norton transformation of Figure 4.11(f) to make the network transformation shown in Figure 4.11(g). The geometric symmetry condition makes it possible to apply the transformation of Figure 4.11(h) to the final network of Figure 4.11(g). In other words, the transformation of Figure 4.11(h) can be used in the bandpass case when the attenuation poles are symmetrical about the center frequency. If we choose to start with a lowpass network, we can make the transformation shown in Figure 4.11(i) and then convert the bridged circuit to its bandpass equivalent by making use of the well-known lowpass-to-bandpass transformations shown in Figure 4.11(j). An example of the use of lowpass element values is described in Reference [10]. In this paper, explicit equations for the coupling coefficients between resonators, in terms of the lowpass element values, are shown for the case of nine resonators.

The next step in the filter design of Figure 4.12 is to use the impedance inverters of Figure 4.11(d) to convert the series-arm series-tuned circuits of

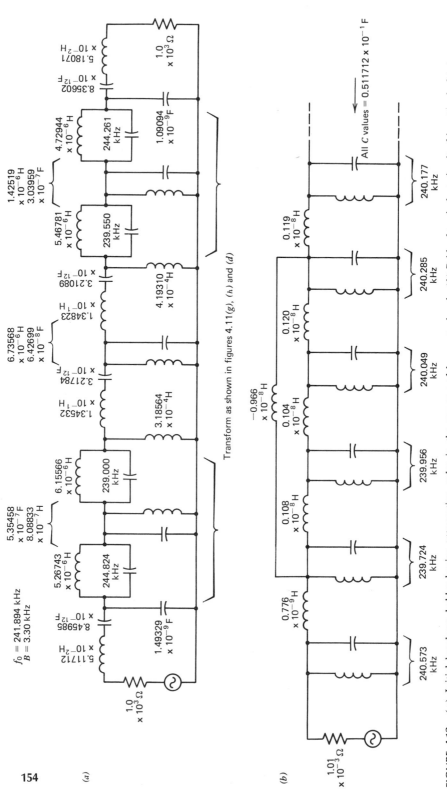

FIGURE 4.12. (a) Initial bandpass ladder having attenuation poles in the upper and lower stopbands. (b) Bridged equivalent circuit of input end of (a).

154

(a)

(b)

4.12(a) and 4.11(h) into shunt-arm parallel-tuned circuits. The inverter imped-ance values are chosen to create equal capacitance (equivalent mass) resona-tors, as shown in the final circuit of 4.12(b). The negative bridging inductance can be converted to a positive inductance and phase inverter by means of the transformation of Figure 4.11(k). Methods of converting the end circuits to circuits that include the transducer and tuning elements will be discussed later in this chapter.

Multiple-Resonator Bridging. The simple ladder to a bridged-ladder net-work transformations of Figure 4.11(h) and (i) are based on the more general Cauer transformation (plus inverters) shown in Figure 4.11(l). This transfor-mation shows that a two-port lowpass network containing n attenuation poles is equivalent to a series-parallel connection of an ($n - 1$) pole two-port and an ideal transformer. Successive applications of this transformation results in a pole-free ladder circuit with n bridging branches. In the bandpass case, each lowpass attenuation pole becomes a bandpass attenuation pole in each stop-band. This results in the circuit configuration and response shown in Figure 4.6(d). Treatment of an even more general network where the bridging wires are crossed is shown in Reference [11].

Summary of Bridging Network Design. Table 4.2 gives us a feeling for the various methods used to obtain bridging-wire electrical equivalent circuits. Most of the methods are also applicable to microwave filter design, and in fact, some were primarily motivated by this application. Lee's work was applied to crystal filter design, but can be used in the design of narrowband mechanical filters with capacitance bridging. Yano's parallel-ladder realization is   dis-cussed in the following section.

At this point, I must suggest caution in first obtaining an electrical design and then rushing into the lab to build it. This is what I did with my first bridging-wire design where I used two single-disk bridging sections in a five-disk filter. I could obtain only a single attenuation pole. When I tried a six-disk filter, where the bridging was from resonator one to resonator three and four to six, it worked, in the sense that I was able to vary the bridging and obtain two distinct poles [13]. My problem was that I did not use an adequate model; this was explained to me 17 years later [23]. What was clear to me was that I could not end one bridging wire and start another on the same resonator. Therefore, we are limited in our bridging wire designs to a maxi-mum of $n/2$ attenuation poles in the double resonator bridging case, where n is the number of resonators.

This example is typical of what is found when attempting to realize the first-order electrical model. Through experience, the effect of parasitics can be reduced or can be included in the design. An example of using parasitic modes of vibration is in the design of disk-wire single sideband filters, where the coupling-wire orientation is used to control the selectivity of the carrier-frequency side of the filter [24].

TABLE 4.2. Summary of Bridging Techniques

SYNTHESIS PAPER (References are in Brackets)	Frequency Plane		Starting Point			Method				Frequency Response			Final Topology		
	Lowpass	Bandpass	Polynomial	Ladder	Bridged Topology	Image Parameter	Insertion Loss	Optimization	Equivalent Network Transformation	Symmetric	Nonsymmetric	Delay Correction	Single-Wire Bridging	Multiple-Wire Bridging	Parallel Ladder
Börner [12]	×				×	×			×	×		×	×		
Johnson [13]		×		×		×			×		×		×		
Kohlhammer & Schüssler [14]	×			×					×	×		×	×		
Kohlhammer [15]	×		×				×			×		×		×	
Künemund [16]	×	×		×					×	×			×		
Wittmann et al. [17]	×		×				×			×	×	×		×	
Temes [18]		×	×				×				×	×		×	
Günther [19], [20]		×			×			×			×	×		×	
Yano et al. [21]		×		×					×		×			×	
Lee [22]		×		×					×	×			×		
Cucchi & Molo [11]		×		×					×	×	×			×	

[a]Adapted, with permission, from *Proc. 1975 IEEE ISCAS*, April 1975, Boston, pp. 313–316. ©1975 IEEE.

Parallel-Ladder Networks

In this section we look at acoustically and electrically coupled parallel-ladder networks having a single resonator in each of the ladder circuits. In the acoustic case, the parallel-ladder is bracketed by resonators imbedded in a larger ladder network. The electrically coupled parallel-ladder is simply a two-resonator filter. Parallel-ladders provide the flexibility inherent in lattice designs, for instance, quartz-crystal lattice filters. This flexibility allows the designer to realize a general-frequency set of attenuation poles, that is, the designer is freed from being limited by frequency symmetry. The electrically coupled case is important where there are spurious-mode or structural limitations caused by using acoustic coupling.

Electrically Coupled Resonators. In this section, we derive the equivalent circuit of the electrically coupled parallel-ladder network from the ladder network of Figure 4.8 (where $n = 2$). We first derive equations for the monotonic stopband response case and then show how variations in electromechanical coupling and transducer capacitance can result in upper and lower stopband attenuation poles. Although the following discussion may seem tedious, it is typical of numerous other mechanical filter derivations and illustrates a number of important transformations.

We derive the design equations from the circuit of Figure 4.8 and the series of network transformations shown in Figure 4.13. The first transformation involves splitting the capacitor C_1' into two capacitors, where one of these, C_0', is transformed along with R_1 into a series equivalent circuit C_0 and R. This transformation, which is shown in Figure 4.37, is basic to the design of narrowband mechanical filters employing piezoelectric ceramic transducers. This transformation is not necessary if the design begins with the network of Figure 4.13(c), which is the parametric-symmetric type filter shown in Figure 4.5. If the value of C_0' is chosen so that its reactance (at center frequency f_0) is equal to the resistance R_1, the electromechanical coupling will be the minimum required value. This fact is be discussed in a later section. After combining L_1' and $-L_{12}$, we bisect the network of Figure 4.13(c) and apply Bartlett's bisection theorem.

Bartlett's theorem states that we can form a symmetrical lattice network of series arm Z_a and diagonal arm Z_b from a symmetrical network (physical symmetry) by bisecting the network and then forming the Z_a and Z_b arms from the short-circuit and open-circuit input impedances. For example, the Z_a arm is found by first removing both the right-hand half of the network in Figure 4.13(c) and the source resistance R and then shorting the terminals b and b'. The remaining network between terminals a and a' is the lattice arm impedance Z_a. The arm Z_b is the open-circuit case, where $L_{12}/2$ is left hanging and therefore can be ignored. Having formed the two lattice arm impedances, we can apply the Norton transformation to the capacitor C_0, to form the pi network shown in Figure 4.13(d). By moving the 2:1 transformer through the parallel-tuned circuits, we obtain the lattice shown in Figure 4.13(e).

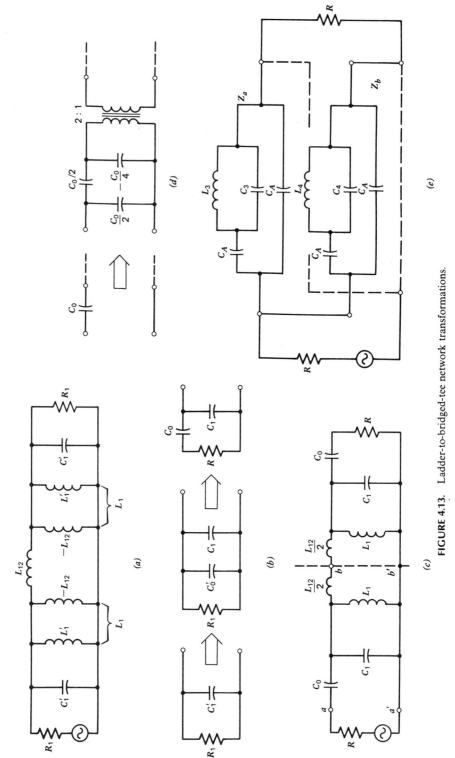

FIGURE 4.13. Ladder-to-bridged-tee network transformations.

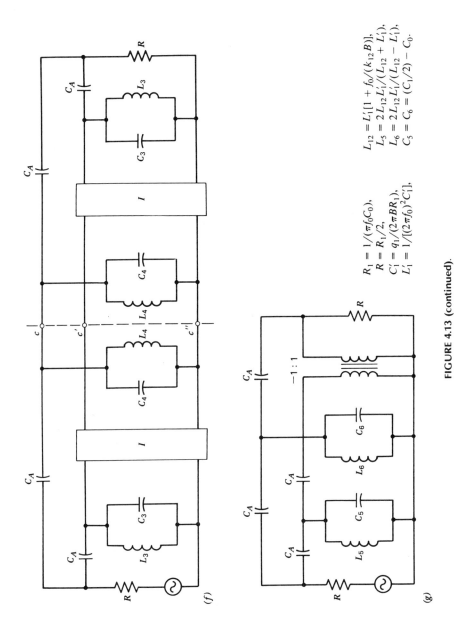

$R_1 = 1/(\pi f_0 C_0)$,
$R = R_1/2$,
$C_1' = q_1/(2\pi B R_1)$,
$L_1' = 1/[(2\pi f_0)^2 C_1']$,

$L_{12} = L_1'[1 + f_0/(k_{12}B)]$,
$L_5 = 2L_{12}L_1'/(L_{12} + L_1')$,
$L_6 = 2L_{12}L_1'/(L_{12} - L_1')$,
$C_5 = C_6 = (C_1/2) - C_0$.

(f)

(g)

FIGURE 4.13 (continued).

159

Rather than trying to show how to transform the lattice to the parallel ladder, we start with the parallel ladder in Figure 4.13(f), and by applying Bartlett's theorem to this network, show that it is equivalent to the lattice. In applying the theorem, it should be noted that the impedance inverter circuit I causes the shorts and opens across its output terminals to conversely look like opens and shorts when viewed from the input terminals. Therefore, when c, c', and c'' are left open, a short appears across the tuned circuit L_3C_3, leaving L_4C_4 as the only tuned circuit in arm Z_b.

Being convinced that the networks in Figure 4.13(e) and (f) are equivalent, we can then combine the impedance inverters to form a phase inverter and combine the parallel-tuned circuits to obtain the parallel-ladder circuit in Figure 4.13(g). The design equations, for the case where minimum coupling is needed, are shown in the figure; f_0 is the center frequency, B is the bandwidth.

To this point, we have assumed that the stopband of our design is monotonic and symmetrical and that we do not use acoustic coupling between the resonators. By making small modifications in the element values, we can realize attenuation poles and allow the design to have some acoustic coupling.

One reason for using a parallel-ladder configuration is to reduce the effect of spurious responses caused by the large coupling wires needed in designs having wide bandwidths. Since the coupling wire is often used as a support wire, it may be helpful to use a small wire or pair of wires for this purpose, but still use the electrical coupling to realize the greater part of the filter bandwidth. This acoustic coupling is represented by the inductance L_{AB} between the two ladders as shown in Figure 4.14. The filter must be designed to a reduced bandwidth by taking into account the additional coupling produced by the coupling wires.

In addition to the acoustic coupling, the parallel-ladder filter of Figure 4.14 shows a capacitor C_c from the input to the output terminals. C_c, which represents an actual capacitor, produces attenuation poles in both the upper and lower stopbands. When the two ladders are essentially identical (except for the frequencies of the resonators), the attenuation poles are nearly symmetrical about the center frequency. When either a static capacitance value C_A or C_B, or an electromechanical coupling value $k_{em2}^2 = C_A/C_5$ or $k_{em1}^2 = C_B/C_6$, is varied, the position of the attenuation poles becomes dissymmetrical. Figure 4.15 shows the result of these changes for the cases where capacitance bridging is present and is not present.

Acoustically Coupled Resonators.   As in the electrically coupled case, the parallel-ladder filter with acoustic coupling allows us to obtain upper and lower stopband attenuation poles at any specified frequency. Since the details of designing this type of filter are lengthy, we simply outline the method used in Reference [25]. A starting point is one of the P-ladder networks shown in Figure 4.16. As the result of a ladder synthesis, this network is imbedded in a complete filter network, part of which is shown in Figure 4.17(a). Successive Norton transformations, first of the series-arm parallel tuned circuit and then

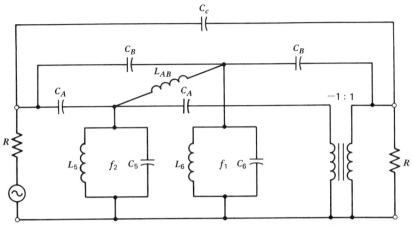

FIGURE 4.14. Parallel-ladder filter with electrical bridging and a combination of electrical and acoustic coupling.

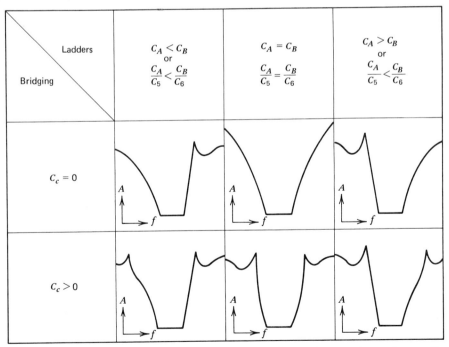

FIGURE 4.15. Frequency response shapes for various static capacitance and capacitance ratio conditions relating to the parallel-ladder filter of Figure 4.14.

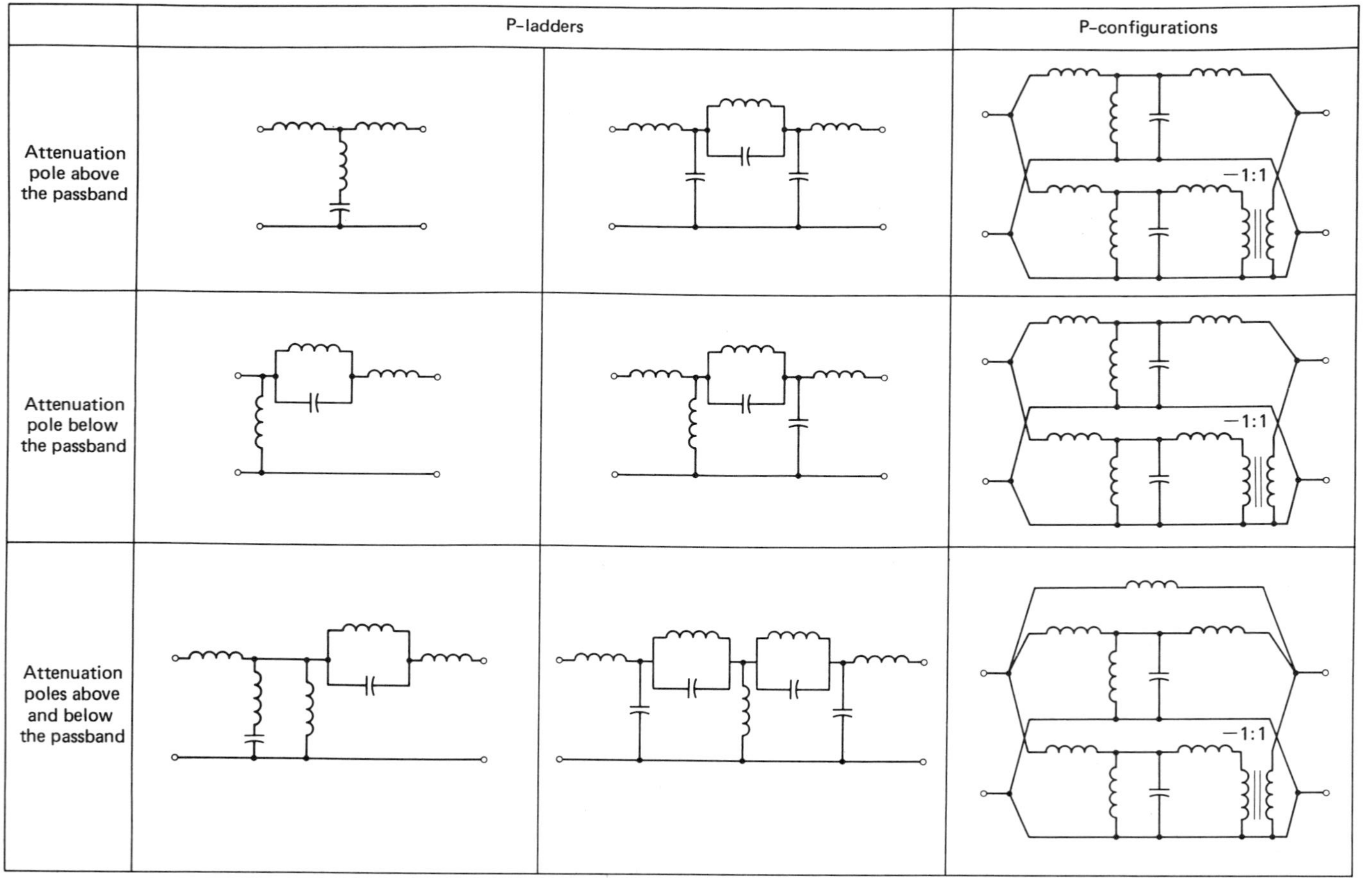

FIGURE 4.16. Conventional ladder and equivalent parallel ladder networks. [Adapted, with permission, from *Proc. 1975 IEEE ISCAS*, Boston, 305–308 (Apr. 1975). © 1975 IEEE.]

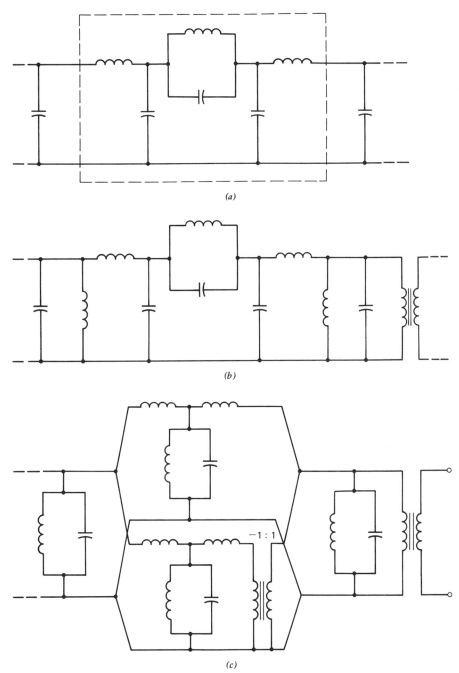

FIGURE 4.17. Ladder to parallel-ladder transformations. [Adapted, with permission, from *Proc. 1975 IEEE ISCAS*, Boston, 305–308 (Apr. 1975). © 1975 IEEE.]

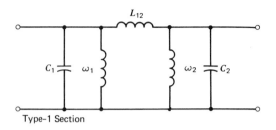

Type-1 Section

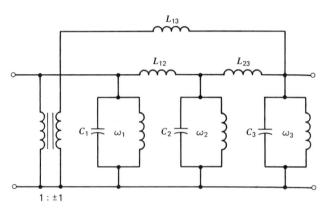

1 : ±1

Type-2 Section use (+) sign
Type-3 Section use (−) sign

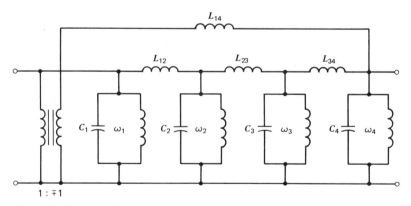

1 : ∓1

Type-4 Section use (−) sign
Type-5 Section use (+) sign

FIGURE 4.18. (*a*) Image-parameter filter sections.

GENERAL EQUATIONS

$$\left.\begin{array}{l} \omega_{-g} = 2\pi f_{-g} \\[4pt] \omega_g = 2\pi f_g \end{array}\right\} \text{Lower and Upper Band Edge Frequencies}$$

$$\Delta\omega_1 = 2\pi B = \omega_g - \omega_{-g} \quad \text{Radian Bandwidth}$$

$$\omega_m = 2\pi f_0 = \sqrt{\omega_g \omega_{-g}} \quad \text{Center Frequency}$$

$$R_I \quad \text{Image Impedance at } \omega_m$$

$$C_k = 1/(\Delta\omega_1 R_I)$$

TYPE-1 SECTION

$$C_1 = C_2 = C_k; \ \omega_1 = \omega_2 = \omega_{-g}; \ L_{12} = 2R_I/(\omega_g + \omega_{-g})$$

TYPE-2 AND TYPE-3 SECTIONS

$$\omega_\infty = \text{Pole Frequency}$$

When: $\omega_\infty > \omega_g \rightarrow$ Type-2 Section, choose upper signs

$\omega_\infty < \omega_{-g} \rightarrow$ Type-3 Section, choose lower signs

$$u = \omega_\infty^2 \mp \sqrt{\left(\omega_\infty^2 - \omega_g^2\right)\left(\omega_\infty^2 - \omega_{-g}^2\right)} \ ; \ v = \mp \frac{\omega_g^2 + \omega_{-g}^2 - 2u}{2} \ ;$$

$$w = \sqrt{\left(\frac{\omega_g^2 - \omega_{-g}^2}{2}\right)^2 - v^2} \ ; \ C_1 = C_3 = C_k; \ C_2 = 2C_k; \ L_{12} = L_{23} = \frac{1}{wC_k}$$

$$L_{13} = \frac{1}{vC_k} \ ; \ \omega_1 = \omega_3 = \sqrt{u - w - v \mp v} \ ; \ \omega_2 = \sqrt{u - w}$$

TYPE-4 AND TYPE-5 SECTIONS
When: Poles are at $\omega_\infty, \omega_{-\infty}$ on $j\omega$-axis, use upper sign

Poles are at $\pm a \pm jb$, use lower sign

$$\text{Restriction:} \left.\begin{array}{l} \omega_\infty^2 + \omega_{-\infty}^2 \\[4pt] 2(b^2 - a^2) \end{array}\right\} = \omega_g^2 + \omega_{-g}^2$$

$$\delta_0 = \omega_g^2 - \omega_{-g}^2; \ \delta_\infty = \left\{\begin{array}{c} \omega_\infty^2 - \omega_{-\infty}^2 \\ \text{or} \\ 4ab \end{array}\right\} \ ; \ X = \left(1 \mp \frac{\delta_0^2}{\delta_\infty^2}\right)^{1/4} \ ; \ C_1 = C_4 = C_k$$

$$C_2 = C_3 = 2C_k; \ L_{23} = \frac{1 + X^2}{\delta_0 C_k} \ ; \ L_{12} = L_{34} = \frac{L_{23}}{X} \ ; \ L_{14} = \pm \frac{2L_{23}}{1 - X^2} \ ;$$

$$\omega_1 = \omega_4 = \sqrt{\omega_{\pm g}^2 - \delta_0 \frac{X \pm 1}{X^2 + 1}} \ ; \qquad \omega_2 = \omega_3 = \sqrt{\frac{1}{2}\left(\omega_g^2 + \omega_{-g}^2 - \delta_0 \frac{1 + X}{1 + X^2}\right)}$$

FIGURE 4.18 (Continued). (b) Design equations. [Adapted, with permission, from *IEEE Trans. Sonics Ultrason.*, **SU-20**, pp. 294–301 (Oct. 1973). © 1973 IEEE.]

of the resultant inverted-L pairs of inductors, leads to the network of Figure 4.17(b). The input and output shunt arm resonators in Figure 4.17(b) are then each split into two resonators of such values to satisfy the following theorem: "given a P-ladder, there exists two such two-ports of type-R (a shunt parallel-tuned circuit) that a cascade connection of one two-port (type-R), the P-ladder, and the other two-port (type-R), in that order, can be transformed into a cascade connection of a P-configuration and an ideal transformer." The P-configuration mentioned in the theorem is one of those shown in Figure 4.16. The results of the P-ladder to P-configuration transformation is shown in Figure 4.17(c).

The circuits of Figure 4.17 realize a single attenuation pole above the passband. A pole below the passband is achieved with similar structures, but as Figure 4.16 shows, a bridging inductor (wire) across the parallel ladder is necessary in order to realize attenuation poles both above and below the filter passband. The mechanical realization of this electrical equivalent circuit is shown in Figure 7.17 of Chapter 7.

Image-Parameter Design

To this point, we have assumed exact synthesis methods based on realizing transfer functions like that of Equation (4.1). Although these methods are exact, under lossless nonparasitic conditions, they have two major drawbacks. The first is that for bridging-wire designs a digital computer is needed. Second, the computer programs often lack flexibility, which is overcome only by writing more sub-programs. The image-parameter design method, shown in Figure 4.7, is not exact but is flexible and can be programmed on a calculator.

Figure 4.18 shows three basic image-parameter sections [20]. The Type-1 section is used to design monotonic-stopband filters or can be used in conjunction with the attenuation pole producing networks of Types 2 to 5. The Type-2 section produces an attenuation pole above the passband, the Type-3 section produces a lower stopband pole. The Type-4 section realizes one pole above and one pole below the passband, whereas the Type-5 section produces symmetrically spaced poles in the complex s-plane, for the purpose of delay correction.

In this book, we do not attempt a detailed discussion of image-parameter theory. Instead, we illustrate the use of image-parameter sections through an example.

Example 4.3.   A Four-Resonator Image-Parameter Filter.   We will choose a very simple example to show how the sections of Figure 4.18 are used. Our specification requires four resonators and a single attenuation pole above the passband. The center frequency is 253.95 kHz, the bandwidth is 3.21 kHz, and the attenuation pole is located at 256.375 kHz. To realize the four-resonator circuit in Figure 4.19, we will simply cascade the Type-1 and the Type-2 sections shown in Figure 4.18(a). Because there is no phase inverter used (i.e., the sign of the secondary turns of the ideal transformer is positive), the transformer can be eliminated. If the sign had been negative (Type-3 section), a transformation like that in Figure 4.11(k) could have been used to eliminate the need for a transformer in the equivalent circuit.

The next step is simply to combine the parallel elements C_2 and C_2' and L_2 and L_2'. If we want the resonators to be equal in size, we can apply the first Norton transformation in Figure 4.11(f) to the coupling inductors L_{12} and L_{34}. The turns ratio is chosen to be equal to $1/(2)^{1/2}$, so that when the

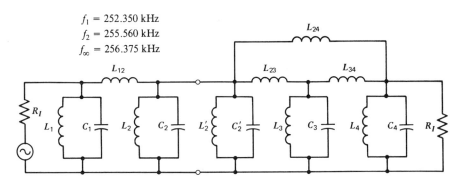

$f_1 = 252.350$ kHz
$f_2 = 255.560$ kHz
$f_\infty = 256.375$ kHz

$R_I = 1.0\ \Omega$ (Arbitrarily chosen)
$C_1 = C_2 = 4.95810 \times 10^{-5}$ F
$L_1 = L_2 = 8.02266 \times 10^{-9}$ H
$C_2' = C_4 = 4.95810 \times 10^{-5}$ F
$L_2' = L_4 = 8.05398 \times 10^{-9}$ H

$C_3 = 9.91620 \times 10^{-5}$ F
$L_3 = 3.98825 \times 10^{-9}$ H
$L_{12} = 6.26705 \times 10^{-7}$ H
$L_{23} = L_{34} = 6.76924 \times 10^{-7}$ H
$L_{24} = 1.65804 \times 10^{-6}$ H

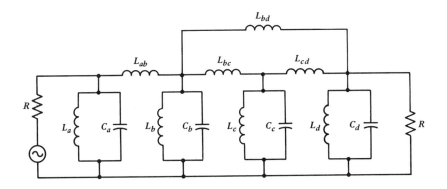

$R = 0.50\ \Omega$
$C_a = C_b = C_c = C_d = 9.91620 \times 10^{-5}$ F
$L_a = 7.9629471 \times 10^{-9}$ H
$L_b = 4.0339108 \times 10^{-9}$ H
$L_c = 3.9980069 \times 10^{-9}$ H

$L_d = 4.0073238 \times 10^{-9}$ H
$L_{ab} = 4.43147 \times 10^{-7}$ H
$L_{bc} = L_{23} = 6.76924 \times 10^{-7}$ H
$L_{cd} = 4.7865755 \times 10^{-7}$ H
$L_{bd} = 1.1724112 \times 10^{-6}$ H

FIGURE 4.19.   Image-parameter design example.

transformers are moved to the outside of resonators 1 and 4, their impedances are reduced to the values of the inside resonators. This results in the terminating resistances being reduced to one-half their original value and a transformer is now associated with the bridging inductor L_{24}. That transformer can be eliminated by the last Norton transformation in Figure 4.11(f).

Optimization

Response improvement, parasitic compensation, design centering, and design for ease of manufacturing can all be realized by means of filter optimization. The optimization may involve the filter impedance or delay, as well as the amplitude. The optimization techniques may be simple trial-and-error methods, the use of well-known optimization algorithms, or a combination of optimization and statistical computer programs.

One of the most obvious uses of optimization techniques is simply to improve the amplitude response. For instance, if an image-parameter design has undesirable ripple at the passband edges, the element values are adjusted so that these variations are reduced or maybe even eliminated. This may result in more ripple in the center of the passband, but that is one of the design trade-offs. Optimization may also be used to distort the filter response to compensate for peaks or roll-off in the users' system.

A second use of optimization is for compensation of parasitic elements, such as finite resonator losses, or for taking into account additional resonator responses (spurious responses). Using an optimization program to fully predistort (compensate) a lossy filter introduces the danger of having an amplitude response that is very sensitive to element value changes. A solution to this problem is a statistical design approach, where the filter sensitivity is included in the optimization [26]. This method, which makes use of Monte Carlo techniques, is known as design centering.

Spurious responses also cause a distortion of the filter response. The adverse effects of the spurious modes can be removed by optimization or, conversely, the spurious modes can be used to improve the filter selectivity, as in the case of disk-wire mechanical filters. Removal of the distortion effects has been accomplished as follows [20]:

1. Design an electrical equivalent prototype filter using image-parameter techniques.

2. Add the equivalent circuit elements corresponding to the spurious modes (see Figure 3.23 of Chapter 3).

3. Calculate the sensitivities of the characteristic function of the modified filter (which can be calculated from the attenuation) to changes in resonator frequencies and coupling values. These become the elements of an $n \times n$ Jacobian matrix.

4. By multiplying the matrix of differences between the actual characteristic-function values and the desired extrema (maximum values of the ripples) by

the inverse of the Jacobian matrix, we find the changes needed in the frequencies and couplings. This procedure is repeated until the error is reduced to the maximum allowable value. In short, this is simply the Newton-Raphson method applied to the characteristic function.

THE PHYSICAL REALIZATION

To this point, we have looked at the electrical equivalent circuits of various network configurations. It is our purpose, in this section, to show methods of realizing those configurations in mechanical (acoustic) and electromechanical forms. In Figure 4.1 we showed that some of this work had to be done prior to the electrical design, but those calculations of resonator sizes, electromechanical coupling, and so on, were only approximations and must be redone on the basis of more exact frequencies and couplings. The physical design process becomes approximately that in Figure 4.20.

Realistically speaking, a new design is a series of loops around Figures 4.1 and 4.20. The looping-back is the result of such things as wire sizes not being available, the appearance of spurious modes near or in the passband, the filter being too weak structurally, the parts not conforming to the factory processes, intermodulation distortion being too high, and on and on. Because of the changes that will probably need to be made, it is helpful to program the design equations to avoid repeated hand calculations. In this next section we discuss the flowchart in Figure 4.20 in more detail.

Resonator Coupling and Bridging

Prior to building a filter, it is necessary to choose the mode of vibration of the transducer resonators and the interior resonators, as well as the means of coupling and bridging. These choices are based on myriad factors, such as:

Center frequency	Dynamic range
Selectivity	Intermodulation distortion
Bandwidth	Microphonic responses
Passband ripple	Spurious responses
Size and strength	Material availability
Cost of production	Processing capability
Stability	Process tolerances

The designer is faced with keeping these factors in mind while choosing from an array of resonator, transducer, coupling, and bridging combinations.

Calculate interior resonator dimensions based on the electrical model resonator frequencies, the resonance equations in Figure 3.19, and the resonator material characteristics in Table 3.5.

↓

Calculate the transducer dimensions based in Equations (2.15), (2.16), (2.17), and (2.19), or those in Figure 3.19.

↓

Calculate coupling-wire and bridging-wire dimensions based on the normalized coupling coefficient k_{ij} of the electrical model, and the coupling-wire equations in Figure 3.32, the resonator equivalent-mass equations in Figure 3.19, and the coupling-wire material characteristics in Table 3.6.

↓

Build pairs or quads of coupled resonators for the purpose of determining the effects of using approximate equations, the effects of variations in material constants, and the effects of spurious modes of vibration. Make the necessary modifications in equations, material constants, or the electrical model.

↓

Build a filter and change frequencies and couplings until the required frequency response is obtained.

↓

Check the filter to the complete specification (shock, vibration, intermodulation distortion, etc.) and against available parts and factory processes. Modify the electrical or acoustical design as is necessary.

FIGURE 4.20.   A physical realization method for mechanical filters.

Some of the possible combinations are shown in Figures 4.21, 4.27, and 4.29. Having made a choice, the designer proceeds with calculations of coupling, bridging, and resonator tuning.

Interior-Resonator Coupling

Figure 4.21 shows a matrix of resonator and coupling-wire combinations. Ignoring the influence of spurious responses, these two-resonator sections behave like the spring-mass systems described in the previous chapter and mathematically can be tested in much the same way. The following mechanical

Resonator / Coupling	Extensional	Torsional	Flexural
Extensional	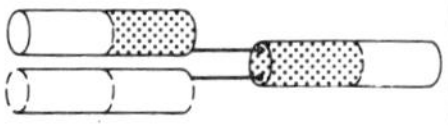 Tanaka, Kokusai, Telefunken, RFT	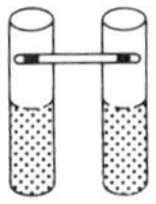 Telefunken, Fujitsu, NEC, Tesla, Oki, Tanaka, NTT, RFT	Rockwell, Kokusai 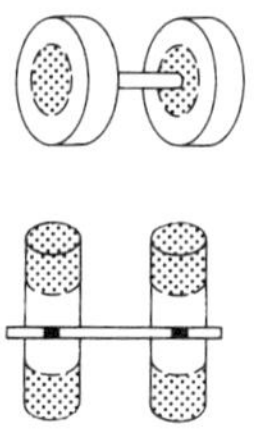 Siemens, Telefunken, Telettra, SEL, Italtel
Torsional		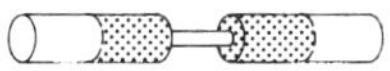 RCA	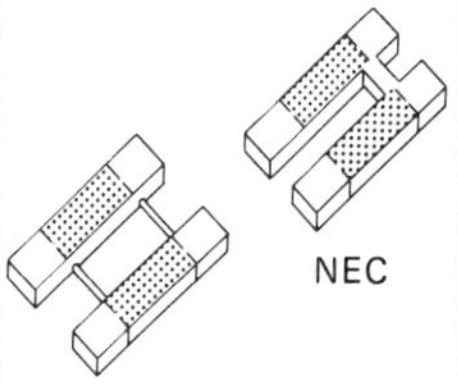 NEC Seiko, Rockwell, Fujitsu, Konno
Flexural	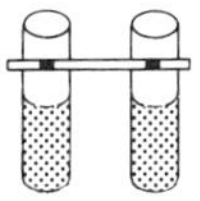 Toko, Telefunken, LTT	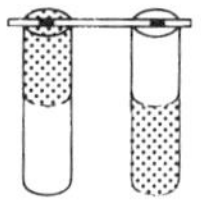 Telefunken	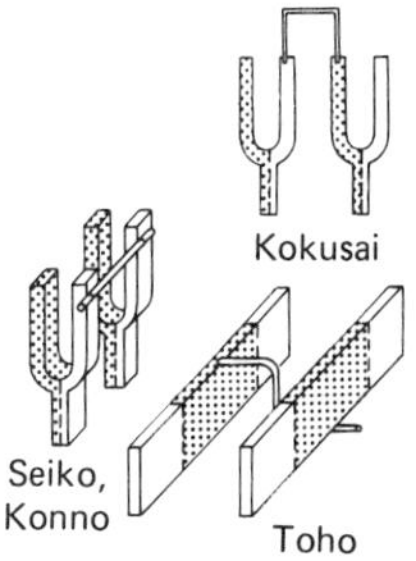 Kokusai Seiko, Konno Toho

FIGURE 4.21. Interior coupled resonator-pairs using extensional, torsional, and flexural resonators and coupling elements. Displacements in the shaded areas are in phase at the lower natural resonance.

equations are also related to the electrical analogy Equations (4.3) to (4.6). First we must determine the resonator dimensions. The arbitrary dimensions, such as the diameter of a torsional-mode resonator, are selected on the basis of the resulting wire dimensions, process tolerances, allowable filter size, spurious responses, and signal level sensitivity. The other dimensions are fixed by the resonator frequency equations, like those of Figure 3.19.

Having chosen and calculated the resonator dimensions, we next find the required coupling-wire stiffness from

$$K_{ij} = k_{ij} \frac{B}{f_0} \sqrt{K_i K_j} = k_{ij} \frac{B}{f_0} K_i \Big|_{K_i = K_j}, \tag{4.9}$$

where k_{ij} is the normalized coupling coefficient between resonators, B is the filter bandwidth, f_0 is the center frequency, and K_i and K_j are the resonator equivalent stiffness. Equivalent stiffness and equivalent mass are related by the equation for resonance frequency written in the form

$$K_i = \left(2\pi f_i\right)^2 M_i. \tag{4.10}$$

The equations for the equivalent masses M_i of thin or slender resonators are shown in Figure 3.19 and are a function of wire position x.

Knowing the required wire stiffness, we can then calculate the series-arm mechanical impedance $Y_{b\,ij}$,

$$Y_{b\,ij} = \frac{K_{ij}}{j2\pi f}. \tag{4.11}$$

Knowing $Y_{b\,ij}$ and the material characteristics E_c, G_c, and ρ_c, we can use the equations of Figure 3.32 to calculate the coupling-wire dimensions.

From the calculated resonator and coupling-wire stiffness, we can calculate the frequency difference between the first and second natural resonances of an identically tuned resonator pair:

$$f_2 - f_1 = \frac{f_r K_{ij}}{2} \left(\frac{1}{K_i} + \frac{1}{K_j} \right), \tag{4.12}$$

where f_r is the resonator frequency. This frequency difference is then compared to the measured value of an actual coupled-pair. The difference can then be used to modify the equivalent-mass and stiffness of the resonators to account for the effect of nonslenderness, spurious modes, and wire deformation due to welding.

A final step in calculating coupling-wire dimensions is that of locating wire resonances that are primarily due to bending modes. If these modes fall near the filter passband, then the wire position, the wire diameter or length, or the resonator size must be varied (in order to change the coupling-wire length or diameter) so the spurious responses can be moved away from the passband. As was discussed in Chapter 3, the equations governing the flexure modes are dependent on the end conditions, that is, dependent on the linear or angular displacements at the weld points. A good first-order approximation is to assume that the weld region does not deform and that the ends of the wires follow the motion of the resonator at the weld points. Under these conditions, the coupling wire may act like the ends are free or fixed (Equation 3.29 applies to both), hinged, or sliding [Figure 3.31(b) and Figure 3.32]. The resonance frequency for both the hinged and sliding case is

$$f_n = \frac{(n\pi)^2}{2\pi\ell^2} \sqrt{\frac{EI}{\rho A}} .$$

(4.13)

Imperfect filter construction should be assumed so as to account for all of the spurious modes that, in practice, will be present in the filter.

Example 4.4.   Extensional-Mode Coupling Between Torsional Rods.   A very popular filter is one with torsional-mode resonators and extensional-mode coupling wires. In this example problem, we calculate the coupling-wire length between resonator welds for a given coupling-wire diameter and resonator size. The two-resonator section, and its dimensions and material constants, is shown in Figure 4.22. The resonator material is Tokin TE-3 and the coupling wire material is Tokin TE-2. The filter bandwidth is 3.4 kHz, the center frequency is 130 kHz, and the coupling coefficient k_{ij} of the section in which we are interested is 0.560. We assume that the resonators are identical, tuned to the same frequency of 130 kHz, and are coupled at the point $x = 0.103$ cm. Having calculated the coupling-wire dimensions, we then study the spurious modes of the wire. But first, let us discuss the more general problem of spurious modes in this type of filter.

Spurious Responses.   Figure 4.23 shows the sources of spurious modes in filters composed of torsional-resonator sections like that in Figure 4.22. These modes have two effects; first, they may produce other responses in what should be the stopband of the filter or even in the passband, and second, they cause variations in the coupling between the resonators, which results in variations of bandwidth, passband ripple, or attenuation-pole frequencies. In general, the filter designer tries to move the responses as far from the passband as possible. As an example, by using the $\ell = (3)^{1/2}a$ approximation discussed in Chapter 3, we can use the curve of Figure 3.5 to find the flexural-mode resonance frequencies of the bars of our example problem. The lowest two frequencies are roughly at 85 kHz and 195 kHz. In other words, the flexural-mode resonances are spread about equal-distance from the torsional-mode frequency of 130 kHz. As the resonator diameter is increased, the lower flexural resonance

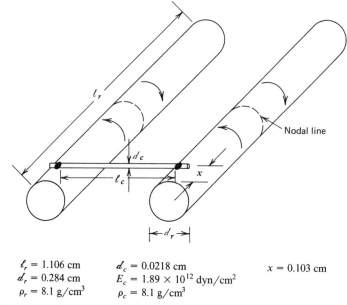

$\ell_r = 1.106$ cm $d_c = 0.0218$ cm $x = 0.103$ cm
$d_r = 0.284$ cm $E_c = 1.89 \times 10^{12}$ dyn/cm^2
$\rho_r = 8.1$ g/cm^3 $\rho_c = 8.1$ g/cm^3

FIGURE 4.22.   Torsional-mode resonators with extensional-mode wire coupling.

moves up in frequency toward the passband, and as the diameter is decreased, the higher resonance moves down toward the passband.

The design process, for dealing with spurious modes, is roughly as follows. First, adjust the resonators' dimensions to position their spurious responses as far from the passband as possible. The coupling wires will have little effect on these modes. Next, the coupling-wire resonances should be analyzed. As stated earlier, the wire resonances are a function of the resonator mode of vibration and the position of the wire on the resonator. This analysis is first done on coupled pairs of resonators. Next, the resonances of the supporting structure are taken into account. In our example problem, the supports are most often located at the dashed line nodes shown in Figure 4.22. Ideally, the supports located at the nodal lines will have no effect on the filter response, but if we take the reality of manufacturing processes into account, we can assume that the support points will not be exactly at the nodes and that resonances in the support structure will affect the filter characteristics. Finally, modes of the entire filter structure must be taken into account. If the filter is quite complex, these modes will be difficult to calculate, and therefore will not be evident until the filter is built. At that point, adjustments may have to be made in coupling-wire, resonator, and support dimensions.

Calculation of Coupling-Wire Length.   Our first step in calculating the coupling-wire length is the calculation of the torsional-resonator equivalent mass, at the point x from the end of the rod. Since the coupling wire axis is

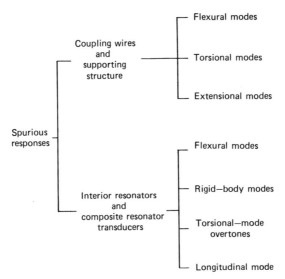

FIGURE 4.23. Sources of spurious modes of vibration in torsional-resonator mechanical filters. (Reprinted, with permission, from *CST '76 of IECE*, Japan, and S. Sugawara, T. Ogasawara, and M. Konno of Yamagata University, Japan.)

perpendicular to the axis of the resonator and is welded on the surface of the rod, we will use Equations (3.20), or Figure 3.19, to find the equivalent mass. Therefore,

$$M_r = M_{eq\,x} = \frac{\rho_r A_r \ell_r}{4\cos^2(n\pi x/\ell_r)} = \frac{8.1\left[\pi(0.284)^2/4\right]1.106}{4\cos^2\left[(1\pi \times 0.103/1.106)\right]} = 0.1544 \text{ gms.}$$

Knowing the equivalent mass of the resonator, we can calculate the equivalent stiffness from Equation (4.10),

$$K_r = (2\pi f_r)^2 M_r = (2\pi \times 1.30 \times 10^5)^2 \times 0.1544 = 1.031 \times 10^{11} \text{ dyn/cm.}$$

Knowing the normalized coupling and the filter's bandwidth and center frequency, we can use the resonator stiffness to find the coupling-wire stiffness. From Equation (4.9),

$$K_c = K_{ij} = K_{ij}\frac{B}{f_0}K_r = 0.560\left(\frac{3.4}{130}\right)1.031 \times 10^{11} = 1.509 \times 10^9 \text{ dyn/cm.}$$

Knowing the coupling-wire stiffness, we can next calculate the mechanical impedance of the wire from Equation (4.11) and then the wire length from the series-arm equations of Figure 3.32:

$$Y_b = \frac{K_c}{j2\pi f} = \frac{1.509 \times 10^9}{j2\pi \times 1.30 \times 10^5} = -j1.848 \times 10^3 \frac{\text{dyn-sec}}{\text{cm}}$$

and

$$\sin \alpha \ell_c = \sin\left(2\pi f \ell_c \sqrt{\rho_c/E_c}\right) = \frac{-j}{Z_0 Y_b} = \frac{-j\pi d_c^2 \sqrt{\rho_c E_c}}{4Y_b}$$

or

$$\ell_c = \frac{1}{2\pi f \sqrt{\rho_c/E_c}} \sin^{-1}\left(\frac{-j\pi d_c^2 \sqrt{\rho_c E_c}}{4Y_b}\right)$$

$$= \frac{1}{2\pi 1.30 \times 10^5 \sqrt{8.1/1.89 \times 10^{12}}} \sin^{-1}\left[\frac{\pi(0.0218)^2 \sqrt{8.1(1.89 \times 10^{12})}}{4 \times 1.848 \times 10^3}\right]$$

$$= 0.5388 \text{ cm.}$$

Next, we look at the coupling-wire spurious modes.

Flexure-Mode Spurious Responses.   In this section we study the effect of coupling-wire spurious modes on inter-resonator coupling. We assume that the construction of the filter is ideal, so that the only flexure-modes present are those due to the resonator applying a bending moment to the ends of the coupling wires. In other words, as the resonator, at the weld point, moves in the direction of the wire axis, it applies a force to the wire at that point, producing a bending moment equal to the force times one-half of the diameter of the wire. At the ends of the wire there is no first-order rotation or displacement in the transverse direction so the wire acts as if it were canti-levered between two solid supports. The resonances of this system are the same as those of the free-free case of Equation (3.29). An analysis of the effects of the spurious modes on coupling is beyond the scope of this book, other than to show the results of analyzing the resonator-pair of our example.

Figure 4.24 shows the effect of varying coupling-wire length on the denor-malized coupling coefficient $k_{ij}B/f_0$ [27]. The resonator and coupling-wire dimensions and characteristics are close to those of Figure 4.22. It is clear that wire lengths, wire diameters and weld positions x should be chosen in such a way that our design is not near the resonance-produced discontinuities. If time permits, the designer will generate coupling coefficient versus coupling-wire length curves, for various wire diameters.

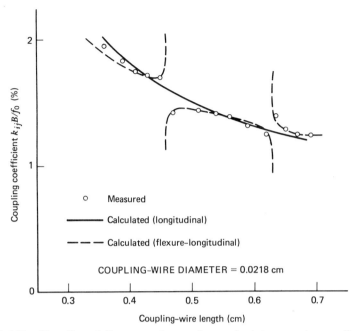

FIGURE 4.24. The effect of flexure spurious modes on the inter-resonator coupling between torsional mode resonators. [Adapted, with permission, from *31st Annual Freq. Control Symp.*, 207–212 (June 1977), and the authors K. Yakuwa, S. Okuda, K. Shirai, and Y. Kasai.]

Let's next look at another very popular but very different type of filter that is used for low-frequency, narrow bandwidth applications.

Example 4.5. Torsional Coupling of Flexural-Mode Bars. In this example we will calculate the dimensions of the two wires used to couple two identical flexural mode bar resonators. Figure 4.25 shows the coupled resonator-pair dimensions and material constants. The resonator material is Ni-Span C (Table 3.5) and the coupling-wire material is Magnetics Inc. 56/44 alloy (Table 3.6). The filter center frequency is 3.825 kHz, the bandwidth is 55 Hz and the coupling coefficient between resonators is 0.71 (which for a two-resonator filter corresponds to 0.1 dB passband ripple). This problem provides an approximate solution for the case where the resonators include the piezoelectric ceramic transducer plates, as shown in Figure 2.1. In our coupling calculations, we assume that the bars are tuned to 3.825 kHz and that the spacing between the inside edges of the welds is 0.254 cm. We also assume that the coupling wires are at the flexural-mode nodal points.

Let's start by finding the equivalent mass (actually the inertia J) of the resonator bars from the thin-bar equation (3.36) and Table 3.2. The nodal point is at $x/\ell = 0.224$ where the normalized angular displacement is 3.90.

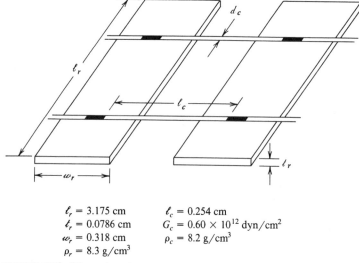

$$\ell_r = 3.175 \text{ cm} \qquad \ell_c = 0.254 \text{ cm}$$
$$t_r = 0.0786 \text{ cm} \qquad G_c = 0.60 \times 10^{12} \text{ dyn/cm}^2$$
$$w_r = 0.318 \text{ cm} \qquad \rho_c = 8.2 \text{ g/cm}^3$$
$$\rho_r = 8.3 \text{ g/cm}^3$$

FIGURE 4.25. Flexural-mode resonators with torsional-mode wire coupling.

Therefore,

$$J_r\bigg|_{x=0.224} = \frac{\rho_r A_r \ell_r}{4(3.90/\ell_r)^2} = \frac{8.3 \times 0.0786 \times 0.318 \times 3.175)}{4(3.90/3.175)^2}$$

$$= 0.1091 \text{ dyn-sec}^2 \text{cm},$$

and from (4.10),

$$K_r = (2\pi f_r)^2 J_r = (2\pi \times 3.825 \times 10^3)^2 0.1091 = 6.302 \times 10^7 \text{ dyn-cm.}$$

From (4.9), for each of the two coupling wires,

$$\frac{K_c}{2} = k_{12} \frac{B}{f_0} \frac{K_r}{2} = 0.71 \left(\frac{55}{3825}\right) 6.302 \times \frac{10^7}{2} = 3.217 \times 10^5 \text{ dyn-cm.}$$

At a low frequency of 3825 Hz, the coupling wire acts like a short line (wire length of less than $\lambda/8$). Therefore, we can solve for the coupling-wire diameters directly from Figure 3.32 and (4.11).

$$d_c = \left(\frac{32\ell_c K_b}{\pi G}\right)^{1/4} = \left(\frac{32 \times 0.254 \times 3.217 \times 10^5}{\pi 0.6 \times 10^{12}}\right)^{1/4} = 0.0343 \text{ cm.}$$

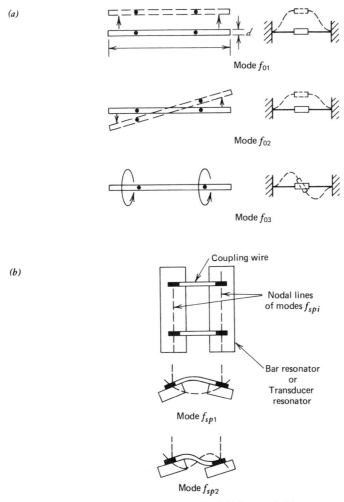

FIGURE 4.26. Rigid-body modes of (*a*) single rigid bars and (*b*) a two-resonator filter. [© 1979 IEEE. Reprinted, with permission, from *Proc. 1979 IEEE ISCAS*, Tokyo, 1068–1071, (July 1979).]

Spurious responses in this filter are also troublesome. The resonances are due to higher-order flexure modes (see Table 3.1), coupling-wire or support-wire flexure modes, and rigid-body modes of the entire structure. What do we mean by the term *rigid-body modes*?

Figure 4.26(*a*) shows an example of three resonance modes of a single rigid bar supported at its fundamental flexural-mode nodal points by four wires. Of these three modes, the one of primary importance when considering the vibrations of two or more resonators is the torsional rigid-body mode f_{03}. Figure 4.26(*b*) shows the two lowest torsional modes. The higher mode f_{sp2}, if the designer is not careful, may be located near the filter passband causing bandwidth variations and reduced stopband attenuation.

Coupling Corrections

We usually do not have time to account for all of the factors that cause the measured coupling to differ from our calculated values. Therefore, based on resonator-pair data, we include a multiplying factor to account for spurious-mode effects, nonideal welding, erroneous material constants, and second-order coupled modes. Having done this and then having built and measured a multiple-resonator filter, we sometimes find that we are still in error. One cause of this error is that wider bandwidth networks are more greatly affected by spurious modes than narrow bandwidth networks. For instance, a 10-resonator disk-wire filter will theoretically have almost twice the theoretical bandwidth of a simple resonator-pair, but in practice, the filter bandwidth may be considerably less than twice the pair-bandwidth value. In a case like this, it is helpful to build a quad of coupled disks. The quad has a theoretical bandwidth of 1.707 (i.e., almost 2.0) times that of a resonator pair [24]. Multiplying factors will then be based on this more accurate data.

Transducer-to-Resonator Coupling

Most of the same principles we developed regarding interior-resonator coupling can be applied to the case of coupling a transducer resonator to an interior resonator. For instance, in the next example we use Equations (4.9) to (4.12) to calculate coupling-wire dimensions and coupled-pair resonance frequency differences; but as a starting point, let's look at the matrix of Figure 4.27, which shows various combinations of transducers and metal resonators.

The transducers in Figure 4.27 were described in Chapter 2 and the resonators and coupling wires were described in Chapter 3; the figure therefore describes how various manufacturers have combined the elements. Just as there are more modes of vibration than the three shown in the figure, there are numerous other ways of coupling the resonators. The coupling wire could be shaped, for instance in the form of a U, in order to change the amount of coupling or the direction of the coupling. Note also that the coupling-wire mode of vibration is another variable, as can be seen in the extensional-extensional, and flexural-flexural cases shown in Figure 4.27. Let's next look at an example of how to calculate the coupling-wire dimensions between a transducer and an interior resonator.

Example 4.6.   Ferrite-Rod Transducer to Disk-Resonator Coupling.   In this example we will calculate the diameter of the coupling wire used between an extensional-mode ferrite transducer and a two-circle, flexure-mode disk resonator; the section is shown in Figure 4.28. The ferrite material is Ferroxcube 7A1 (Table 2.3), the coupling wire is Magnetics Inc 56/44 (Table 3.6), and the disk resonator material is Ni-Span C (Table 3.5). The coupling coefficient k_{tr} between the resonators is 0.700, the filter bandwidth is 3.840 kHz, and the

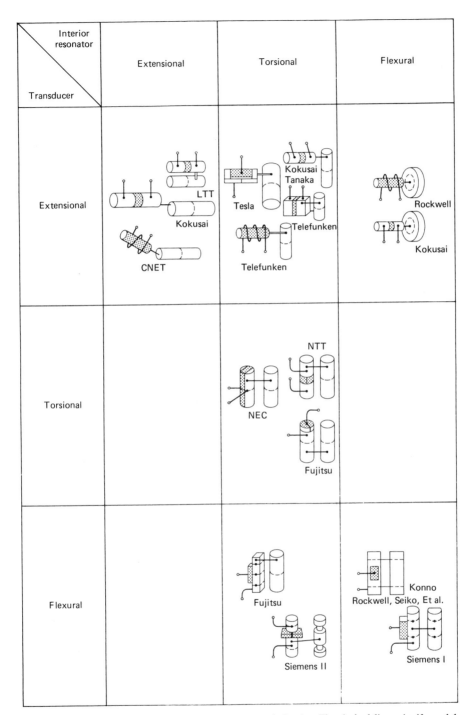

FIGURE 4.27. Transducer and interior resonator coupled pairs. The dashed lines signify nodal lines. True flexure nodes are shown as points.

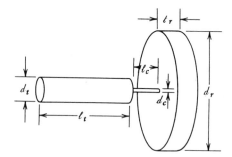

$\ell_t = 1.065$ cm,

$d_t = 0.175$ cm,

$E_t = 1.5 \times 10^{12}$ dyn/cm^2,

$\rho_t = 5/25$ g.cm^3,

$\ell_c = 0.295$ cm,

$E_c = 1.54 \times 10^{12}$ dyn/cm^2,

$\rho_c = 8.2$ g/cm^3,

$d_r = 1.466$ cm,

$\ell_r = 0.236$ cm,

$\mu = 0.312$,

$\rho_r = 8.3$ g/cm^3.

FIGURE 4.28. Wire coupling between an extensional-mode ferrite transducer rod and a flexure-mode metal alloy disk.

filter center frequency is 254.0 kHz. In calculating the coupling, we tune the resonators to 251.0 kHz.

First we calculate the equivalent-stiffness values of the ferrite rod and the disk. The equation for the rod is (3.14),

$$M_t = \frac{\rho_t A_t \ell_t}{2} = 5.25\pi \left(\frac{0.175}{2}\right)^2 \times \frac{1.065}{2} = 0.6724 \times 10^{-1} \text{g}$$

and

$$K_t = (2\pi f)^2 M_t = (2\pi 0.251 \times 10^6)^2 \times 0.6724 \times 10^{-1}$$

$$= 1.672 \times 10^{11} \text{ dyn/cm.}$$

The disk equivalent mass can be estimated by using Figure 3.12, although the diameter-to-thickness ratio of our example and the figure are not the same. Note that Poisson's ratio is also slightly different. Because we have no other data available, we must be aware of the fact that the calculation of disk equivalent mass may be in error by about 10 percent. Therefore, from the z-direction curve of Figure 3.12 and its intersection with the ordinate,

$$M_r = \left(\frac{M_{eq}}{M_{st}}\right) M_{st} \simeq 0.123 \rho_r \ell_r \pi \left(\frac{d_r}{2}\right)^2$$

$$= 0.123 \times 8.3 \times 0.236\pi \left(\frac{1.466}{2}\right)^2 = 0.4067 \text{ g}$$

and

$$K_r = (2\pi f)^2 M_r = (1.577 \times 10^6)^2 \times 0.4067 = 1.011 \times 10^{12} \text{ dyn/cm.}$$

In calculating the coupling-wire diameter, we start from Equation (4.9),

$$K_{tr} = k_{tr}\frac{B}{f_0}\sqrt{K_t K_r} = 0.700\left(\frac{3.840}{254}\right)\sqrt{(1.672 \times 10^{11})(1.01 \times 10^{12})}$$

$$= 4.351 \times 10^9 \text{ dyn/cm.}$$

From (4.11) we can write

$$Y_{b\,tr} = \frac{K_{tr}}{j2\pi f} = \frac{4.351 \times 10^9}{j2\pi(251 \times 10^3)} = -j2.759 \times 10^3 \text{ dyn-sec/cm.}$$

From Figure 3.32, for extensional-mode coupling,

$$\alpha\ell = 2\pi f \ell_c \sqrt{\rho_c/E_c} = 2\pi \times 251 \times 10^3 \times 0.295\sqrt{8.2/1.54 \times 10^{12}} = 1.074 \text{ rad}$$

and

$$Z_0 = \frac{-j}{Y_{b\,tr}\sin \alpha\ell} = \frac{1}{2.759 \times 10^3 \times 0.8789} = 4.124 \times 10^{-4} \text{ cm/dyn-sec.}$$

And finally, from the equation for Z_0,

$$d_c = \left(\frac{4}{\pi Z_0 \sqrt{\rho_c E_c}}\right)^{1/2} = \left(\frac{4}{\pi \times 4.124 \times 10^{-4}\sqrt{8.2 \times 1.54 \times 10^{12}}}\right)^{1/2}$$

$$= 2.94 \times 10^{-2} \text{ cm.}$$

The frequency spacing between the natural resonances, from (4.12), is

$$f_2 - f_1 = \frac{fK_{tr}}{2}\left(\frac{1}{K_t} + \frac{1}{K_r}\right) = 3.806 \times 10^3 \text{ Hz.}$$

With regard to spurious responses, the coupling wire is carefully centered on the disk resonator in order to reduce the coupling between the ferrite transducer and the diameter-mode spurious responses shown in Table 3.4. The idea of coupling to a spurious-response nodal point is not unique to end resonators; it is used on interior resonators as well. In each case, the narrowed bandwidth of the coupled sections does not match well with the terminating resistance, causing a reduction in the amplitude of the spurious response. But let's move on to the subject of realizing bridging topologies.

Realizing Bridging-Wire Circuits

In this section, we study mechanical realizations of the electrical bridged networks discussed earlier in this chapter and summarized in Figure 4.29. As a means of understanding the concepts involved in acoustic bridging, let us first look at the four-resonator section and its frequency response, which are shown in Figure 4.30.

In the torsional-resonator quad of Figure 4.30(a), adjacent resonators are directly coupled by one-quarter wavelength wires. The two end resonators are coupled by a three-quarter wavelength wire. At the attenuation-pole frequencies $f_{\infty 1}$ and $f_{\infty 2}$ in Figure 4.30(b), the amplitude of the force contribution of the bridging wire is equal to that of the one-quarter wavelength wire attached to the same end resonator, but the contributions are opposite in phase, causing the signal to be cancelled. At each attenuation pole the out-of-phase condition is the result of the bridging wire acting as a compliance and a phase inverter, the phase inverter being derived from the greater than one-half wavelength and less than a full wavelength condition of the wire. At $f_{\infty 1}$ the adjacent resonators are in-phase, but because we are outside of the passband edge f_1 the main resonator-to-resonator signal is attenuated. When the amplitude of the bridging force and main coupling force are equal, the resultant vibration is zero (i.e., an attenuation pole occurs). Above the filter passband, at $f_{\infty 2}$, we have the same condition resulting from the two center resonators each adding a 180 degree phase shift. By adding a bridging element to the spring-mass systems in Figure 3.34 and then by applying superposition and observing the forces acting on the last resonators, the concepts discussed above should become more clear. If the bridging wire is between one-half and a full wavelength, please remember that if one-half of the wire is stretched, the other half is compressed. Also, below and above the lowest and highest natural resonances, the phase relationships of the displacements, when bridging is added, are essentially, but not exactly, those shown in Figure 3.34. Therefore, a study of the applied forces on the last resonators is the most profitable way to analyze the pole-producing behavior of these sections.

Conditions for Realizing Attenuation Poles. Study of a three-resonator system shows that simple noninverted bridging results in an attenuation pole above the filter passband, whereas phase inversion produces a pole in the lower stopband. In general, simple bridging across an even number of resonators results in a symmetric frequency response, whereas bridging across an odd number of resonators results in an attenuation pole being produced in the lower stopband or the upper stopband, but not in both. These concepts are summarized in Table 4.3.

The number n in Table 4.3 corresponds to the total number of resonators in the bridged section or the number of resonators that are bridged. Phase inversion of $-1:1$ results when the total number of phase inversions in the bridging and the coupling under the bridging is an odd number. This same rule

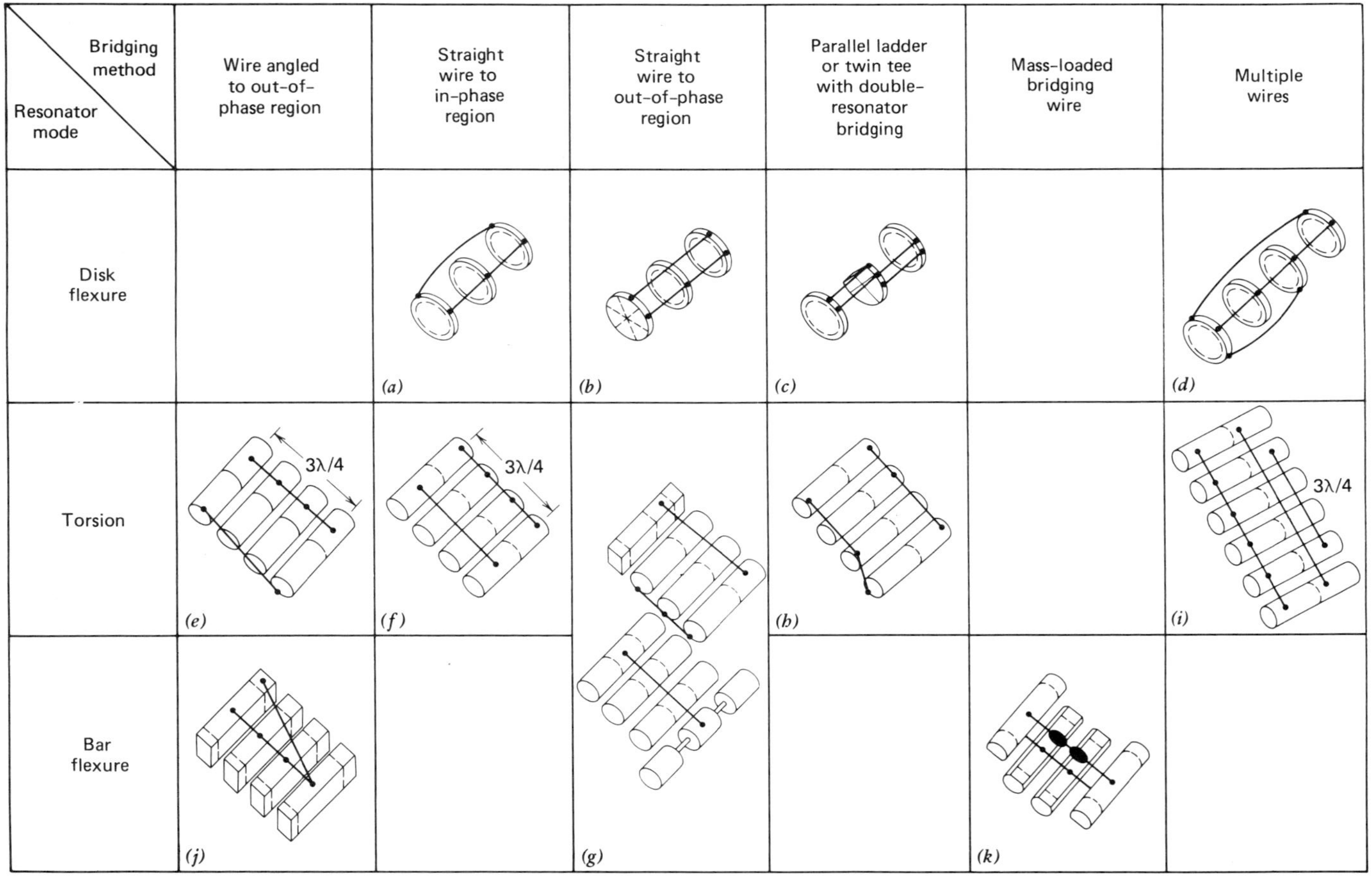

Bridging method / Resonator mode	Wire angled to out-of-phase region	Straight wire to in-phase region	Straight wire to out-of-phase region	Parallel ladder or twin tee with double-resonator bridging	Mass-loaded bridging wire	Multiple wires
Disk flexure		(a)	(b)	(c)		(d)
Torsion	(e) 3λ/4	(f) 3λ/4	(g)	(h)		(i) 3λ/4
Bar flexure	(j)				(k)	

FIGURE 4.29. Various bridging-wire configurations for realizing attenuation poles of the frequency response. [Adapted, with permission, from *Proc. 1975 IEEE ISCAS*, Boston, 313–316 (Apr. 1975). © 1975 IEEE.]

185

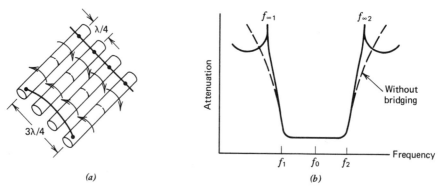

FIGURE 4.30. (*a*) Bridging across two resonators, and (*b*) the resultant frequency response. [© 1975 IEEE. Reprinted, with permission, from *Proc. 1975 IEEE ISCAS*, Boston, 313–316 (Apr. 1975).]

applies to electrical bridging across the mechanical filter structure when we recognize that the two transducers (because each acts as an impedance inverter) produce a phase inversion under the bridging capacitor or inductor. These concepts are important for the filter users who may want to exert additional control over the stopband behavior of the filter.

To this point, we have looked at the conditions for realizing attenuation poles at various locations. Let us now study methods of realizing physical structures, such as those that were shown in Figure 4.29.

Physical Realizations of Bridging Networks. Figure 4.29 showed various configurations for realizing attenuation poles above and below the filter passband and in the complex *s*-plane. In those cases where phase inversion is necessary, it is accomplished by either inverting the phase of the signal through the wire or by coupling to out-of-phase regions on the resonators. Means of doing this are outlined in Figure 4.31.

The most straightforward phase-inversion method shown in Figure 4.31 is to simply increase the bridging-wire length to where it acts as a phase inverter. Problems with this method are most pronounced at low frequencies where the wire may become excessively long. In this case, masses can be added to the

TABLE 4.3. Attenuation Pole Locations for Different Numbers *n* of Bridged Resonators and Phase Conditions

n	Phase	Attenuation Pole Location
Even	$-1{:}1$	$j\omega$-Axis—above and below passband
Even	$+1{:}1$	Complex—right and left half-planes
Odd	$-1{:}1$	$j\omega$-Axis—below filter passband
Odd	$+1{:}1$	$j\omega$-Axis—above filter passband

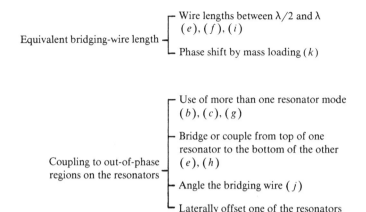

Equivalent bridging-wire length
— Wire lengths between $\lambda/2$ and λ
  $(e), (f), (i)$
— Phase shift by mass loading (k)

Coupling to out-of-phase regions on the resonators
— Use of more than one resonator mode
  $(b), (c), (g)$
— Bridge or couple from top of one resonator to the bottom of the other
  $(e), (h)$
— Angle the bridging wire (j)
— Laterally offset one of the resonators

FIGURE 4.31.   Phase inversion methods for realizing attenuation poles. Bracketed letters refer to Figure 4.29.

wire to obtain a spring-mass lowpass filter having greater than 180 degrees phase shift at the bandpass filter attenuation-pole frequencies [11]. The mass-loading method involves additional manufacturing costs, which possibly can be avoided if the main coupling and the bridging can be connected to out-of-phase regions of the first or last resonators in the bridged section, or if phase inversion can be accomplished under the bridging, as in Figure 4.29(h). The term *out-of-phase regions* refers to portions of the bridging-section resonators that are vibrating in opposite directions, at the lowest natural frequency of the undamped section. To clarify these concepts, let's look at an example.

Example 4.7. Extensional-Mode Coupling of a Flexural to a Torsional Resonator.   In this example we look at the bridging section shown in Figure 4.32, where the bridging from the flexural-mode end resonator to the torsional-mode center (4th) resonator is accomplished through the extensional-mode bridging wire [27]. This is like a section of the telephone channel filter of Figure 7.20. The thickness of the flexural-mode end resonator is increased slightly to approximate the actual ceramic/metal composite resonator transducer. Note that at the lowest natural resonance, all regions connected to the main coupling wires are in phase, whereas the bridging wire is connecting out-of-phase portions of the first and last resonators. In this way, phase inversion is achieved and finite upper and lower stopband attenuation poles result (see Figure 7.21).

As we have done in previous sections, let us calculate the coupling-wire diameter for a normalized coupling coefficient, which in this example is 0.200. Since we have no tables in this book for the equivalent mass of a second-mode flexural resonator, it was necessary to calculate the equivalent mass from Equations (3.30), (3.31), and (3.33). The resulting equivalent mass is $M_f = 0.164$

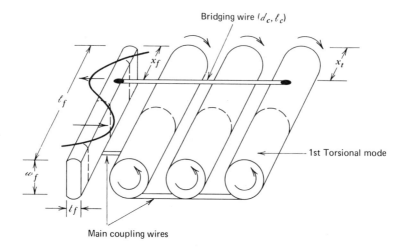

Flexural	$\ell_f = 1.067$ cm	$\ell_c = 1.643$ cm
transducer-resonance	$w_f = 0.29$ cm	$E_c = 1.89 \times 10^{12}$ dyn/cm²
characteristics	$\ell_f = 0.111$ cm	$\rho_c = 8.1$ g/cm
	$\rho_f = 8.1$ g/cm³	$x_f = 0.345$ cm
		$x_t = 0.364$ cm

Torsional resonator characteristics are shown in figure 4.22

FIGURE 4.32.   Four-resonator bridging section with a phase-inverting end (transducer) resonator. The directions of motion in various regions, at the lowest natural resonance, are shown by arrows.

g at $x_f = 0.345$. The torsional equivalent mass can be found from the data of Example 4.4 with $x = x_t = 0.364$. The results of these calculations are that $M_t = 0.542$ g. From (4.10),

$$K_f = (2\pi \times 1.30 \times 10^5)^2 \times 0.164 = 1.094 \times 10^{11} \text{ dyn/cm}$$

$$K_t = (2\pi \times 1.30 \times 10^5)^2 \times 0.542 = 3.616 \times 10^{11} \text{ dyn/cm}$$

and (4.9), where $B = 3.4$ kHz,

$$K_{ft} = 0.2 \times \frac{3.4}{130.0} \times \sqrt{(1.094)(3.616) \times 10^{22}} = 1.040 \times 10^9 \text{ dyn/cm}$$

and, therefore, from (4.11),

$$Y_{b\,ft} = \frac{1.040 \times 10^9}{j2\pi \times 130 \times 10^3} = j1.274 \times 10^3 \text{ dyn-sec/cm}$$

and finally, from Figure 3.32,

$$d_c = \left[\frac{4|Y_{b_{ft}}| \sin \left(2\pi f_0 \ell_c \sqrt{\rho_c/E_c}\right)}{\pi \sqrt{\rho_c E_c}} \right]^{1/2}$$

$$= \left[\frac{4 \times 1.274 \times 10^3 \times \sin \left(2\pi \times 1.30 \times 10^5 \times 1.643\sqrt{8.1/1.89 \times 10^{12}}\right)}{\pi \sqrt{8.1 \times 1.89 \times 10^{12}}} \right]^{1/2}$$

$$= 0.0121 \text{ cm}.$$

Resonator Tuning

The next step in the physical realization of a mechanical filter is that of determining new resonator frequencies based on both the frequencies of the electrical equivalent circuit and the effect of the coupling wires and support wires on the frequencies after welding. In other words, we must preadjust the frequency of each mechanical resonator so that, after the wires are attached, the frequency of the shunt tuned circuit is equal to f_j of Figure 4.8(a) or ω_2 of Figure 4.18, and so on.

In terms of the mechanical network, the electrical resonator becomes the spring-mass tuned circuit of Figure 4.33(a) surrounded by two coupling-wire equivalent circuits. We assume for purposes of illustration that the coupling wires are short extensional-mode lines (see Figure 3.32). In previous sections of this chapter, we discussed the calculation of the resonator equivalent mass and stiffness and the coupling-wire and bridging-wire stiffness K_b, where K_b corresponds to the series-arm inductances in Figure 4.8(a) and the series-arm stiffnesses shown in Figure 4.33(a). Now our task is to determine the shunt element values of the wires and their effect on the resonator frequencies.

From Figure 4.33(a), it is apparent that if we simply attach the wires to the resonators, the resonators will be mistuned by the shunt masses of the wires. Therefore, we must remove some mass from M_j by an amount equal to the sum of all of the shunt wire masses attached in parallel with resonator j. (Remember that parallel masses in the mobility analogy are added and subtracted like parallel capacitors in an electrical circuit.) Our resultant mechanical circuit becomes that of Figure 4.33(b) where the sum of the three inside masses are equal to M_j. The resonator is now tuned to f_j' where

$$f_j' = \frac{1}{2\pi} \left[\frac{K_j}{(M_j - \Sigma M_a)} \right]^{1/2} \tag{4.14}$$

and where the masses M_a correspond to the shunt masses of each of the wires connected to the resonator.

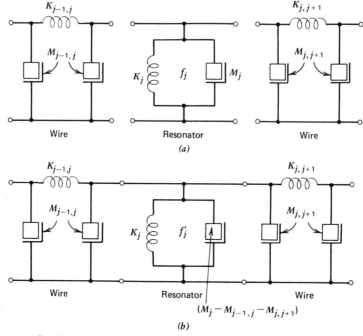

FIGURE 4.33. Coupling-wire and resonator equivalent circuits: (*a*) before welding and adjustment of the resonator frequency to take into account the wire shunt masses, and (*b*) the pretuned resonator and attached coupling wires.

Referring to the shunt-arm equivalent circuit values of Figure 3.32, we see that short-wire extensional-mode and torsional-mode element values act like inertia elements (ω is in the denominator of Y_a). In quarter-wavelength flexural lines the shunt elements act like stiffness and must be added to the resonator spring constant (since stiffness is analogous to reciprocal inductance, the spring constants are additive).

In addition to the coupling wire effects, we must also take the resonator supports into account. For instance, if the end of the left-hand wire in Figure 4.33(*b*) is clamped to a rigid support, the stiffness $K_{j-1,j}$ is shorted to ground and must be added to the resonator stiffness K_j. If the support-wire length is short compared to a wavelength, the wire stiffness has a greater effect on resonator frequency than the wire mass, thereby increasing the resonator's frequency after welding.

Because we are dealing with narrowband networks there is little difference, when using the general equations, if we choose the shunt impedance Y_a to act like a spring by multiplying Y_a by $j\omega$ or to act like a mass by dividing Y_a by $j\omega$. Although the choice is often made on the basis of the sign of Y_a at f_0 ($+j$ corresponds to a mass, $-j$ corresponds to a spring), a better approximation is when both the slope $dY/d\omega$ and sign are considered. For instance, a positive mass corresponds to the case where the sign of Y is $+j$ and the slope is positive, a negative mass corresponds to the sign of Y being $-j$ and the slope of Y is negative.

Example 4.8   Frequency Shift Calculations.   In this example, we calculate the resonator frequency shifts, due to mass and inertia loading of the coupling wires, of the previous four examples. In the calculation of the frequency mistuning of the flexural bar, torsional coupling-wire case we will assume that the coupling wire is attached to the nodal point on the side of the resonator bar rather than on the top surface of the bar. If we choose to account for the fact that the wire is one-half of the bar thickness ℓ_r from the nodal point, we must add a factor $J = (m/2)[(\ell_r + d_c)/2]^2$ (where m is the wire static mass) to the resonator moment of inertia to find the mistuning. This factor turns out to be more than 20 times the inertia of the coupling wire about its axis.

Table 4.4 summarizes these calculations and shows the wide variations of frequency shifts relative to the filter bandwidth. For example, at $f = 3.825$ kHz the coupling wire located at the nodal point only mistunes the resonator 0.0045 percent of the bandwidth, whereas the transducer resonators in the last two cases are shifted by 11 and 44 percent. The numbers shown are due to a single wire either entering or leaving a resonator. The more coupling and bridging wires entering and leaving a resonator, the greater the frequency shift. The effect of multiple wires is simply additive, as implied by Equation (4.14).

Not included in these calculations is the effect of the welded portion of the coupling wire. A good approximation is to simply add the static mass of the welded part of the wire to the resonator equivalent mass. Also not included is the effect of the resonator support structure.

Transducer Design Considerations

In Chapter 2 we looked at various electromechanical transducer configurations, and in the first part of this chapter we looked at electrical equivalent circuits of entire filters, including the transducers. The objective of this section is to integrate these concepts through practical advice and applications. Let's start with a discussion of electrical tuning.

The intermediate-band design concept, which uses a coil to tune the static capacitance of a piezoelectric transducer, generally leads to the most stable design, but the drawbacks of this kind of filter are higher cost, greater size, and reduced reliability, when compared with an inductorless design. There are only minor disadvantages in the case of a magnetostrictive transducer filter, because the transducers can be tuned with a capacitor. Therefore, magnetostrictive transducer filters are designed as intermediate-band or wideband filters, whereas piezoelectric ceramic transducer filters are designed with and without inductors.

TABLE 4.4.— Resonator Frequency Shifts due to Mass and Inertia Effects of a Single Coupling Wire. The Wire Dimensions Were Calculated in Examples 4.4 – 4.7

Figure Number and Frequency, f	Resonator Mass M_1 (g) (or J_1)	M_2 (g) (or J_2)	Shunt Element Equation from Figure 3.32	$Y_a\left(\frac{dyn\text{-}sec}{cm}\right)$ (or J_a)	$M_a = \frac{Y_a}{J\omega}(g)$ (or J_a)	$\Delta f_1(H_z)$ $=\frac{fM_a}{2M_1}$	$\Delta f_2(H_z)$ $=\frac{fM_a}{2M_2}$	Bandwidth $B(H_z)$	$\frac{\Delta f_1}{B}(\%)$	$\frac{\Delta f_2}{B}(\%)$
Figure 4.22 (130 kHz)	0.1544	0.1544	(1) $Y_a = \dfrac{j\tan\frac{\alpha\ell_c}{2}}{Z_0}$	$j228$	0.279×10^{-3}	117	117	3400	3.45	3.45
Figure 4.25 (3.825 kHz)	0.1091	0.1091	(2) $J_a = \dfrac{\pi d_c^4 \rho_c \ell_c}{64}$	—	[a] 0.142×10^{-6}	0.0025	0.0025	55	0.0045	0.0045
Figure 4.28 (254 kHz)	0.06724 Ferrite	0.4067 Disk	Same as (1) above	$j1440$	0.913×10^{-3}	1690	281	3840	44.01	7.32
Figure 4.32 (130 kHz)	0.164 Flexure bar	0.542 Torsional rod	Same as (2) above	$j780$	0.954×10^{-3}	378	114	3400	11.13	3.36

[a] In the case where the coupling wire is welded to the top surface of the resonator bar, the moment of inertia value is 3.066×10^{-6} g.

Piezoelectric Transducer Designs

Let's start this section by looking at the advantages of each of the three different design types:

Narrowband	Intermediate-band	Wideband
Smaller size	Highly stable	Can achieve
Lower cost of parts and labor	Greater transducer choices	maximum bandwidth
		Greater selectivity
Higher reliability	Reduced nonlinearities	
Higher percentage automation production		

Transformer impedance matching

Spurious response rejection

Balanced input and output

Electrical tuning for minimizing ripple

The advantages of the narrowband design method are directly related to the absence of input and output tuning inductors or transformers. The advantages of the intermediate-band design are based on: (1) the electromechanical coupling coefficient requirements are low, resulting in high Q, stable transducers; and (2) the coil or transformer can be used for balance (eliminating a balanced-modulator transformer), for impedance matching, for increased stopband selectivity (reducing spurious responses), and for reducing passband ripple due to interior resonator mistunings or miscouplings. The wideband design has the advantage of two additional resonators' selectivity plus the advantages listed in (2). Unfortunately, our choice of transducer type may be limited by the filter bandwidth.

Electromechanical Coupling Requirements.   The most binding limitation relating to the three transducer types is the achievable electromechanical coupling coefficient. As an example, for very large bandwidths the intermediate-band design, because of coupling limitations, becomes a wideband design. If stray capacitance is high, or if temperature compensation is necessary, the achievable bandwidth of each design is further limited, as shown in circuits 2 and 4 of Figure 4.34. Note that the minimum k_{em} is a function of the passband-response shape through the normalized q in the narrowband case and through the electrical circuit/mechanical circuit end normalized coupling coefficient k_{12} in the wideband and intermediate-band cases. The intermediate-band design problem is that of increasing k_{12} (which reduces the sensitivity of the response to electrical tuning changes) to the point where the required higher k_{em} results in equal electrical and mechanical tuned-circuit

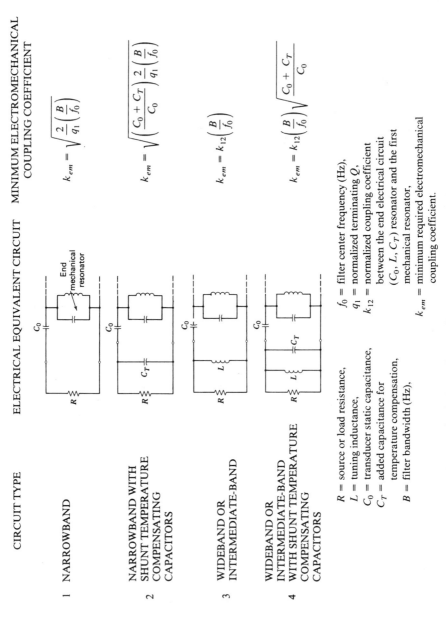

FIGURE 4.34. Minimum electromechanical coupling coefficient requirements for different piezoelectric transducer configurations.

stabilities. In the wideband case, k_{12} is simply the handbook value for the required response shape. In the intermediate-band design, k_{12} is a derived parameter that indicates to the designer how much overcoupling there is between the electrical and mechanical tuned circuits.

If in the manufacturing process the transducers are to be depoled for the purpose of achieving an exact value of electromechanical coupling, then this lower value must be taken into account when using Figure 4.34.

Knowing the filter bandwidth and center frequency, as well as the normalized k_{12} and q_1 values, we can use Figure 4.34 to find the minimum k_{em} requirement. Knowing k_{em}, we must then choose a transducer type from an array, such as those shown in Figure 2.18. A rough starting point in the selection process is recognizing that the -31 extensional-mode transducers are limited to $k_{em} \leq 15\%$, whereas the thickness, the -33, and the -15 modes can achieve 15 to 50 percent values of k_{em}. Let's look at an example.

Example 4.9. A Torsional-Mode Transducer.   In this example we look at methods of realizing transducers for a 3.4 kHz bandwidth filter at 130 kHz. Although stability and the other advantages of the coil-tuned designs are important, we assume that size, cost, and reliability win out and we choose to use a narrowband transducer. We also assume that the stray capacitance is not negligible and that some temperature compensation might be necessary, so we choose $(C_0 + C_T)/C_0 = 1.2$. Assuming a low passband-ripple normalized q_1 value of 0.82, we can find the required k_{em} value from Circuit Type 2 of Figure 4.34,

$$
k_{em} = \sqrt{\frac{C_0 + C_T}{C_0}\left(\frac{2}{q_1}\right)\frac{B}{f_0}} = \sqrt{1.2 \times \frac{2}{0.82} \times \frac{3.4}{130}} = 0.277.
$$

This high coupling coefficient eliminates the composite -31 mode transducers. Since size is very important, we limit ourselves to torsional-mode or flexural-mode resonators, which can be stacked in parallel as shown in Figure 4.27. Only the pairs, in 4.27, designated NEC, NTT, and Siemens II are capable of realizing a 0.277 coupling coefficient. Of these, we choose the NTT for investigation.

The torsional resonator we are investigating is shown in Figure 4.35. The coupling coefficient is

$$
k_{em} = \left[\frac{64}{\pi^4} \times \frac{\ell_t}{\ell_2} k_{15}^2 \cos^2\left(\pi\frac{U}{\ell_t}\right)\sin^2\left(\frac{\pi}{2} \times \frac{\ell_2}{\ell_t}\right)\right]^{1/2},
$$

where

$$
\ell_t = \ell_1 + \ell_2 + \ell_3 \quad \text{and} \quad U = \left(\frac{\ell_t}{2}\right) - \left(\ell_1 + \frac{\ell_2}{2}\right).
$$

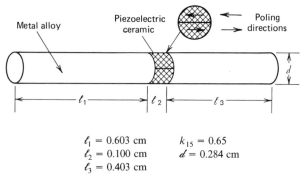

$$\ell_1 = 0.603 \text{ cm} \qquad k_{15} = 0.65$$
$$\ell_2 = 0.100 \text{ cm} \qquad d = 0.284 \text{ cm}$$
$$\ell_3 = 0.403 \text{ cm}$$

FIGURE 4.35. Torsional-mode transducer for realizing high electromechanical coupling coefficients.

Because the thickness of the ceramic is small, we assume that the resonator length ℓ_t is approximately that of the resonators in Figure 4.22, that is, 1.106 cm. To obtain maximum coupling, the ceramic should be centered on the bar ($\ell_1 = \ell_3$), although this is not always desirable if the resonator support is also at the center. Let's assume that the support is at the center and the dimensions are those shown in Figure 4.35. We choose TDK-61A, which has a k_{15} of 0.65 (see Table 2.6), as our transducer material. We may have to use a higher coupling material such as Tokin NEPEC-8 ($k_{15} = 0.78$) if this is not adequate.

Substituting the numbers from Figure 4.35 into Equation (4.14), we obtain

$$U = \left(\frac{1.106}{2}\right) - \left(0.603 + \frac{0.10}{2}\right) = -0.100$$

$$k_{em} = \left[\frac{64}{\pi^4} \times \frac{1.106}{0.100}(0.65)^2 \cos^2\left(\frac{\pi(-0.10)}{1.106}\right) \sin^2\left(\frac{\pi}{2} \times \frac{0.10}{1.106}\right)\right]^{1/2} = 0.238.$$

Using the $k_{15} = 0.78$ material, we obtain a k_{em} of 0.342.

We must now choose whether to increase the thickness of the -61A material or use the NEPEC-8 ceramic with a decreased thickness. Having made this choice, we should check the value of C_0 to be sure that its reactance is a practical value at 130 kHz. From Equation (2.22) and a relative permittivity of 1400 for the NEPEC-8 material,

$$C_0 = \frac{\pi}{4} \times \frac{d^2}{\ell_2}\left(\frac{\varepsilon_{22}^S}{\varepsilon_0}\right)\varepsilon_0 = \frac{\pi}{4} \times \frac{(0.284)^2}{0.10} \times 1400 \times 8.85 \times 10^{-14} = 78.5 \text{ pF},$$

or a reactance of 15.59 kΩ at 130 kHz. Using a thinner ceramic will reduce the reactance and also reduce the coupling to the point where the reactance and the terminating resistance are roughly equal in magnitude. A terminating resistance less than 15 kΩ should not present the circuit designer any problems. It is at this point in the design process that the filter designer must consider the effects of temperature and aging.

Designing for Constant Terminations. Having realized the desired transducer characteristics at room temperature, we must next implement ways of maintaining these characteristics over the specified temperature range. Table 2.1 showed the circuit parameters governing transducer performance, their function, and those external variables (such as temperature and aging) that affect these parameters. In this section we simplify the design problem by considering only three transducer characteristics: mechanical resonance frequency, electromechanical coupling, and transducer static capacitance, and two circuit parameters: mechanical resonator tuning, and terminating resistance.

An excellent method of relating the transducer and circuit parameters is described in Reference [28] and involves essentially the following:

1. Write equations that relate the final circuit model to the original circuit before transformations. An example of the final circuit is Circuit Type 4 in Figure 4.34; the original circuit may simply be the terminating resistance and a parallel tuned circuit.

2. Next find the derivatives of the tuning and terminating Q values of the original circuit with respect to the three transducer parameters.

3. Finally, write a set of linear equations that relate the total shift in the tunings or the terminating Q to the sum of the shifts due to variations of the transducer parameters.

Having a linear set of equations relating the transducer variables to the circuit parameters, the designer then "zeros" the circuit parameter changes (thus keeping the frequency response of the filter constant) by the choice of materials, electrical circuit tuning frequency, metal heat treatment, temperature compensating capacitors, and circuit transformations. This technique can be used for both narrowband filters and filters with tuning coils. Let's illustrate the method through the least complex circuit, which is the Type 1 in Figure 4.34. To simplify the problem we ignore the resonator losses.

Figure 4.36(a) shows the transformations involved in working backward from the final circuit, which contains the transducer parameters and the terminating resistance, to the original circuit, which is simply a resonator and a parallel resistance. The transformations between the series elements, R_s and C_s, and the parallel values, R_p and C_p, are shown in Figure 4.37. These equations are only valid at a single frequency, but in the design of narrowband filters, the approximation holds very well over the entire filter passband. These are a very important set of transformations and are used elsewhere in this book.

Applying the transformations in Figure 4.37 to the circuits in Figure 4.36(a), we obtain

$$\omega_a = \omega_1 \left[1 - \frac{1}{2r(1 + x^2)} \right] \tag{4.15}$$

and

$$Q_a = \frac{R_p}{X_{C_1}} = \left(x + \frac{1}{x} \right) r, \tag{4.16}$$

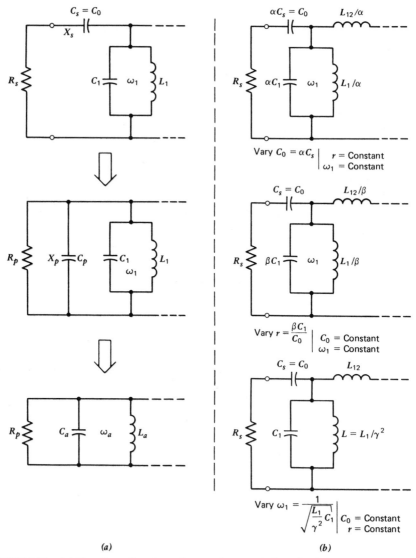

$$(a) \qquad\qquad\qquad (b)$$

FIGURE 4.36. (a) Transformations from the transducer equivalent circuit back to the prototype circuit, and (b) transducer variation conditions.

where the nominal values for x and capacitance ratio r are

$$x = R_s \omega C_s = \frac{1}{Q_s}$$

$$r = \frac{C_1}{C_s} = \frac{1}{K_{em}^2}.$$

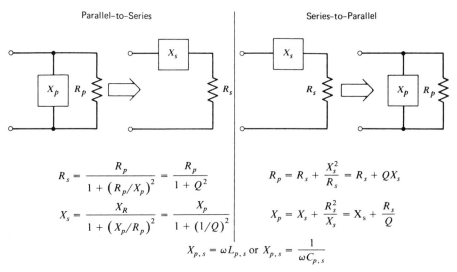

FIGURE 4.37. Parallel-to-series and series-to-parallel transformations at frequency ω which are useful for narrowband circuits.

The networks in Figure 4.36(b) show the conditions for varying C_0, r and ω_1. If we differentiate Equations (4.15) and (4.16) with respect to x, r, and ω_1, and then multiply both sides of the equation by the differential of the independent variable and then multiply and divide both sides by the nominal values, we obtain expressions like:

$$\frac{d\omega_a}{dx}\bigg| = \frac{\omega_1 x}{r(1+x^2)^2} \quad \text{or} \quad \frac{\delta\omega_a}{\omega_a} \simeq \left(\frac{x^2}{r(1+x^2)^2}\right)\frac{\delta C_0}{C_0}$$

$$\omega_1 = \text{Constant}$$

$$r = \text{Constant.}$$

From these equations we can write equations for the total frequency shift and change in Q_a (termination),

$$\frac{\delta f_a}{f_a} = \left[1 - \frac{1}{2r(1+x^2)}\right]\frac{\delta f_1}{f_1} + \left[\frac{1}{2r(1+x^2)}\right]\frac{\delta r}{r} + \left[\frac{x^2}{r(1+x^2)^2}\right]\frac{\delta C_0}{C_0}$$

$$(4.17)$$

and

$$\frac{\delta Q_a}{Q_a} = \left[\frac{x^2-1}{x^2+1}\right]\frac{\delta C_0}{C_0} + \frac{\delta r}{r}.$$

$$(4.18)$$

The design problem is to set

$$\frac{\delta f_a}{f_a} = 0 \quad \text{and} \quad \frac{\delta Q_a}{Q_a} = 0.$$

Let's look at an example to illustrate the foregoing.

Example 4.10. Analysis of a Low-Frequency Filter. In this example, we will analyze the stability of an already-designed filter and look for ways of improving its performance. The filter is a narrowband, two-resonator type like that shown in Figure 7.34. The center frequency is 13.6 kHz, the bandwidth is 18 Hz, the composite transducer-resonator coupling coefficient $k_{em} = 0.05948$, the terminating resistance is 18 kΩ, and the static capacitance is 300 pF. The transducer material is TDK-61A. From this data,

$$r = \frac{1}{k_{em}^2} = \left(\frac{1}{0.05948}\right)^2 = 282.6$$

and

$$x = R_s \omega C_s = 18 \times 10^3 (2\pi 13.6 \times 10^3)(300 \times 10^{-12}) = 0.461.$$

Substituting r and x values into (4.17) and (4.18), we obtain

$$\frac{\delta f_a}{f_a} = (0.9985)\frac{\delta f_1}{f_1} + (1.459 \times 10^{-3})\frac{\delta r_0}{r_0} + (0.5115 \times 10^{-3})\frac{\delta C_0}{C_0}$$

and

$$\frac{\delta Q_a}{Q_a} = (-0.6495)\frac{\delta C_0}{C_0} + \frac{\delta r}{r}.$$

The $\delta C_0/C_0$ and so forth terms can be expressed as temperature or aging coefficients in parts per million (ppm) or percent, or can be fabrication or assembly tolerances, or can be expressed as changes due to signal level, and so on. For this example let us consider only the variations due to temperature. From Table 2.6 for TDK-61A material, $\delta r/r = -2 \times \delta k_{em}/k_{em} = +1000$ ppm/°C and $\delta C_0/C_0 = +1600$ ppm/°C (the capacitance change is proportional to the dielectric constant change). The resonators were designed to have a zero frequency vs. temperature slope at room temperature, so $\delta f_1/f_1 = 0$. Therefore,

$$\frac{\delta f_a}{f_a} = 0 + (1.459 \times 10^{-3})1000 \text{ ppm}/°C + (0.5115 \times 10^{-3})1600 \text{ ppm}/°C$$

$$= 2.277 \text{ ppm}/°C,$$

and

$$\frac{\delta Q_a}{Q_a} = (-0.6495)1600 + 1000 = -39.2 \text{ ppm} / °C.$$

For a temperature change of $+60°C$,

$$\delta f_1 = \frac{2.277 \times 60 \times 13,600}{1,000,000} = 1.858 \text{ Hz}$$

and

$$\frac{\delta Q_a}{Q_a} = \frac{-39.2 \times 60 \times 100 \text{ (percent)}}{1,000,000} = -0.235\%.$$

Since $\delta Q / Q_a \simeq 0$, we should simply change the heat treatment of the alloy bar so that the composite resonator has a slope of -2.277 ppm/°C. If $\delta Q_a/Q_a$ had been too negative, we would have reduced the size of the piezoelectric ceramic, which would have resulted in a larger value of x and a reduction in the magnitude of $\delta Q_a/Q_a$ [this is not obvious but can be seen by studying the transformations of Figure 4.36(a)].

Magnetostrictive Transducer Designs

Most magnetostrictive-transducer filters utilize an extensional-mode ferrite transducer rod which is wire-coupled to an iron-nickel alloy resonator. Examples of this type of section are shown in Figure 4.27; exceptions are direct attachment of alloy wires or ferrite rods to flexural-mode disks. The indirect attachment simplifies the design because the electromechanical coupling coefficient is that of the ferrite rod, magnet, and coil assembly; that is, the design of a composite resonator is not involved. The minimum requirement on the transducer coupling coefficient is that of the wideband or intermediate-band design as expressed in Figure 4.34,

$$k_{em} = k_{12}\frac{B}{f_0}, \tag{4.19}$$

where

$$k_{12} > k_{tr},$$

and where k_{tr} is the coupling between the ferrite rod and the metal resonator. The design of this type of filter is similar to that shown in Figure 4.3, but with an additional set of zero quads. One set corresponds to the electrical/mechanical k_{12} coupling and the other to the coupling between the ferrite transducer and the first metal resonator. The design problem therefore is (1) determining a

k_{tr} that is high enough to reduce the sensitivity of the filter to changes in the mechanical frequency of the transducer, but not so high that the electrical tuning becomes critical; and (2) realizing the transducer/interior-resonator coupled pair in a way similar to Example 4.6.

In the case of directly attaching a ferrite to a disk, the diameter of the ferrite is adjusted to balance the sensitivity of the filter to the temperature shift of the resonator frequency and the sensitivity of the filter to changes in the electrical tuning frequency.

REFERENCES

1. R. A. Johnson, "The design and manufacture of mechanical filters," in *Proc. 1976 IEEE ISCAS*, München, 750–753 (Apr. 1976).
2. E. Christian, *LC-Filters: Design Testing and Manufacturing*. New York: Wiley, 1982.
3. H. Betzl, "Ein Beitrag zur Berechnung von eingliedrigen Quarz-Brückenbandpässen mittlerer Bandbreite nach der Betriebsparametertheorie," *Frequenz*, **19**, 206–209 (June 1965).
4. A. I. Zverev, *Handbook of Filter Synthesis*. New York: Wiley, 1968.
5. ITT Staff, *Reference Data for Radio Engineers*. Indianapolis: Sams, 1974.
6. Y. Peless and T. Murakami, "Analysis and synthesis of transitional Butterworth-Thomson filters and bandpass amplifiers," *RCA Review*, 60–94 (Mar. 1957).
7. H. J. Orchard and G. C. Temes, "Filter design using transformed variables," *IEEE Trans. Circuit Theory*, **CT-15**(4), 385–408 (Dec. 1968).
8. G. Szentirmai, "Interactive filter design by computer," *Circuits and Systems*, **12**(5), 1–13 (Oct. 1978).
9. R. A. Johnson, "Mechanical bandpass filters," in *Modern Filter Theory and Design*, G. C. Temes and S. K. Mitra, Eds. New York: Wiley, 1973.
10. K. Sawamoto, S. Kondo, N. Watanabe, K. Tsukamoto, M. Kiyomoto, and O. Ibaraki, "A torsional-mode pole type mechanical channel filter," in *Modern Crystal and Mechanical Filters*, D. F. Sheahan and R. A. Johnson, Eds. New York: IEEE Press, 1977.
11. S. Cucchi and F. Molo, "Bridging elements in mechanical filters," in *Modern Crystal and Mechanical Filters*, D. F. Sheahan and R. A. Johnson, Eds. New York: IEEE Press, 1977.
12. M. Börner, "Mechanische Filter mit Dämpfungsspolen," *Arch. Elek. Übertr.*, **17**, 103–107 (Mar. 1963).
13. R. A. Johnson, "A single-sideband disk-wire type mechanical filter," *IEEE Trans. Component Parts*, **CP-11**, 3–7 (Dec. 1964).
14. B. Kohlhammer and H. Schüssler, "Berechnung allgemeiner mechanischer Koppelfilter mit Hilfe von äquivalenten Schaltungen aus konzentrierten elektrischen Schaltelementen," *Wiss. Ber. AEG-Telefunken*, **41**, 150–159 (1968).
15. B. Kohlhammer, "Ein neuartiges Entwurfsverfahren zur Synthese von mechanischen Filtern und von Gyrator-Filtern," *Wiss. Ber. AEG-Telefunken*, **43**, 170–177 (Fall, 1970).
16. F. Künemund, "Dimensionierung überbrückter Bandpässe mit Dämpfungsspolen," *Frequenz*, **24**(6), 190–192 (1970).
17. K. Wittmann, G. Pfitzenmaier and F. Künemund, "Dimensionierung reflexionsfaktor- und laufzeitgeebneter versteilerter Filter mit Überbrückungen," *Frequenz*, **24**, 307–312 (Oct. 1970).
18. G. C. Temes, "Asymmetrical loss-pole mechanical filter," U.S. Patent 3,725,828 (Apr. 1973).

19. A. E. Günther, "Electromechanical filters: satisfying additional demands," in *Proc. IEEE Int. Symp. Circuit Theory*, 142–145, (Apr. 1973).

20. A. E. Günther, "High-quality wide-band mechanical filters, theory and design," *IEEE Trans. Sonics Ultrason.*, **SU-20**, 294–301 (Oct. 1973).

21. T. Yano, T. Futami, and S. Kanazawa, "New torsional mode electromechanical channel filter," in *Proc. 1974 European Conf. on Circuit Theory and Design*, London, 121–126, (July 1974).

22. M. S. Lee, "Equivalent network for bridged crystal filters," *Electronic Letters*, **10**(24), 507–508 (Nov. 1974).

23. A. E. Günther and K. Traub, "Precise equivalent circuits of mechanical filters," *IEEE Trans. Sonics Ultrason.*, **SU-27**, 236–244 (Sept. 1980).

24. R. A. Johnson, "Electrical circuit models of disk-wire mechanical filters," *IEEE Trans. Sonics Ultrason.*, **SU-15**, 41–50 (Jan. 1968).

25. Y. Nakauchi, "A synthesis of parallel ladder circuits using equivalent transformation techniques," in *Proc. 1975 IEEE ISCAS*, Boston, 305–308 (Apr. 1975).

26. J. Trnka and V. Sobotka, "Respecting certain manufacturing and technological problems in the design of mechanical filters," in *Proc. 1979 IEEE ISCAS*, Tokyo, 1080–1081 (July 1979).

27. K. Yakuwa, S. Okuda, K. Shirai, and Y. Kasai, "128 kHz pole-type mechanical channel filter," *Proc. 31st Annual Freq. Control Symp.*, 207–212 (June 1977).

28. T. Ashida, "Design of piezoelectric transducers for temperature stabilized mechanical filters," *Elect. and Commun. in Japan*, **57-A**(5), 10–17 (1974).

Chapter Five _____

MECHANICAL DESIGN

In Chapter 4 we looked at the design of mechanical filter resonators, transducers, and coupling elements in terms of a desired frequency response and stability, and available materials. Not included were subjects often lumped under the category "mechanical design," which include designing for factory processes, support and package design, and design in light of performance criteria such as shock and vibration. A complete design matrix includes:

Design Evaluation Criteria	Component Parts	Factory Processes
Electrical performance	Resonators	Materials
Sensitivity and stability	Transducers	Fabrication
Environmental tests	Coupling wires	of parts
Size requirements	Bridging wires	Tuning and
Production costs	Electrical	adjusting
Reliability	components	Assembly
Patent status	Supports	Testing
	Packaging	

Looking at the design evaluation criteria, we see that it leans toward items of concern to the filter user. Therefore, an understanding of the material presented in this chapter is helpful in evaluating the strengths and limitations of mechanical filters when considering their use in a system. We don't have space to cover all subjects in the design matrix but we look closely at the most important, starting with size requirements.

DESIGNING FOR MINIMUM SIZE

When the physical size of components is important, for instance in telephone equipment, system designers have often turned to mechanical filters [1]. The size of a mechanical filter resonator is often smaller than the capacitor in an equivalent *LC* network, and even smaller than the resonator sections of most active filters. The ability to realize small resonators is made possible by the inherent high Q and the linearity at high signal levels of the resonator materials. In this section we will look at various resonator shapes and filter configurations, as well as fundamental limitations on resonator size.

Resonator Shapes and Modes of Vibration

One of the first items a mechanical filter designer must choose is a resonator mode that will be compatible with the packaging requirements. At low frequencies, flexure modes must be used, because extensional or torsional resonators become too long. At very low frequencies, like a few hundred Hz to 3 kHz, the flexural-mode resonator will most often be in the shape of a tuning fork, which is simply a folded bar resonator. A single resonator having input and output transducers can have a Q greater than 250, at a frequency of 300 Hz, and can be packaged in a volume of 0.6 cm³. At frequencies as high as 32.768 kHz, tuning forks using ZnO transducers have been designed with Q's of over 30, 000 and with a volume of 0.003 cm³ [2].

Figure 5.1 shows the range of frequencies over which mechanical filters using flexural, torsional, and extensional resonators have been used. Although the frequency ranges could be extended on the high end, there are practical limitations with regard to manufacturing tolerances, and physical limitations with regard to nonlinear behavior. The low-frequency end of the range is simply limited by the length of the resonator, or in the case of the flexure modes, how thin the resonator can be made and still be useful as a filter element. With the exception of the tuning fork, the frequency equations pertaining to these resonators are found in Chapter 3.

Mass Loading

Resonator size can be increased for ease of manufacturing by using higher order modes. Conversely, resonator lengths can be reduced by using composite

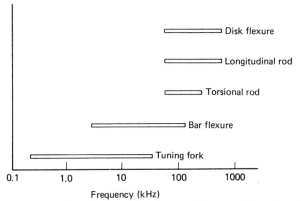

FIGURE 5.1.   Frequency ranges of various resonators used in manufactured mechanical filters.

resonator shapes like shown in Figure 3.18. The basic concept used in these dumbbell or mass-loaded designs is that of using necked-down regions to make the resonator act more like a spring-mass system. This technique has been used in the 64 to 108 kHz frequency range to build small telephone channel filters [3], and at 128 kHz, to realize a second-mode torsional resonator used for phase inverting purposes (Figure 7.15).

Multimode Resonators

Another method of reducing filter size is to use multiple mode (multimode) resonators. One example is the three-prong tuning fork, which has two funda-

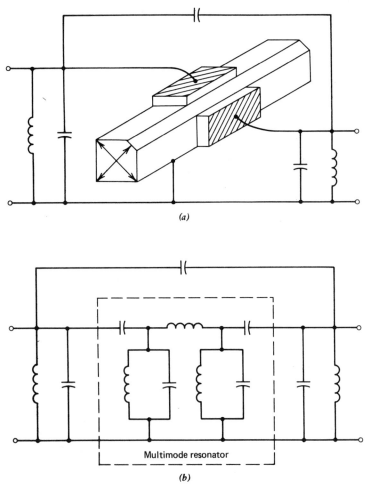

(a)

(b)

FIGURE 5.2. A multimode filter that employs electrical tuning and capacitive bridging. [© 1978 IEEE. Reprinted, with permission, from *Proc. 1978 IEEE ISCAS*, New York, 330–335 (May 1978).)

mental resonances, and another is the chamfered flexural-mode bar filter shown in Figure 5.2 [4]. The principle of operation of the chamfered filter is related to the fact that certain resonator shapes cause the natural modes of a resonator to be degenerate. By degenerate we mean, for example, that a square or round bar will vibrate at a single fundamental frequency around any axis in the plane of the cross section. This degeneracy is removed when the cross section is perturbed causing the resonator to vibrate about the two axes corresponding to the highest and lowest natural frequencies. For instance, the filter resonator in Figure 5.2(a) will vibrate in the direction of the arrows when one of the corners is chamfered. Used in conjunction with two transducers, this resonator provides two natural modes in a single bar for greater selectivity. An advantage of this type of design, over wire coupling, is that the multimode filter is free of spurious modes near the passband so that the fractional bandwidth can be as great as 15 percent. The tuning coils, which are needed in wide-bandwidth designs, can be used to increase the filter selectivity. The input-to-output capacitor provides additional selectivity by causing attenuation poles to be generated in the upper and lower stopband.

Configurations for Reduced Filter Size

As was mentioned in the previous example, additional stopband selectivity can be obtained by generating attenuation poles. Additional selectivity means that fewer resonators are needed and the filter size is reduced. In Chapter 4 we looked at various methods of wire bridging, but we were not concerned with filter size reduction. A design that combines both bridging and what we call folding is shown in Figure 5.3(a) [5]. The folding reduces the length by almost one-half and provides a method for multiple resonator bridging similar to that used in microwave filter circuits. The resonators in the folded design vibrate in an extensional mode, and the coupling is flexural. Other methods of folding have also been used, including the zig-zag design shown in Figure 5.3(b) [3]. In addition to the folded designs, space reduction has been realized by using a common transducer between two filters, as shown in Figure 7.19.

Limitations on Size Reduction

There are three basic limitations to size reduction beyond that of the resonator wavelength. These are: manufacturing tolerances, physical strength, and frequency and Q shifts due to the drive level. Let's start with the drive level.

Drive Level Sensitivity

Both resonance frequency and mechanical Q vary with changes in the signal power applied to the filter [6], [7]. These variations are determined by the size, vibration mode, and material of the resonator. For a particular mode of

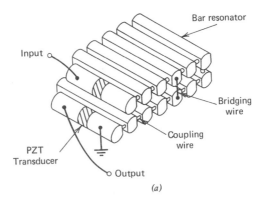

(a)

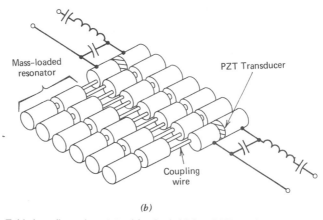

(b)

FIGURE 5.3. Folded configurations (a) with wire bridging (LTT), and (b) a zig-zag design with mass-loaded resonators (Kokusai Electric). [© 1978 IEEE. Reprinted, with permission, from *Proc. 1978 IEEE ISCAS*, New York, 330–335, (May 1978).]

vibration and material, the amount that the resonator volume can be reduced depends upon the Q and frequency shift allowed by the filter's frequency response specification. The shift in resonance frequency with signal level is usually the more important factor, therefore curves of only frequency shift versus energy-density level are shown in Figure 5.4. The curves show that there is more than one order of magnitude difference between each material, in terms of frequency change for a prescribed volume. This difference is closely related to resonator Q. Typical Q values are: Q (ferrite-core inductor and mica capacitor) = 200, Q (PZT-ceramic transducer) = 2,000, Q (Fe-Ni alloy) = 20,000, and Q (AT-cut quartz) = 200,000. A significant fact relating to resonator Q is that the nominal value of the Q of acoustic resonators is relatively independent of volume, whereas in LC-circuit inductors, the Q drops substantially with decreasing volume.

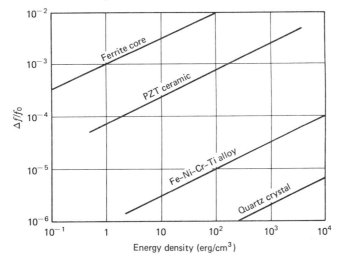

FIGURE 5.4. Frequency change versus energy density for various resonator materials. [© 1978 IEEE. Reprinted, with permission, from *Proc. 1978 IEEE ISCAS*, New York, 330–335, (May 1978).]

An example, which is discussed in Reference [8], is the determination of the minimum volume of a metal alloy resonator used in a bending mode at 48 kHz. The filter bandwidth is 3.250 kHz and the allowable frequency shift is 0.02 percent for a maximum input power level of -10 dBm. The calculations show that the minimum resonator volume is 0.0299 cm^3.

Manufacturing Considerations

It is necessary that the mechanical-filter designer look at limitations on building small filters in a factory environment. Some of the factors that must be considered are the machining of parts, resonator tuning, and wire-to-resonator welding. Machining of small parts involves variations in flatness, parallelism, and perpendicularity, which lead to welding variations, and in turn, resonator coupling variations. Also, deviations in resonator volume (and therefore in mass) result in coupling variations, which become important in the manufacturing of low passband-ripple telephone channel filters.

With regard to tuning, small resonators limit the methods that can be used for tuning when the sensitivity of the resonance frequency to material removal exceeds the tolerances of the grinder, drill, or laser. It is, therefore, difficult to tune a very small resonator to the better than 50 ppm allowable frequency shift needed in high-performance filter manufacturing. Another difficulty in the manufacture of very small mechanical filters is the ability to obtain consistent welding because of variations of wire diameter, straightness, and positioning.

The foregoing fabrication limitations are basically a function of the state-of-the-art of machining, material removal, and automation processes and should not be viewed as absolute limitations. In the case of channel filters, in the late 1970s the advancements in laser tuning allowed the production of small, extremely tight tolerance filters on an automated production line basis.

Physical Strength

Sometimes reductions in size are not possible because of a loss of structural strength. This causes problems in fabrication when hand operations are involved, but is not a major problem when assembly equipment is automated. Hand operations, as opposed to shock and vibration effects, which are scaled down with resonator size, impart fixed forces that cause bent support wires and coupling wires as well as broken bonds between ceramic and metal plates. Strength problems can often be circumvented by supporting each resonator at its nodal points, as opposed to using designs where the coupling wires are also used as the resonator support.

DESIGNING A RESONATOR SUPPORT SYSTEM

When designing a mechanical filter support system, the basic goal is to find means for the transducers, resonators, and coupling wires to vibrate freely, or at least vibrate in a predictable and consistent manner. Ideally, we would like to float the filter in a vacuum, but since this is not possible, we must look for ways of supporting it with wires, brackets, or soft mounts. Generally, the support system is composed of a primary support, a secondary support, and sometimes a tertiary support, as well as the filter case and the external mounting. The support system is also responsible for isolating the filter elements from external shock and vibration, which can cause both damage and unwanted microphonic responses. Figure 5.5 shows a block diagram of a resonator and its support system.

The most important characteristics of the elements in Figure 5.5 are the driving-point mechanical impedances, Y_r and Y_s, of the resonator and support at their point of attachment, the transmission of an external acoustic vibration of velocity v from the equipment to the resonator, and acoustic coupling to other resonators.

The Resonator Impedance

The effect of the support system on a resonator is dependent on the mechanical impedances Y_r and Y_s, shown in Figure 5.5. Y_s acts like a parallel impedance across the support terminals of the resonator. If the impedance of the resonator, looking into the support point, is Y_r, then the combined mechanical impedance of the resonator plus the support is $Y_r + Y_s$ (remembering that mechanical impedances add like electrical admittances). Therefore,

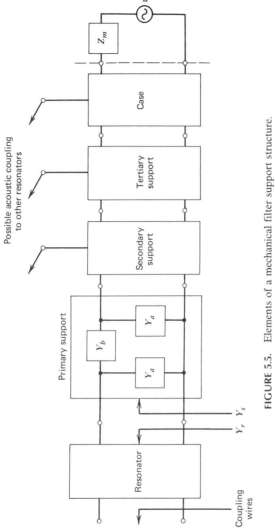

FIGURE 5.5. Elements of a mechanical filter support structure.

the larger we make Y_r, the less is the effect of the support impedance Y_s on the resonator's resonance frequency and Q. Ideally, we would attach the supports to the resonator nodes, causing Y_r to be infinite. It is possible to approximate this condition, in the case of torsional- and extensional-mode resonators, but in bending-mode resonators, regardless of the position of the support, some angular or linear motion is present. In other words, there is no point on a flexural-mode resonator where all impedances are infinite.

Of the different resonator types, the highest mechanical impedance is achieved at the nodal point of a torsional-mode resonator. This is because there is zero lateral expansion or compression at the nodal point, and it therefore acts as a true node. If the support is not attached exactly at the node, or has a finite width, the impedance Y_r cannot be infinite but can only approach infinity. The nodal points of an extensional-mode resonator are not true nodes in the radial direction (from the axis of the bar) but do act like high impedance points, in contrast to the flexural-mode support points.

Figures 2.19, 3.4, and 3.19 show the nodal point positions of flexural-mode bar resonators. The nodal points on the sides of these resonators are nodes of motion in the y and x directions but are not nodes of torsion. In this case, the input impedance Y_r is related to angular rather than linear motion and is therefore a function of the torsional inertia J_{eqx} of Figure 3.19. If the resonator is wire-supported on the top or bottom surfaces, directly above or below the nodal point, an additional set of terminals may be added to our multiport resonator to take into account the effect of the supports on the motion in the x-direction. Alternatively, a transformer, in cascade with a support equivalent circuit representing the effect of the support on the linear motion, can be connected in parallel across the torsional-mode terminals. The equivalent circuit for this example is shown in Figure 5.6. Note that as the resonator thickness approaches zero, the turns n_2 approach zero and the mechanical impedance $Y_{rf} = Y_{rt}/n_2^2$ as viewed from the flexural-mode wire equivalent circuit, becomes infinite. Any Δx errors in positioning the support will result in y-direction displacements, making it necessary to add another parallel flexural-mode equivalent circuit. In addition, if a resonator is supported by more than one wire, the equivalent circuits of the additional wires must also be added in parallel.

In the case of disk-wire filters, it is difficult to support each individual disk at its nodal point. Therefore, the coupling wires act as internal supports, as well as supports to a secondary support structure. In other words, the filter is supported at a low-impedance point, making the consistency of the support structure very important. Figures 7.1 and 7.24 show two mechanical filters that are supported by the coupling wires. The main wires are extended on the input and output ends to a tubular support structure. Advantages of this type of support are: (1) simplicity, because no additional wires or welding processes are needed; and (2) the impedance of the one-circle and two-circle modes is constant around the circumference of the disk.

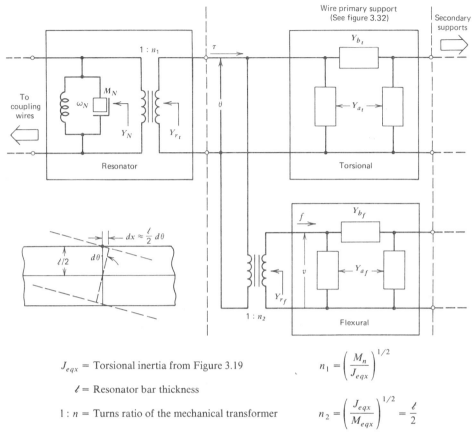

J_{eqx} = Torsional inertia from Figure 3.19 $n_1 = \left(\dfrac{M_n}{J_{eqx}}\right)^{1/2}$

ℓ = Resonator bar thickness

$1 : n$ = Turns ratio of the mechanical transformer $n_2 = \left(\dfrac{J_{eqx}}{M_{eqx}}\right)^{1/2} = \dfrac{\ell}{2}$

FIGURE 5.6. Parallel equivalent circuit effects of a support wire on a resonator supported at a point having both torsional motion and linear motion in the x-direction (i.e., support welding to the top or bottom of the resonator bar.)

Let's now summarize the major points in this section.

1. Each direction of motion of the resonator at the point or region of attachment of the support, must be accounted for.

2. Corresponding to each linear or angular motion is an equivalent mechanical resonator impedance.

3. Attached to the resonator terminals is a support equivalent circuit corresponding to each motion.

4. Mechanical transformers allow us to transform angular motion to linear motion and vice versa.

Resonator Support Structures

In this section we look at various types of resonator support structures corresponding to the block diagram in Figure 5.5. We categorize the supports in terms of their primary element: a wire, a bracket, or a soft foam or silicone material. We also study combinations of primary and secondary supports and the advantages and disadvantages of metals, hard plastics, and soft low-Q materials. Let's start with wire primary supports.

Wire Supports

The most common method of supporting flexural-mode resonators is with small diameter wires. Figure 5.7 shows a variety of primary wire-support methods using secondary supports such as wires and brackets, solder baths and plastic posts, and soft silicone elastomers. The wire provides a consistent low-impedance isolation element between the resonator and the secondary support. A hard secondary support provides a consistent impedance to the wire if spurious modes are not nearby, but does not absorb external shock. A plastic mount is excellent in terms of not introducing spurious modes but does not provide shock and vibration isolation. The silicone support does not provide an exact impedance, but unlike the other secondary supports, it reduces the effects of shock and vibration.

As a means of realizing a very-low-impedance support, it is possible to shape the wire in the form of an "S" or a "U" or the simple "L" shown in Figure 5.8(a). With the wire welded to the node of motion in the x and y directions, the motion transmitted to the support is in the form of a torque τ, and an angular velocity $\dot{\theta}$. At the bend of the "L", the torque τ_2 acts as a bending moment M_2 applied to the vertical leg, which is terminated in a hard support. The wire support, therefore, acts like the cascade connection of a torsional element and a flexural element, as shown in Figure 5.8(b). From Figure 3.32 we can calculate the Y_a and Y_b arm impedances, and from Equation (3.70), having set $f_1 = v_1 = \dot{\theta}_2 = v_2 = 0$, we can calculate the flexural input impedance Y_f, where, for a circular wire cross section,

$$Y_f = \frac{EI\alpha}{j\omega_0 \ell_2} \times \frac{1 + Cc}{Sc + Cs} = \frac{Ed^4\alpha}{j128 f_0 \ell_2} \times \frac{1 + Cc}{Sc + Cs} \qquad (5.1)$$

and

$$k = 2 \frac{\omega_0^2 \rho^{1/4}}{d^2 E}, s = \sin\alpha, S = \sinh\alpha, c = \cos\alpha, C = \cosh\alpha, \alpha = k\ell_2,$$

where E is Young's modulus, d is the wire diameter, f_0 is the center frequency of the filter, and k is the propagation constant of Equation (3.28)[9]. Let's apply these concepts to a practical problem.

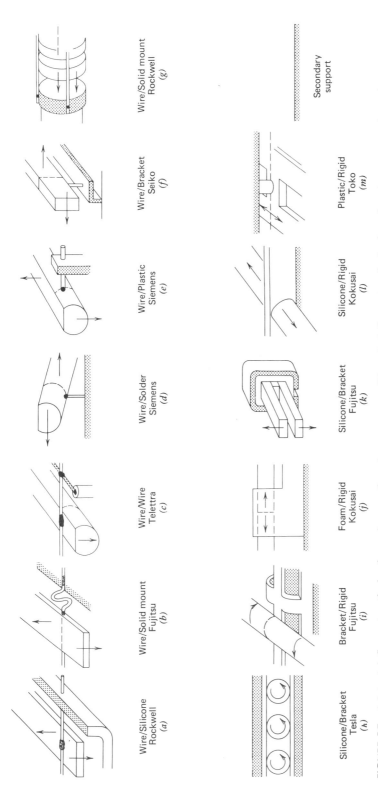

FIGURE 5.7. (*a*) to (*g*) Support methods where wires are the primary support. (*h*) to (*m*) Support methods where the primary support is a soft mount or a bracket.

Wire/Silicone
Rockwell
(*a*)

Wire/Solid mount
Fujitsu
(*b*)

Wire/Wire
Telettra
(*c*)

Wire/Solder
Siemens
(*d*)

Wire/Plastic
Siemens
(*e*)

Wire/Bracket
Seiko
(*f*)

Wire/Solid mount
Rockwell
(*g*)

Silicone/Bracket
Tesla
(*h*)

Bracket/Rigid
Fujitsu
(*i*)

Foam/Rigid
Kokusai
(*j*)

Silicone/Bracket
Fujitsu
(*k*)

Silicone/Rigid
Kokusai
(*l*)

Plastic/Rigid
Toko
(*m*)

Secondary
support

215

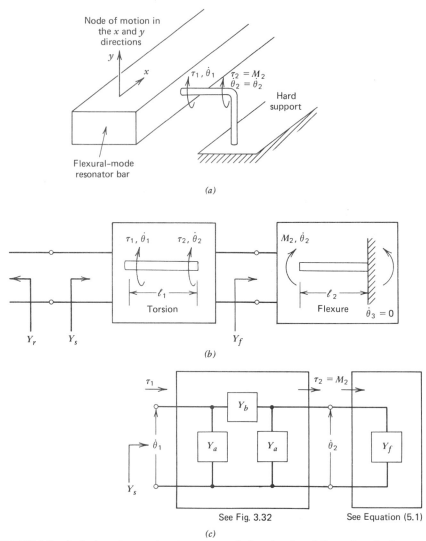

FIGURE 5.8.   An L-shaped support system composed of torsional- and flexural-mode elements.

Example 5.1.   Support Systems for a 12.080 kHz Pilot-Tone Filter.   This example is roughly based on the pilot-tone filter design shown in Figure 7.27. We assume, for ease of calculation, a rectangular cross section for the fundamental flexural-mode bar. The resonator dimensions are $\ell_r = 3.34$ cm, $w_r = 0.43$ cm, and $\ell_r = 0.27$ cm. The length ℓ_1 of the torsional arm of the support wire is 0.32 cm and the diameter d is 0.045 cm. The resonator material is Thermelast 4002 and the coupling wire is Thermelast 5409. The center frequency is 12.080 kHz. In this example, we vary the length, ℓ_2, of the flexural arm of the support and note its effect on the frequency of the resonator.

Let us first calculate Y_a and Y_b of Figure 5.4 from the torsional-mode equations of Figure 3.32.

At 12.08 kHz the wire acts like a short line. The torsional impedances are

$$Y_a = \frac{j\pi\omega_0\ell_1\rho d^4}{64} = \frac{j\pi \times 2\pi12.08 \times 10^3 \times 0.32 \times 8.16 \times (0.045)^4}{64}$$

$$= j0.0399 \text{ dyn-cm-sec}$$

and

$$Y_b = \frac{\pi d^4 G}{j32\omega_0\ell_1} = \frac{\pi(0.045)^4 \times 0.653 \times 10^{12}}{j32 \times 2\pi \times 12.08 \times 10^3 \times 0.32} = -j10.82 \text{ dyn-cm-sec.}$$

Let's look at five flexural arm lengths, $\ell_2 = 0.0, 0.2635, 0.4941, 0.6148$, and 0.6232 cm. From Equation (5.1),

$$k = 2\left[\frac{\omega_0^2\rho}{d^2 E}\right]^{1/4} = 2\left[\left(\frac{2\pi \times 12.08 \times 10^3}{0.045}\right)^2 \frac{8.16}{1.79 \times 10^{12}}\right]^{1/4} = 3.795$$

and therefore,

$$\alpha = k\ell_2 = 0.0, 1.0, 1.875, 2.333, \text{ and } 2.365.$$

Again, from (5.1),

$$Y_f = \frac{Ed^4(\alpha)}{j128f_0\ell_2}\left[\frac{1 + Cc}{Sc + Cs}\right] = \frac{1.79 \times 10^{12}(0.045)^4}{j128 \times 12.08 \times 10^3}\left[\frac{1}{\ell_2}\right](\alpha)\left[\frac{1 + Cc}{Sc + Cs}\right]$$

$$= \frac{4.747}{j\ell_2} \times (\alpha)\left[\frac{1 + Cc}{Sc + Cs}\right]$$

or

ℓ_2 (cm)	0	0.2635	0.4941	0.6148	0.6232
α	0	1.0	1.875	2.333	2.365
Y_f (dyn-cm-sec)	$-j\infty$	$-j17.05$	$j0$	$j10.910$	$j\infty$

The values of ℓ_2 were chosen to show that Y_f can vary drastically for small changes in ℓ_2. At $\ell_2 = 0$, the flexural arm acts like a rigid support; at $\ell_2 = 0.4941$ cm, the flexural arm acts like an impedance inverter, causing the rigid support to look like a zero impedance (free end) to the torsional arm; and at $\ell_2 = 0.6232$, the flexural arm again acts like a rigid support. Because Y_a has a very small value, we ignore its effects; therefore, the support impedances Y_s

(see Figure 5.8) can be found from

$$Y_s = \frac{Y_f Y_b}{Y_f + Y_b}$$

or

ℓ_2 (cm)	0	0.2635	0.4941	0.6148	0.6232
Y_s (dyn-cm-sec)	$-j10.82$	$-j6.619$	$j0$	$j\infty$	$-j10.82$

Note that at $\ell_2 = 0.6148$ the wire looks like an infinite impedance.

The equivalent stiffness of the resonator, from Figure 3.19 and Table 3.2 [at $x/\ell_r = 0.224$, $J_{eq\,r} = \rho A \ell_r/(4(-3.90/\ell_r)^2)$], is

$$K_r = \omega_0^2 J_{eq\,r} = \frac{(2\pi f_0)^2 \rho \ell_r\, w_r \ell_r}{60.84}$$

$$= \frac{(2\pi \times 12.08 \times 10^3)^2 8.3 \times 0.27 \times 0.43 \times 3.34}{60.84}$$

$$= 34.02 \times 10^8 \text{ dyn-cm.}$$

The stiffness of the support for $\ell_2 = 0$ is

$$K_s = j\omega_0 Y_s = 2\pi \times 12.08 \times 10^3 \times 10.82 = 8.21 \times 10^5 \text{ dyn-cm.}$$

The frequency shift in this case, due to the stiffness of a single support wire, is approximately

$$\Delta f\Big|_{\ell_2=0} \simeq f_0 \Delta K/2K = f_0 K_s/2K_r = \frac{12.08 \times 10^3(8.21 \times 10^5)}{2(34.02 \times 10^8)} = 1.46 \text{ Hz,}$$

whereas, for $\ell_2 = 0.4941$ cm, since the frequency shift is proportional to the wire stiffness ratios,

$$\Delta f\Big|_{\ell_2=0.4941 \text{ cm}} = \frac{0}{10.82} \times 1.46 = 0 \text{ Hz.}$$

In the case where $\ell_2 = 0.6148$, that is, at the frequency where $Y_f = Y_b$, the support wire "clamps" the resonator, causing an attenuation pole.

We must remember that for ease of explanation we ignored the shunt element (Y_a) effects, which contribute a small amount of resonator frequency shift.

Let's next compare the L-shaped support method to one consisting of a straight support wire attached to a rubberlike mount. Figure 5.9 shows both

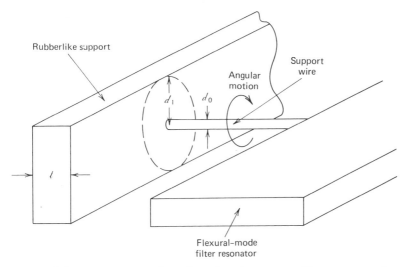

FIGURE 5.9. Soft support structure for a flexural-mode resonator. The support consists of a straight torsional-mode wire penetrating an amount ℓ into a rubberlike support.

the support wire, which is butt-welded to the nodal point of the resonator, and the elastomer support structure. We can assume the following: that the wire is rigid and does not slip, the rubber grommet can be approximated by a circular disk of diameter d_1, and the grommet is clamped along its outer boundary. We also assume that there is only torsional motion of the support wire. The equation relating the spring constant K_e of the elastomer support, to the wire penetration ℓ, the wire diameter d_0, the disk diameter d_1, and the shear modulus of the elastomer G is

$$K_e = \frac{\pi \ell\, G}{\left(\dfrac{1}{d_0} \right)^2 - \left(\dfrac{1}{d_1} \right)^2}. \qquad (5.2)$$

We assume a support diameter d_1 of 0.25 cm, a wire diameter d_0 of 0.045 cm (as in the previous example), a penetration depth ℓ of 0.25 cm and a shear modulus G of 5.2×10^7 dynes/cm^2. The given shear modulus is that of natural rubber, which is about five times as stiff as neoprene and equivalent silicones. Substituting these values into Equation (5.2), we obtain the torsional stiffness of the elastomer:

$$K_e = \frac{\pi \times 0.25 \times 5.2 \times 10^7}{(1/0.045)^2 - (1/0.025)^2} = 8.55 \times 10^4 \text{ dyn-cm.}$$

From the previous example where the resonator stiffness K_r is 34.02×10^8 dyn-cm,

$$\Delta f \simeq \frac{f_0 K_e}{2 K_r} = \frac{12.08 \times 10^3 \times 8.55 \times 10^4}{2 \times 34.02 \times 10^8} = 0.15 \text{ Hz.}$$

In this example we ignored the minor effects of the support-wire mass and compliance. Also we assumed that the support wire is located at the nodal point of the resonator. If the support is positioned off of the nodal point, the wire will be driven in flexure, causing the elastomer support to present a higher impedance, which results in a greater frequency shift than the 0.15 Hz calculated.

DESIGNING FOR HIGH RELIABILITY

In this section we look at ways that reliability is built into the mechanical filter design. The design of reliable filters involves interaction with the filter user so that the size is great enough and the bandwidth is wide enough, that the physical structure will stand up under shock and vibration. In addition, the filter designer must use materials and factory processes that result in reliable parts. And finally, the designer must be clever in his choice of the physical structure of the mechanical filter, its support elements, and its packaging.

Failure Rates and Failure Mechanisms

We now look at two different types of failures: (1) catastrophic failures, where the filter simply does not operate, and (2) the degradation mode, where failures are due to the filter drifting out of the specification.

Catastrophic Failures

The most common catastrophic failures in ceramic transducer filters involve broken lead wires, bad solder bonds between the lead wires and the ceramic transducer plating, and broken or loose coupling wires [10]. There are also a small number of failures where the insulation resistance breaks down. In the case of ferrite-transducer filters, or ceramic-transducer filters with tuning coils, open coil leads are also a possible source of failure. The use of epoxy, to bond ceramic transducers to metal coupling wires or to bars, disks, or plates, is an additional source of failures when the bond is subjected to harsh atmospheric conditions.

It is, in fact, the environment that is responsible for most failures. Operation at room temperature, in moderate humidity, with no shock and vibration, and with normal operating signal levels, results in very few failures. In other words, the filter rarely fails because of the fact that it is a device with vibrating elements. The normal operating signal vibration levels are too low to cause fatigue of the wires, metal resonators, ceramics, ferrites, or the solder or epoxy bonds.

In the case of bar flexural-mode filters, the failure rate is related to center frequency and bandwidth; the lower the center frequency or the narrower the bandwidth, the higher the failure rate. For example, a very narrow-bandwidth filter will require small coupling wires and even small-diameter wire supports (in order to reduce resonator frequency variations). This weaker structure is more susceptible to shock or vibration damage. It is the responsibility of the designer to insure that the filter design is compatible with the environmental specifications; this may involve negotiating with the filter user in regard to the performance specifications.

Typical of what we can expect with regard to failure rate, as a function of the time the filter has been in service, is shown in Figure 5.10. The curve is based on data compiled by Fujitsu Ltd. on ten years of field operation. The filter is a two-resonator, ceramic-transducer, telephone signaling filter with input and output tuning coils and capacitors. Note that the infant mortality is considerably greater than the failure rate after about two years of service. From four years to an estimated nine years, the failure rate is level; the failures in this region are called chance failures. There seems to be no indication of wear-out failures, that is, the curve does not have the up-swing of the well-known bathtub failure curves. The failure rate is in Fit, or failures per 10^9 hours per filter. The reciprocal of the failure rate is the mean time between failures MTBF.

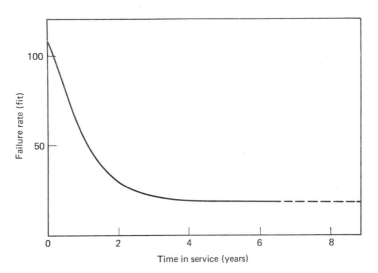

FIGURE 5.10.   Failure rate as a function of field operation time for a two-mechanical resonator, two-coil, telephone signaling filter of Fujitsu Ltd. One Fit equals 10^{-9} failures per hour per filter. [Adapted, with permission, from *FUJITSU*, **24** (1), 172–179 (1973).]

Failures Resulting from Aging

More common than catastrophic failures are failures due to the filter drifting outside of its specification. Types of failures can be as numerous as the specific electrical parameters, but the majority are due to frequency shifts of the electrical and mechanical resonators. These frequency shifts cause both center frequency changes and passband ripple variations. Other sources of failure are changes in the electromechanical coupling and the inter-resonator coupling and bridging. In some cases the aging rate is a function of operating temperature, but in others, for instance the aging of ceramic transducers, the drift is simply a function of time.

Designing for High Reliability

In designing to prevent catastrophic failures, the following must be observed: (1) materials and purchased parts must be well chosen, (2) the designer must take into account variations in the manufacturing process that could lead to structural or electrical failures, and (3) bonding processes must be chosen with reliability in mind. The following are some rules-of-thumb regarding reliability. Parenthetically, not all of these are universally agreed upon.

Material and Parts
Avoid outgassing, corrosion-producing elastomers (such as early types of RTV's).
Use noncorrosive flux.
Test every batch of ceramic and ferrite transducers for plating strength.
Use Litz wires rather than single-strand lead wire on ceramic transducers.
Avoid polyester foam supports, which may only have a few years' life.
Avoid unproven materials, parts, and vendors.

Processes and Design
Use solder bonds, rather than epoxy bonds, when possible.
Tightly control time and temperature in processes where there is soldering to the transducer plating.
Minimize hand work, automate processes when possible.
Design welds for strength, as well as consistency.
Give ruggedness priority at the expense of cost, size, or even performance.
Make sure that enough cleaning steps are included in the manufacturing process, so as to eliminate the chance of metal particle shorts.
Run environmental tests beyond the specifications.

With regard to filter aging, the best prevention against the filter drifting outside of its specification is to choose parts that exhibit a low rate of aging and to choose processes that minimize aging. For instance, a PZT ceramic transducer material can be chosen for its aging characteristics, although this may involve sacrificing Q or electromechanical coupling or temperature characteristics. The transducer-to-metal bond also involves a decision regarding aging. Should solder or epoxy be used, and what kind, and what should be the thickness of the bonding layer, and what kind of pre-aging is necessary?

Pre-aging is one means of insuring against excessive drift. This usually involves some pre-aging of the parts before assembly and aging of the entire filter before final testing and shipment. The aging may be in the form of temperature cycling, burn-in at an elevated temperature, driving the parts with a high amplitude signal, or simply storing the parts at room temperature for a specified period of time. In the case of accelerated aging, care must be taken to insure that the aging process does not damage the parts through excessive temperature extremes or vibration levels. If this care is taken and long-life materials and parts are used, the aging of a mechanical filter will decrease with time. This will make it possible to choose a pre-aging scheme that will guarantee compliance with the specification during the desired lifetime of the filter.

INTERNAL INPUT-TO-OUTPUT FEEDTHROUGH

Both the mechanical filter designer and the filter user must prevent input signals from bypassing the filter and adding to the filtered output signal. This subject is discussed in Chapter 9, in regard to preventing external feedthrough, whereas in this chapter, we are concerned with feedthrough within the filter package. Input-to-output feedthrough can involve both mechanical and electrical signals. In either case, the result is reduced, or noncontrolled stopband attenuation. Let's first look at mechanical feedthrough.

Mechanical Feedthrough

There are two types of mechanical feedthrough. One is the superposition of rigid-body modes on the desired mechanical vibrations. This was briefly discussed in the previous chapter. The rigid-body modes involve a spring-mass interaction between the coupling and support wires (which act as springs) and the resonators (which act as masses or rotational inertia elements); an example was shown in Figure 4.26. Each rigid-body mode can be modeled by a parallel-tuned circuit that is transformer-coupled in a parallel connection to the mechanical input and output of the filter. This model is shown in Figure 5.11(a). Figure 5.11(b) shows the effect of rigid-body modes on the frequency response of a four-resonator flexural-mode mechanical filter. Note that these

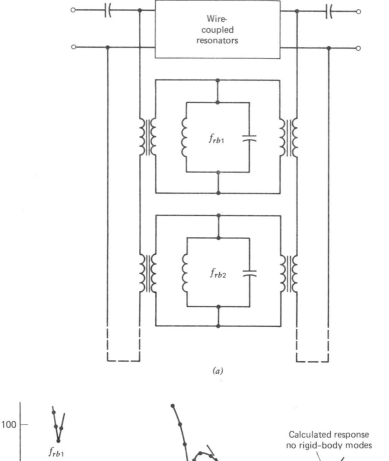

(a)

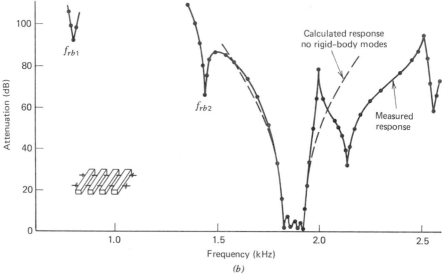

(b)

FIGURE 5.11. (a) Rigid-body mode feedthrough model. (b) The effect of rigid-body modes on the frequency response of a flexural-mode filter. (Courtesy of Fujitsu, Japan.)

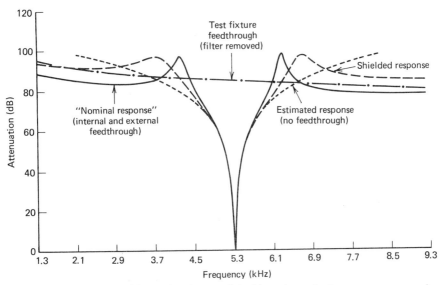

FIGURE 5.12. The effect of internal and external feedthrough on the frequency response of a two-resonator flexural-mode mechanical filter. (Courtesy of Rockwell International, USA.)

unwanted responses have high-Q peaks and dips that tend to detract from, rather than improve, the stopband frequency response.

Although there are numerous methods of reducing the effects of rigid-body modes, one of the best methods is to mount each resonator at its nodes with a wire terminated in a solid support. In this way, the rigid-body modes are raised in frequency and reduced in number over the frequency bands of interest. A disadvantage in this method is the possibility of transmission through the supports, as was shown in Figure 5.5.

One method of reducing input-to-output transmission through the supports is to mount the primary wire support in a low-Q secondary support, such as the solder bath shown in Figures 5.7(d) and 7.13, or in plastic. If rigid-body modes are not a problem, the wires can be mounted in an elastomer, such as silicone. Parenthetically, silicone mounts have been used to isolate the vibration of two signaling filters enclosed in the same package.

Electrical Feedthrough

In this section we look at magnetic and electric-field coupling between input and output circuits and ways of reducing these types of electrical feedthrough. Magnetic coupling is normally not a problem in intermediate-band piezoelectric-ceramic transducer filters, because the tuning coils are well shielded by the ferrite cup cores. Capacitive (electric-field) coupling is only a problem in two-resonator designs where the transducers are close to one another.

Figure 5.12 shows response curves of a low-frequency mechanical filter having two transducers of the type shown in Figure 2.1. The short-dash curve is the estimated monotonic response if no internal or external feedthrough is present. The dash-dot curve shows the feedthrough of the test fixture with the filter removed from the circuit. The long-dash curve shows the filter response when a grounded-conductor shield is placed between the two transducers. Note that the stopband attenuation poles are close to the cross-overs of the test fixture and monotonic response curves. Near those points, the non-bypassed and the feedthrough signals cancel. The solid curve is designated "nominal response" and is the result of having both test fixture and transducer feedthrough. Because this filter was being designed to meet an 80 dB stopband attenuation specification, and had to be automatically tested, it was necessary to spread the resonators apart to reduce the feedthrough on the high-frequency side of the passband.

A method of estimating the stopband attenuation change due to separating adjacent transducers is to use the equation for the capacitance between two plates whose major surfaces (the electrodes) lie in the same plane,

$$C_{ft} \propto \ln\left(\frac{2w + d}{d}\right) \tag{5.3}$$

where d is the distance between the nearest edges of the transducers and w is the electrode widths. As an example, the ratio of the capacitances of two transducers having widths w of 0.25 cm and spacings of 0.1 cm and of 0.5 cm is

$$\frac{C_{ft(d=0.1)}}{C_{ft(d=0.5)}} = \frac{\ln[2(0.25) + 0.1/0.1]}{\ln[2(0.25) + 0.5/0.5]} = \frac{2.58}{1}.$$

A 2.5/1 reduction in the feedthrough capacitance results in an 8 to 10 dB improvement in stopband attenuation outside of the attenuation-pole frequencies.

Feedthrough in magnetostrictive-transducer filters was briefly discussed in reference to the coil support (shield) tube shown in Figure 2.11. The support tube acts as a shield, preventing the input alternating magnetic flux field from coupling to the output coil if (1) the support is a high permeability magnetic material, such as a ferrite, or (2) the support is highly conductive, as in the case of brass or aluminum. The amount of shielding is a function of the thickness and length of the shield, the shield's distance from the coil, the distance between input and output coils, and the amount of additional shielding, such as the use of end-caps (see Figure 2.11). The effects of magnetic feedthrough are similar to those shown in Figures 5.12 and 9.10.

PACKAGING

Packaging involves supporting the filter structure, providing a means of getting the electrical signals in and out of the filter, and isolation of the filter from both its outside environment and from the package enclosure itself. The package must also be compatible with the user's space limitations and printed circuit board layout. Let's first look at the effect the enclosure has on the mechanical resonators, not through the support structure, but through the surrounding air.

Air Resonances

In the frequency range of 20 to 75 kHz, there is an interaction between air and flexural-mode mechanical resonators that causes variations in insertion loss and passband ripple. At these frequencies, the spacing between resonators, or the distances between the resonators and the enclosure cover (or base) are on the order of an acoustic (air) wavelength. Assuming that the wavelengths are large enough so there is little heat flow, the compressions and expansions are adiabatic and the equation for the velocity of the wave becomes $V = (B/\rho)^{1/2}$ where B is the adiabatic bulk modulus of the air (or gas) and ρ is the density. Also, the acoustic velocity in a gas is proportional to the square root of temperature (in degrees Kelvin); therefore, from reference tables we find for air that

$$V \simeq 2.0 \times 10^3 \sqrt{T_c + 273.16} \text{ cm/sec.} \tag{5.4}$$

Therefore, the velocity of an acoustic wave in air at 20°C is

$$V \simeq 2.0 \times 10^3 \sqrt{20 + 273.16} = 3.4 \times 10^4 \text{ cm/sec.}$$

At 10 kHz, a quarter wavelength ($\lambda/4$), in air and at room temperature, is 0.86 cm; at 50 kHz, $\lambda/4 = 0.17$ cm; and at 455 kHz, $\lambda/4 = 0.019$ cm. At frequencies below 10 kHz, distances between enclosure surfaces and the resonators are usually smaller than a quarter wavelength, and therefore, variations of resonator frequency and Q are small. Above 20 kHz, and to about 75 kHz, resonance effects become pronounced, but at high frequencies the spacings are on the order of many quarter wavelengths and the variations tend to "damp-out." Let's look at an example.

Resonator Damping at 43 kHz

Figure 5.13 shows the variation of the amplitude of the input admittance zero (impedance pole), of a two-terminal resonator, as a function of the distance of the resonator bar from a hard reflective surface. The resonator vibrates in a fundamental flexure mode, and the displacement is in a direction perpendicular to the hard reflecting surface. The ΔV minimums correspond to high Q points, and the $\Delta V = 0$ value is the amplitude where $d = \infty$.

The solid curve of Figure 5.13 is for the case where there is no damping material, whereas the dashed curve corresponds to the use of a 0.25 cm layer of 25 pores/cm polyether-polyurethane foam between the resonator and the

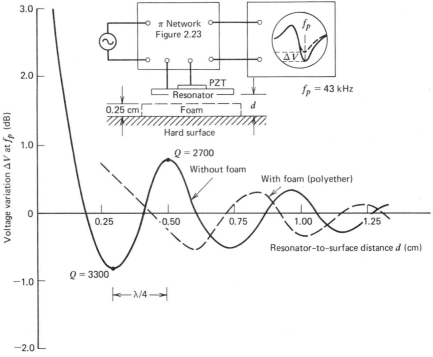

FIGURE 5.13. Variation of amplitude ΔV at impedance-pole frequency f_p as a function of the spacing d between a composite flexural-mode resonator and a hard reflection surface. Measured values at 43 kHz.

reflecting surface. Use of the foam shifts the first zero crossing an amount about equal to the thickness of the foam and reduces the amplitude of the peaks and dips. In the no-foam case, the resonator mechanical Q variation between the first dip and the first peak is from $Q = 3300$ to $Q = 2700$; the frequency shift between the dip and the peak is less than 1 Hz.

In Figure 5.13, a single surface was shown. In most practical cases, two surfaces are involved and the effects are additive. For example, if the resonator of Figure 5.13 is between a hard support ($d = 0.5$ cm) and a foam support ($d = 0.8$ cm), the Q is reduced ($\Delta V = +1.2$ dB). Conversely, when the resonator is mounted between two hard supports at $d = 0.30$, the Q is improved to 3500.

Reducing the Effects of Shock and Vibration

We are concerned with external shock and vibration because of both damage to the filter and the generation of microphonic responses. Many mechanical filters are hard mounted and rely on the strength of the resonator supports for protection against damage. Other filters are soft mounted with rubberlike materials or foams. An ideal structure, for shock protection, has a strong primary wire support system or bracket which is suspended in a soft mount. The improved shock resistance from using a large wire or bracket support must be balanced against the potential resonator mistuning effects.

With regard to vibration, the designer must take into account both the amplitude and the frequency of the vibration. Since most vibration specifications place greatest emphasis on low frequencies, it is often an advantage to stiffen the support system in order to move the structural resonances above the specification limit. If this is not practical, a damped (low Q) support will reduce the amplitude at the resonance frequency. Although there are lower Q materials, neoprene and the silicones have Q values in the range of 3 to 5 and will maintain their elastic and damping characteristics for more than 20 years.

As estimate of the fundamental resonance frequency of a wire/resonator structure mounted to a soft support can be made by first applying a force to the resonator assembly and measuring its deflection. Then, from Hook's law, we can calculate the spring constant K. From the resonator dimensions and density, we can find the mass of the resonator assembly. Knowing the spring constant and mass, we can calculate the resonance frequency. In this type of calculation, we assume that the coupling-wire and support-wire stiffness is much greater than that of the soft mount. Let's look at an example.

A mass of 275 gm is applied to a flexural-mode resonator structure supported by a soft mount, such as that shown in Figure 7.38. The measured displacement x is 0.061 cm and the total volume of the two Ni-Span C resonators is 0.19 cm^3.

From Hook' law and the fact that $F = mg$,

$$K = \frac{F}{x} = (275 \text{ gms} \times 980 \text{ cm/sec}^2)/0.061 \text{ cm} = 4.42 \times 10^6 \text{ dyn/cm}.$$

From the equation of the resonance frequency of a simple spring-mass system,

$$f = \frac{1}{2\pi}\sqrt{\frac{K}{M}} = \frac{1}{2\pi}\sqrt{\frac{4.42 \times 10^6 \text{ dyn/cm}}{0.19 \text{ cm}^3 \times 8.3 \text{ gm/cm}^3}} = 266 \text{ Hz}.$$

The resonance frequency can be raised or lowered by varying the stiffness (the shore or durometer value) of the mount. This is especially helpful if the g-level, as a function of frequency, is known at the point where the filter is mounted in the equipment. The filter resonance can simply be moved away from the structure resonance.

Enclosure Design

In this section we look at the design of the mechanical filter cover and base assemblies. More specifically, we consider various cover and base materials and sealing methods. Although filters with cylindrical shape enclosures (with terminal pins at the ends) are still being manufactured, most filters are designed in a box-shaped package with terminals compatible with a printed circuit board. The number of terminals normally ranges from three (a common input/output ground) to six (two grounds and floating coils), although some filters have as many as 10 terminals. Also, mounting studs may be included on the base if there is some question regarding the ability of the terminals to withstand shock or vibration.

Enclosure Materials

Mechanical filter enclosures are made of metal or plastic, or combinations of both. In the case of piezoelectric transducer filters, metal enclosures can be magnetic or nonmagnetic. Because of magnetic field considerations, the magnetostrictive-transducer filters usually use nonmagnetic materials, such as brass or copper with tin or nickel plating. The type of metal and plating largely depends on the sealing method.

Plastic enclosure materials are of two types, thermo set or thermo plastic. Both should be flame retardant, but the thermo set material will withstand higher terminal-pin soldering temperatures without melting. The advantages of the thermo plastic covers and bases are that they can be manufactured cheaper and that they can be used with ultrasonic or heat-stake sealing methods.

The choice of a plastic or a metal enclosure is primarily dependent on whether a hermetic seal is needed. If the hermetic seal is not needed, the plastic will usually be chosen because of its lower cost.

Sealing

Mechanical filters are built with materials that are considered noncorrosive under normal atmospheric conditions. Therefore, hermetic sealing is needed only under conditions of high humidity and freezing temperature, or where the filter is immersed in a liquid, as in a printed circuit board wash. The hermetic seal only needs to pass a simple immersion test (no bubbles when submerged in hot water) to prevent damage caused by heavy condensation. Although mechanical filters are usually hermetically sealed in a dry nitrogen atmosphere, room atmospheric conditions are often sufficient.

The most common method of hermetically sealing a metal package is with solder, although cold welding is sometimes used on small parts. Plastic packages can be sealed with epoxy, by ultrasonic bonding, by heat staking, or through use of a snap-on construction. If it is necessary to keep moisture out of the package, a post-coating can be used. Snap-on and metal-tab construction are used on metal-plastic combination packages.

REFERENCES

1. R. A. Johnson and K. Yakuwa, "Miniaturized mechanical filters," in *Proc. 1978 IEEE ISCAS*, New York, 330–335 (May 1978).

2. S. Fujishima, H. Nonaka, T. Nakamura, and N. Nishiyama, "Tuning fork resonators for electronic wrist watches using ZnO sputtered film," *1st Meeting on Applications of Ferroelectric Materials in Japan* (1977).

3. K. Sawamoto, T. Yano, K. Yakuwa, Y. Koh, and M. Konno, "Electromechanical filters developed in Japan, Part 2: Channel EM Filters," *L'Onde Electrique*, **58**(6–7), 482–487 (June–July 1978).

4. M. Konno and Y. Tomikawa, "An electromechanical filter consisting of a flexural vibrator with double resonances," pp. 157–166 of *Modern Crystal and Mechanical Filters*, Ed. by D. F. Sheahan and R. A. Johnson. New York: IEEE Press, 1977.

5. H. H. Ernyei, "Filtre de voie mécanique miniaturisé LTT," *L'Onde Electrique*, **58**(2), 128–135 (Feb. 1978).

6. K. Yakuwa and S. Okuda, "Design of mechanical filters using resonators with minimized volume," *Proc. 1976 IEEE ISCAS*, München, 790–793 (Apr. 1976).

7. Y. Tomikawa, K. Sato, A. Otawara, and M. Konno, "Consideration on small size of low frequency vibrators," *J. Acoust. Soc. Japan*, **31**(5) 318–332 (May 1975).

8. K. Yakuwa, S. Okuda, K. Shirai, and Y. Kasai, "128 kHz pole-type mechanical channel filter," *Proc. 31st Annual Symp. on Freq. Cont.*, Atlantic City, 207–212 (June 1977).

9. K. Nagai and M. Konno, *Electromechanical Resonators and Applications*. Tokyo: Corona, 65 (1974).

10. K. Yakuwa, S. Okuda, Y. Kasai, and Y. Katsuba, "Reliability of electromechanical filters," *Fujitsu*, **24**(1), 172–179 (1973).

Chapter Six

MANUFACTURING

Because of the maturity of mechanical filter production technology, it is possible to make firm statements regarding the manufacturing techniques that are workable and the conditions under which these techniques are valid. We therefore concentrate on methods now being used, basic concepts and strategies as well as process details. This chapter should be of benefit to the filter designer and manufacturing specialist and also to the filter user who is concerned with such things as reliability as a function of basic processes, limitations of the technology, and long-term costs and manufacturability.

GENERAL PRODUCTION CONCEPTS

The production method used to build a specific filter type, of course depends on a multitude of factors. Some things to be considered are: production volume, filter specifications, availability and cost of capital equipment, operator skills, standardization and utilization of the equipment for other filters, factory designer skills and priorities on their time, and the ability to maintain automated equipment. Consideration of these factors determines the basic filter design and the production strategies to be used. Often the production method is simply what is available or is based on intuition; but when time and cost allow, the methods can be determined in a more analytical way.

Manufacturing Strategy

A formal means for attacking the method-of-production problems is as follows:

1. Determine preliminary tolerances on the circuit-parameter values of the electrical circuits, transducers, resonators, and coupling elements. Although the

analysis may start with L's and C's, it should result in a set of tolerance values for frequencies, Q's, equivalent resonator masses, coupling-wire compliances, coupling coefficient, and static capacitance (or inductance). This can be done with a Monte Carlo frequency-response analysis.

2. Determine the necessary tolerances on dimensions and material constants in order to be able to hold the circuit-parameter tolerances.

3. Determine the machines, materials, processes, measurements, parts, and people-skills necessary to satisfy step (2).

4. Look at methods of making corrections after machining or after assembly. These may involve sorting, trimming, and tuning and are designed to reduce the cost of step (3).

5. Each one of the preceding steps is dependent on the others, so trade-offs may be made of the type where coupling tolerances are relaxed and tuning tolerances are tightened in order to reduce the cost of a weld machine or a centerless grinder.

Mechanical filter production lines tend to fall into the following categories which are related to the above analysis.

1. Tight tolerances on dimensions, materials, and processes. This type of production line requires few correction steps.

2. Looser tolerances but many correction loops. For instance: depoling ceramic-metal composite transduces to obtain a tight electromechanical coupling tolerance rather than attempting to hold tight tolerances on ceramic and metal dimensions, ceramic coupling coefficient, and the attachment process. These correction loops may also be in macro form, where a measurement is made farther down the production line, then this information is fed back, and a correction in the process is made. For example, the filter bandwidth is measured after welding and assembly. This information is then fed back to the welder who can increase or decrease pressure or voltage, as is necessary, to force the response into the specified frequency limits.

3. Allow loose tolerances throughout and make most of the corrections after final assembly of the filter (minus the cover). This requires either very skilled operators or very complex auto-correction machines. As an example, resistors and capacitors can be added to the input and output electrical circuits to compensate for variations in transducer coupling-coefficient and static capacitance (or inductance).

Process Flow Charts

The production methods described in the previous section can be illustrated by the use of actual process flow charts. The first chart, Figure 6.1, shows the steps in manufacturing a 50 kHz telephone channel filter (the filter's performance characteristics are described in Chapter 7) [1]. It is clear from the figure

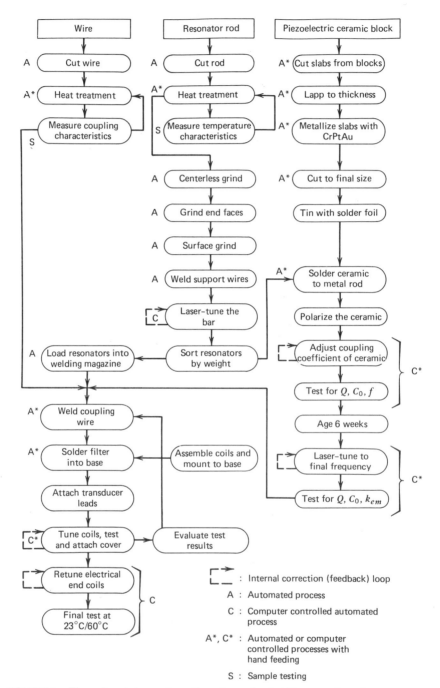

Wire	Resonator rod	Piezoelectric ceramic block

A — Cut wire
A — Cut rod
A* — Cut slabs from blocks

A* — Heat treatment
A* — Heat treatment
A* — Lapp to thickness

Measure coupling characteristics
S — Measure temperature characteristics
A* — Metallize slabs with CrPtAu

S

A — Centerless grind
A* — Cut to final size

A — Grind end faces
Tin with solder foil

A — Surface grind

A — Weld support wires
A* — Solder ceramic to metal rod

C — Laser-tune the bar
Polarize the ceramic

Sort resonators by weight
C* — Adjust coupling coefficient of ceramic } C*

A — Load resonators into welding magazine
Test for Q, C_0, f

A* — Weld coupling wire
Age 6 weeks

A* — Solder filter into base
Assemble coils and mount to base
C* — Laser-tune to final frequency } C*

Attach transducer leads
Test for Q, C_0, k_{em}

C* — Tune coils, test and attach cover
Evaluate test results

C* — Retune electrical end coils } C

Final test at 23°C/60°C

⌐→⌐ : Internal correction (feedback) loop
A : Automated process
C : Computer controlled automated process
A*, C* : Automated or computer controlled processes with hand feeding
S : Sample testing

FIGURE 6.1. Flexural-mode bar and coupling-wire mechanical filter manufacturing process steps. [Adapted, with permission, from Proc. IEEE, **67**(1), 102–108 (Jan. 1979). ©1979 IEEE.]

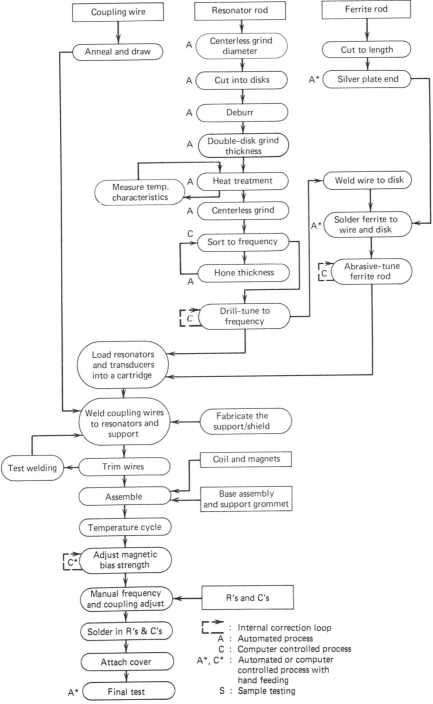

FIGURE 6.2. Disk-wire mechanical filter manufacturing process steps. The transducers are magnetostrictive ferrites. (Courtesy of Rockwell International.)

that this production line is highly automated and makes considerable use of computer control. This degree of automation and use of computers was possible because there was basically a single filter being manufactured. The required capital investment was possible because the filter was used in a high-volume application. The philosophy used in the development of this filter production line was to use materials and fabrication techniques that had the largest possible tolerances. These variations were then corrected or compensated by tuning, depoling, sorting bars, and adjusting coupling-wire parameters. The final result was a filter that met a tight specification with a high yield and required no trimming of resonators and wires after assembly. Most of the steps in this process flow chart will be discussed in subsequent sections of this chapter.

Figure 6.2 shows the process steps in manufacturing a 256 kHz, disk-wire channel filter (performance characteristics are described in Chapter 7). This filter used less automation because of a lower production volume and the availability of experienced, highly skilled people. Also disk-wire welding requires both translation and rotation of the parts, which results in assembled filters that require manual frequency and coupling adjustment. Because manual adjustment and the training of operators to perform this task is necessary, tolerances on all processes are made less stringent resulting in time and cost savings in the pre-assembly steps.

RESONATOR MANUFACTURING

The frequency and the frequency stability of a mechanical filter are determined primarily by its resonators. Therefore, mechanical filter manufacturers have put considerable effort into the development of both resonator materials and resonator manufacturing processes. In this section we summarize the most important aspects of this work. It is assumed in subsequent discussions that the resonator material is purchased in a cold-worked form and is ready for fabrication.

Material Considerations

In Chapter 3 we looked in detail at the performance characteristics of resonator materials. Therefore, in this chapter, we look only at characteristics related to production, such as machinability, weldability, heat size, cost, and lead time.

With regard to machinability, most iron-nickel alloys are difficult to machine because they are tough and ductile. The cutting edges of tools must be kept sharp or excessive work-hardening will take place. The work-hardening will not only decrease the rate of material removal but will cause the tools to dull even further. The cutting speeds should be relatively low and ample cutting fluid used. The resonator alloys lend themselves well to grinding or abrasive cutting, although the ductility of the materials may cause burrs, which must be removed as shown in Figure 6.2. Whether cutting or grinding, care should be taken to prevent a temperature rise that will cause excessive precipitation hardening. In using abrasive cutting to fabricate disks from rods, the surface temperature will cause some local precipitation hardening, but this material is removed during a subsequent grinding operation and the resonator's frequency versus temperature characteristics remain controllable. Let's next look at weldability.

Constant-modulus mechanical filter resonator materials are not easily weldable, so great care must be taken to develop and maintain good welding techniques. Small percentage changes in weld energy can make the difference between a good and bad weld. Selection of a Be-based alloy rather than Cr-based Elinvar alloy may reduce the sensitivity to weld energy changes, but other factors such as the welding machine characteristics, the coupling-wire material, resonator dimensional tolerances, and surface conditions can mask any gains resulting from choosing the most weldable material. Also, possible safety hazards in using Be must be considered.

Small changes in the chemical composition of the constant-modulus alloys result in relatively large changes in the temperature characteristics and in the elastic constants, namely Young's modulus and the shear modulus. In producing constant-modulus alloys, it is difficult to maintain an exact ratio of the elements. Thus each batch (heat, melt) of material is different, and coupling-wire compliances and heat-treatment temperatures may have to be adjusted. Therefore, batch size is a consideration in choosing a material. If the filter characteristics are not tight and if the alloy manufacturer is careful in his processing, then batch-to-batch differences may not be a problem.

Batch size, quality, process methods, and chemical composition all affect the material price. Therefore, the user will find wide differences when shopping for a resonator alloy. The user should also be aware that delivery time for a new heat of material may be very long.

Fabrication and Dimensional Considerations

In this section we look at methods of fabricating resonators, but first we briefly analyze the effect of dimensional variations on frequency and equivalent mass. These effects are best illustrated by an example.

Example 6.1. A 254.5 kHz disk resonator has a diameter of 1.4656 cm and a thickness of 0.2286 cm. The material is Ni-Span C and the disk vibrates in a two-circle flexural mode. We would like to know what tolerances to place on the diameter d and thickness t to insure that the resonator frequency falls between ± 600 Hz from 252.5 kHz. This will allow the disks to be sorted into six trays for further processing, the six trays each having a ± 100 Hz spread. We assume for this analysis that the elastic constants will not vary due to heat-treatment variations. We will also split the allowable frequency tolerance between the diameter and thickness variations. In other words, we will allow ± 300 Hz variation for the diameter and ± 300 Hz variation for the thickness.

Since for flexural-mode disks the frequency is governed by the proportionality $f \sim \ell/d^2$ we can write

$$\Delta d \simeq \left(\frac{f_1 - f_2}{2f_2}\right)d = \pm\left(\frac{300}{2 \times 252.5}\right)1.4656 = \pm 8.7\ \mu\text{m}$$

and

$$\Delta \ell = \pm\left(\frac{f_2 - f_1}{f_1}\right)\ell = \pm\left(\frac{300}{252.5}\right)0.2286 = \pm 2.7\ \mu\text{m}.$$

This example could also have been used to specify tolerances on flexural-mode bar resonators where length ℓ is substituted for diameter d.

The equivalent mass of the disk is proportional to its volume V so we can write:

$$\frac{\Delta M_{eq}}{M_{eq}} = \frac{\pm \Delta V}{V_1} = \pm\frac{V_2 - V_1}{V_1} = \pm\frac{d_2^2 \ell_2 - d_1^2 \ell_1}{d_1^2 \ell_1} = \pm\left[\left(\frac{d_2}{d_1}\right)^2 \frac{\ell_2}{\ell_1} - 1\right]$$

$$= \pm\left[\left(\frac{1.46647}{1.46560}\right)^2\left(\frac{0.22887}{0.22860}\right) - 1\right] = \pm 0.237\%.$$

The percentage change in equivalent mass defines the limits over which the coupling between resonators can vary due to mass variations. If a 1 percent maximum variation of coupling must be held, the above mass change will remove about 25 percent from the allowable variation. If this is not acceptable, the disks must be sorted by weight.

Machining to a Specified Diameter

Centerless grinding is the most common method of machining a rod or disk to a specified diameter. The word *centerless* is used to describe this grinding process because there is no attempt to maintain the original center. The drawing of Figure 6.3 shows the essential elements of the process.

Centerless grinding involves the use of two wheels, a grinding wheel and a regulating wheel. The grinding wheel removes material from the rod and forces and rod against the support and regulating wheel, whereas the regulating wheel both rotates the rod and feeds it past the grinding wheel. The rubber-bonded regulating wheel rotates at about one-tenth of the speed of the grinding wheel and is tilted 3 to 8 degrees to cause both the rod rotation and movement in the axial direction. The peripheral speed of the bar is the same as the regulating

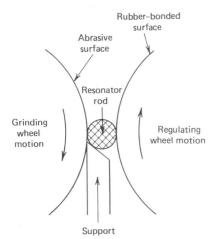

FIGURE 6.3.   Centerless grinding technique.

wheel while the rate of feed is determined by the rod diameter and the regulating wheel speed and angle of tilt.

Dimensional tolerances on the rod or disk diameter range from ± 1 micron to ± 10 microns depending on the machine and the diameter of the work. A tolerance of ± 3 microns is typical for high quality 3 to 4 mm diameter resonators used in telephone channel filters.

Machining to a Specified Thickness

There are a number of methods of machining a resonator to a specified thickness; these include surface grinding, double-disk grinding, honing, and lapping. The essential characteristics of each of these processes are shown in Figure 6.4 and Table 6.1. In all of these processes, material is removed by grinding.

Figure 6.4(a) shows the surface grinding process. The resonators are held to a table by means of a magnetic chuck. The table is moved under the grinding wheel at a speed and rate of material removal that will not cause excessive heating. The table movement is controlled automatically in factories where a large number of resonators are being processed. Aluminum oxide and diamond abrasives are used as the grinding material and tolerances as low as ± 2 microns can be held, but more typical is ± 10 microns.

Figure 6.4(b) illustrates both the double-disk grinding and honing processes. Through use of either a metallic or a nonmetallic carrier, the part is moved past a much-higher-speed grinding wheel or two grinding wheels in the double-disk case. The spacing between the two wheels determines the amount of material removal when using the double-disk grinder. The amount of material removal by the hone is determined by the amount of pressure between the resonator and the grinding wheel, or when the material is removed at a sufficient rate, by a stop. The hone is designed with a finer grinding surface than the double-disk grinder in order to remove smaller amounts of material, and therefore it can be used to either hold tighter tolerances or to tune resonators. Typical thickness tolerances in honing are ± 1 to ± 2 microns and, in the case of the double-disk grinder, ± 5 microns.

In lapping, the abrasive particles are in a liquid binder between the lap surface and the resonators. As shown in Figure 6.4(c), grinding takes place as the carrier moves the parts past the stationary upper and lower lap surfaces.

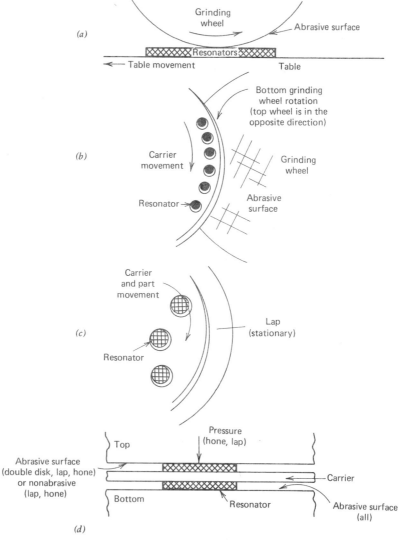

FIGURE 6.4. Methods of reducing resonator thickness: (a) surface grinding, (b) hone or double-disk grinding (top grinding wheel not shown), (c) lapping, (d) side view of hone, double-disk grinding, or lapping equipment.

TABLE 6.1. Characteristics of Machines Used for Material Removal in the Thickness Dimension

	Number of Surfaces Ground at One Time	Grinding Material Mounting	Constraints on Material Removal	Motion for Cutting Action	Resonator Positioning	Resonator Movement
Surface grinding	1	An abrasive wheel or abrasive is hardmounted to the wheel periphery	Position of wheel surface to the part	Wheel rotation	Magnetic chuck	Resonator is moved under the wheel
Double-disk grinding	2	Abrasive wheel	Position of the wheel (disk) surfaces to the parts	Grinding wheel rotation in opposite directions	Carrier held	Carrier rotation
Honing	1 or 2	Abrasive hard-bonded to flat plate(s)	Time, pressure rate of feed, and abrasive or position of a stop	Wheel (disk) rotation	Carrier held	Carrier rotation
Lapping	1 or 2	Abrasive in a liquid binder	Time, pressure rate of feed, and abrasive material	Part rotation in the carrier	Carrier held	Resonator is always in contact with the lap but follows the carrier

Lapping is a relatively slow process which depends on pressure, time, the lapping compound, and the carrier speed, but low-micron tolerances can be held. Figure 6.4(*d*) shows a side-view perspective of the double-disk grinding, honing, and lapping processes.

Machining to a Specified Length

One of the simplest ways to cut a bar or rod into sections is by shearing. Tolerances of ± 50 to ± 100 microns can be held using this process, and perpendicularity and length tolerances of ± 10 microns can be achieved by a follow-up grinding on one or both end surfaces.

An often used process for cutting rods into sections or disks is to use a very narrow abrasive cutoff wheel. The wheel rotates at a high speed, and through application of pressure it saws through the material; this is called abrasive cutting. As in the case of most of the grinding wheels discussed, aluminum oxide is an effective abrasive for cutting constant-modulus alloy resonators. Tolerances for this cutting process are on the order of ± 50 microns so a subsequent grinding step is usually required. A means of avoiding a second step is to use a cutting tool or series of tools. Using this method, tolerances of ± 10 microns can be held.

Heat Treatment

Heat treatment of Elinvar-type materials is performed for two reasons: the first is to relieve stresses in the material to reduce aging, and the second is to modify the resonator temperature-coefficient of frequency characteristics. A stress-relief step is necessary in all cases and involves raising the temperature of the part to a value just under that where precipitation hardening takes place. In the case of Ni-Span C, this is about 400°C. The material should be held at this temperature for 4 hours.

Regarding heat treatment to adjust the temperature coefficient, it is necessary to first make a series of tests to establish a proper heat-treatment temperature and time. This usually involves fixing the time and plotting the ambient-temperature versus frequency characteristics for different heat-treatment temperatures. An example of this type of curve is shown in Figure 3.27 of Chapter 3. Heat-treatment temperature ranges from a nominal value of 510°C for Sumispan EL-3 to 750°C for a special Durinval, and the suggested heat-treatment time ranges from 30 minutes to 4 hours.

In conducting tests and in subsequent production heat treating, because of positional temperature gradients, it is necessary to establish rigid process controls on the position of the resonators in the furnace. The lot size is also important because of time-temperature gradients in heating and cooling the parts. It is also important that the furnace be able to maintain a temperature tolerance compatible with the temperature-coefficient of frequency tolerance of the filter. In this regard, let us look at an example.

Example 6.2. In this example we determine the allowable heat-treatment temperature tolerance for a disk-resonator filter at 250 kHz. We are concerned with both the average variation of frequency shift and the disk-to-disk variations. If all resonators shift equally, the result is simply a center-frequency change of that amount, but a disk-to-disk variation will cause increased passband ripple. Choosing an allowable spread of ± 10 Hz at 60°C we can determine from Figure 3.27 the tolerance that must be held on the furnace temperature. We assume a heat-treatment temperature of 580°C which equalizes the frequency shift at 0°C and $+60$°C. The sensitivity of frequency shift to heat-treatment temperature, at 580°C, is approximately 5 ppm/°C. Calculating the heat-treatment temperature tolerance for a ± 10 Hz allowable frequency shift we obtain

$$\Delta T = \frac{1°C}{5 \text{ ppm}} \times \frac{\pm 10 \text{ Hz}}{250,000 \text{ Hz}} = \pm 8°C.$$

Most heat treatment is done in either vacuum or hydrogen furnaces. The vacuum furnace has the advantage of leaving little oxidation. Resonators heat treated in a hydrogen atmosphere may need to have the oxidation ground from the weld surface, although in some cases, additional grinding steps are necessary regardless of whether there is an oxidation problem.

Resonator Tuning

Some of the aspects of resonator tuning we discuss in this section are methods of detecting the resonance frequency, various means of frequency adjustment, and tuning strategies.

Frequency Detection: Drive and Pickup

In looking at the subject of frequency detection, we will examine both drive and pickup sensors and external detection circuits such as oscillators.

The most widely used sensor is the magnetostrictive drive and pickup shown in Figure 6.5(a). When dealing with flexural-mode resonators the coil is placed near one of the stressed surfaces, as is shown. The coil is first pulsed with a direct current causing magnetic flux lines to penetrate the metal resonator. Most of the lines take the shortest return path and therefore are close to the bottom surface near the coil, which causes only that portion of the metal to be magnetized. When the resonator is excited with an alternating magnetic field from the coil, the bottom surface expands and contracts, causing flexure. Conversely, flexural vibrations cause an alternating magnetic field to be produced by the magnetically biased portion of the bar; this induces a voltage across the coil. For extensional- or torsional-mode resonators, the coil should surround the resonator so that the flux distribution is uniform through its cross section to prevent spurious modes from being excited. In order to excite the

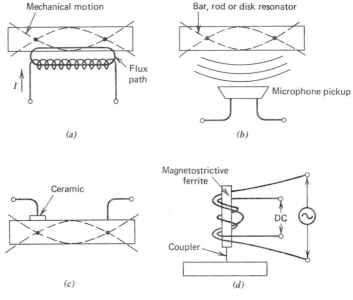

FIGURE 6.5. Frequency detection devices: (*a*) magnetostrictive drive and pickup, (*b*) microphone pickup, (*c*) piezoelectric ceramic transducer drive and pickup, and (*d*) magnetostrictive composite-resonator drive and pickup.

torsional mode, a circumferential magnetic bias is generated by applying a strong direct current along the axis of the resonator.

A commonly used sensor for filters, in the frequency range of 50 kHz and below, is the microphone pickup shown in Figure 6.5(*b*). A principal advantage of the microphone is that shielding is not needed between it and the magnetostrictive driver when used in a two-port resonator oscillator.

A third drive and pickup method is the use of a ceramic transducer bonded to the resonator, as shown in Figure 6.5(*c*). The ceramic may be part of the filter's input or output composite resonator or it may be added solely for the purpose of tuning, but it remains part of the filter. The final circuit, Figure 6.5(*d*), shows a method of measuring the frequency of a magnetostrictive composite resonator. In this case, a bias magnet or a second coil must be used to simulate magnetic-bias conditions in the actual filter.

Detection Circuits

Having looked at drive and pickup means, let us next consider various detection circuits. Three common circuits are the oscillator, the pulse-and-measure, and the stepped-frequency circuit. These three circuits can interact with either the driving-point (input) impedance of the drive/resonator circuit or the transfer function of a drive/resonator/pickup circuit.

The oscillator circuit can be designed to use the variation of phase and amplitude of the driving-point impedance in its feedback loop or use the transmission characteristics (transfer function) of a two-transducer circuit. The driving-point impedance can be that of the circuits of Figure 6.5(a), (c) or (d). An example of the transfer-function method is to use the magnetostrictive drive of Figure 6.5(a) on one side of the bar with the microphone pickup shown in Figure 6.5(b) on the other side of the bar. The output of the oscillator circuit is a direct reading of the resonance frequency of the resonator.

A second method of directly obtaining the resonance frequency is to pulse the resonator at a frequency near its natural resonance. This will result in the resonator ringing at its natural resonance frequency, after the drive is removed. The natural resonance signal, by returning through the drive circuit or from a separate pickup transducer, is then gated into a fast-count frequency counter.

A third method of detection is to use a generator or synthesizer to step through a frequency range, which includes the resonance frequency of the resonator, while at the same time monitoring the amplitude or phase of either impedance or transmission. This is a somewhat slower method than the other two but is more reliable in terms of consistently measuring the desired mode rather than spurious modes. In cases where the drive circuit has a very low coupling coefficient, the resonator can be pulsed and then the amplitude of the ringing signal is measured. The pulse frequency is then changed and the amplitude measured again. This is done until the maximum output, which corresponds to the resonance frequency, is found.

Frequency Adjustment

Having discussed means for measuring resonance frequency, let us now discuss adjustment methods. Table 6.2 lists tuning techniques that have been used by mechanical filter manufacturers. To this list could be added hand tuning methods such as sanding and filing or adding solder, but it is difficult to assign tolerances to these techniques. The largest frequency shifts are obtained by grinding and the smallest by sandblasting, drilling, or laser trimming. As an example, if we measure a resonator to be 8000 ppm higher than the final resonance frequency, we can grind on its major frequency controlling surface, and without repeated measurements, lower the frequency to within ± 800 ppm of the goal. If we start at close to 2500 ppm from the tuned frequency we could grind a major surface, but this would not be as accurate as grinding on a smaller surface such as an edge, where we see from Table 6.2 that we can obtain an accuracy of ± 10 percent of the starting frequency deviation. The more accurate methods such as drilling or laser trimming can be used, but these often cause excessive distortion of the part, such as deep holes, and may be too time consuming. The most accurate methods involve a feedback loop where the difference between the resonator frequency and the goal is measured, and then by removing successively smaller amounts of material the resonator is tuned.

TABLE 6.2.   Frequency Adjustment Tolerances from Ten Times the Values Shown[a]

Process	Tolerances without Feedback (Noniterative)	Tolerances with Feedback
Grinding a major surface	± 800	—
Grinding a minor surface	± 400	—
Edge grinding	± 250	± 100
Sandblasting	± 150	± 75
Drilling	± 100	± 40
Laser trimming	—	± 20

[a] Values are given in parts per million (ppm).

Example 6.3.   In this example we illustrate an automated method of tuning 128 kHz torsional-mode resonators. The resonators are used in low passband-ripple telephone channel filters which require precise tuning of less than ± 4 Hz. In Chapter 3 it was shown that the resonator equivalent mass at the end circumference of a torsional-mode rod is equal to one-quarter of its total (static) mass. Also, because frequency changes are related to the square root of stiffness divided by mass, we find for small tuning changes that

$$\frac{\Delta f}{f_0} = -2\left(\frac{\Delta M}{M}\right)\left(\frac{r}{a}\right)^2, \tag{6.1}$$

where M is the total mass of the resonator rod, ΔM is the removed mass, a is the radius of the resonator rod, and r is the point of the mass removal on the end surface. Figure 6.6(a) shows experimental frequency shift versus shot position results which correlate with the fact that Δf is a function of r^2 as shown in Equation (6.1). This data was obtained from a 30 W, YAG-Nd laser operating in a pulsed mode at 0.5 J/pulse. The shot position is determined by rotation of the laser lens. Figure 6.6(b) shows how the frequency shift per pulse varies with laser power [2].

On the basis of the frequency shift per pulse as a function of both radius and power, the tuning algorithm shown in Figure 6.7 has been developed. The end surfaces are first ground to where the starting frequency is within the laser tuning range. Laser tuning, from start to end, takes from 6 to 12 seconds, with coarse and fine tuning taking about an equal amount of time. Typical rates of material removal for both automated laser tuning and automated drill tuning are from 40 to 80 Hz/sec.

Tuning Strategies

In the preceding example, a laser tuning strategy was described where there were two laser-beam settings: coarse tune and fine tune. This system can

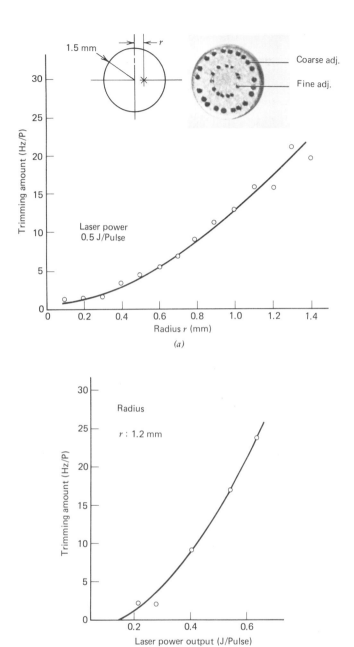

FIGURE 6.6. Torsional-resonator frequency shift as a function of (*a*) the distance of the laser shot from the axis of the rod, and (*b*) the laser output power. [© 1979 IEEE. Reprinted, with permission, from *Proc. IEEE*, **67**(1), 115–119 (Jan. 1979).]

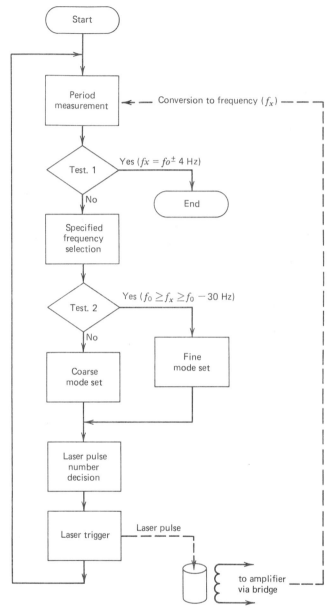

FIGURE 6.7. Method of tuning torsional-mode resonators used in telephone channel filters. [© 1979 IEEE. Reprinted, with permission, from *Proc. IEEE*, **67**(1), 115–119 (Jan. 1979).]

operate in a single laser-shot and frequency-measurement mode, or it can take a number of shots per measurement, the number of shots being proportional to the goal minus the actual frequency difference. Using the proportional method, more care must be taken to insure that there is not excessive frequency overshoot, but this method is usually faster.

Proportional methods can also involve setting an adjustment time, a number of rotations, a pressure, or a displacement, each being proportional to the desired frequency shift. An example is the abrasive tuning (sandblasting) of a ferrite transducer rod. The frequency difference is measured and the Al_2O_3 powder jet is turned on for a time that is proportional to 60 percent of the average time needed to completely tune the resonator. The frequency is again measured and the time again set to 60 percent of the time to completely tune the resonator. This process is repeated until the resonance frequency is within the specification. The percentage of total time is a function of the repeatability of the mass removal. Some algorithms are designed to take the condition of both the tuner and the part into account by calculating the rate of frequency change as a function of time (or displacement, rotations, pressure) for the first removal cycle. This value is then used for the second cycle after which it is updated and used for the third cycle, and so on.

Automatic Feed and Support Methods

When large numbers of resonators must be tuned, it is economical to develop automatic methods of feeding, holding, and tuning the parts. The tuning can be controlled by hard-wired circuits, microprocessors, minicomputers, calculators, and even large digital computers. The feeding of parts to the frequency adjustment equipment is usually by means of vibration trays although methods such as gravity feed are sometimes used. The tray vibrations cause the parts, which have been literally dumped into the tray, to move on their own to the tuner. At that point they fall into a slot, are rammed into position, or are grabbed by their support wires and the tuning begins. In order to prevent tuning-frequency errors, the resonators should either be supported at nodal points or across an entire surface with a fibrous pad. The torsional resonator of Example 6.3 provides an illustration. The rod has two sets of nodes on the surface: one is the axial nodal line appearing as nodal points at the ends, the other is the nodal plane across the center of the bar which appears as a circumferential nodal circle. The rod is vertically supported by a pivot at one end and by knife edges around the circumference.

COUPLING-WIRE FABRICATION

Mechanical filter bandwidth is determined primarily by the wire-coupling between adjacent resonators. For this reason, the characteristics of the coupling-wire material and the wire fabrication are very important. In the manufacturing of high-performance channel filters, the coupling tolerance must be held to less than ± 1 percent. This means that the product of the elastic modulus and the cross-sectional area must be held to a value considerably less than ± 1 percent because of other variations such as wire position, resonator mass, and weld characteristics.

Coupling-Wire Processing

The coupling wire is usually processed from a spool or reel, and after processing it is either cut into short lengths or remains on a reel, ready for welding. The processing involves a series of annealing and cold-working steps. Annealing is done by heating the wire, in the case of simple iron-nickel alloys, or by heating and then quenching the wire, in the case of the precipitation-hardenable materials. The annealing is then followed by cold work (drawing) to a percentage of the annealed cross-sectional area. In order to obtain consistent wire characteristics, the design of the drawing die, the rate of drawing, and the percentage of cold work in each pass must be determined for each type of coupling-wire material.

The percentage of cold work done to the completed wire is determined by the required hardness and strength of the part. If a small amount of weld deformation of the wire is required for the most consistent welds, then a large amount of cold work should be done to the wire. If the wire strength needs to be high, again the wire should have a large amount of cold work, on the order of 80 to 90 percent.

Following the final cold-working step, the precipitation-hardenable materials can be heat-treated to adjust the value of the elastic modulus. Since the coupling is proportional to the product of E or G and the cross-sectional area A, the heat treatment may involve adjusting the elastic modulus to make the EA product constant [1], [3]. The heat treatment will cause the temperature coefficient of the elastic modulus to vary, but this results in only a small change in filter bandwidth over temperature.

Coupling-Wire Measurements

As stated before, we are concerned with variations of the elastic modulus and cross-sectional area. Of course, the area can be found from a micrometer measurement of the diameter, but the value of E or G is more difficult to determine. Of most importance is the relative value of E or G from batch to batch. Because density variation due to cold work or heat treatment is very small, we can measure the relative value of Young's modulus by cutting a wire to a fixed length and measuring its extensional-mode resonance frequency f. From Equation 3.11, we can write

$$E = 4\ell^2 f^2 \rho \qquad (6.2)$$

or

$$\frac{\Delta E}{E} = \frac{2\Delta f}{f}.$$ (6.3)

PIEZOELECTRIC CERAMIC TRANSDUCERS

In this section we deal with the fabrication of PZT ceramic transducer plates and the attachment of the plate to metal bars, disks, or rods to form the composite resonators discussed in Chapter 3. In addition, we discuss methods of coupling coefficient adjustment.

Fabrication of Ceramic Transducer Plates

We do not provide a highly detailed description of the fabrication of PZT plates because these parts can be purchased from vendors that make the material, cut it, plate it, and polarize it. We assume in our discussion that we start with a nonpoled block of material. The following description is based largely on that of Reference [1].

The PZT material arrives from the vendor in the form of slabs, for example, 50 mm by 30 mm by 12 mm. Using mill saws, these slabs are cut into plates and then are lapped to thicknesses on the order of 0.8 mm. The plates are cemented together and sawed and lapped to the desired final thickness and length. The plates are separated and are plated on both major surfaces with a conductive, solderable material, for instance Cr-Pt-Au. The top layer of the conductive plating is usually gold or silver, and the plating process can be that of sputtering, evaporative plating, or even silk screening.

The polarization process can take place either before mounting the ceramic plate to the metal portion of the composite resonator or afterward. As mentioned in Chapter 2, the process is performed at a temperature below the Curie temperature (sometimes at only a slightly elevated temperature) in a 10 to 50 kV/cm field and in an oil bath to prevent flashover. In the case of torsional-mode transducers, the plates are polarized, the plating is removed and the transducer is replated on a major surface 90 degrees from the original plated surface.

Ceramic-to-Metal Bonding

Methods of bonding the ceramic plates to metal bars, rods, or disks are almost as numerous as there are mechanical filter manufacturers. For instance, the bonding materials include solders and epoxys. The solders may be in foil or paste form and are of many compositions, as discussed in Reference [4]. Because the temperature coefficients of Young's modulus of the solders and the epoxys are highly negative and influence the temperature characteristics of the resonator, it is necessary to tightly control the thickness of the bonding material. This can be done in the case of solder bonds by a combination of pressure and by either controlling the solder foil thickness or controlled removal of excessive solder from a pretinned metal surface.

Also important in the bonding process is the preparation of the metal surfaces. With epoxy bonds, the surfaces must be very clean. Preparation for soldering may involve gold plating the metal or tinning the metal with solder and organic acid flux. The tinning may involve only a thin layer of solder or the carefully controlled amount of solder necessary for the ceramic-to-metal bond. In either case, a thorough cleaning step for the removal of the flux is necessary.

Heating the composite structure is necessary to cure the epoxy or to melt the solder. As was discussed in Chapter 2, if the ceramic has been polarized, care must be taken to insure that the temperature does not approach the Curie temperature of the PZT material. In the case of solder bonding, even if the ceramic has not been polarized, time and temperature should be minimized to prevent diffusion of the plating into the solder, thus destroying the ceramic-to-metal bond. Heating methods for the construction of soldered composite resonators include ovens and RF-induction coils. With regard to solder-bonding lead wires with a soldering iron, extreme care must be taken with regard to the temperature of the iron, its wattage, and the time the iron is in contact with the ceramic/wire junction.

Adjusting Frequency and Coupling

Generally, the frequency adjustments made on a composite metal/ceramic transducer are performed in the same way as on the interior resonators. These methods include grinding, sanding, laser tuning, and drill tuning. Because the sensitivity of the filter to variations in the end resonator tuning is low, sometimes a less accurate method is used (such as edge chamfer grinding in the place of laser tuning).

An interesting method of tuning the second flexural-mode transducer used in the filter of Figure 7.20, is to remove material across the bottom metal surface with a laser [5]. Figure 6.8 shows how the amount of tuning varies with the position of the laser shots. The material removal is usually confined to negative frequency shift regions where the center of the resonator is used for fine tuning and the ± 2.5 mm region for rough tuning, although from the curve we can see that it is possible to raise the resonator frequency by trimming the ends.

The coupling coefficient can be adjusted by either of two methods, depolarizing or the removal of plating. In either case, the static capacitance of the transducer varies, as a result of the process, and therefore must be monitored. The depoling method involves applying a bias voltage to the ceramic that corresponds to the deviations of the coupling coefficient and the static capacitance from their nominal values [1]. A small computer (or large calculator) and

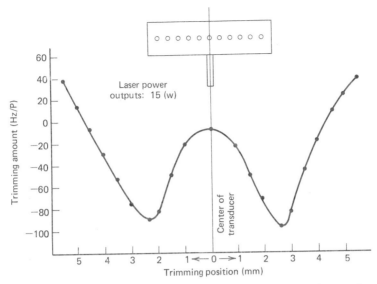

FIGURE 6.8. Frequency trimming of a flexural-mode (first overtone) composite transducer as a function of the laser shot position. See Figure 7.20. [© 1979 IEEE. Reprinted, with permission, from *Proc. 1979 IEEE ISCAS*, Tokyo, 1072–1075 (July 1979).]

equipment for measuring the resonator's reactance can be used to measure Q, static capacitance, electromechanical coupling, and the open-circuit mechanical resonance frequency (f_p of Figure 2.23). The computer is also used to calculate the depoling voltage. A disadvantage in this process is that there is a settling time for the transducer parameters to stabilize. The parameters are then remeasured, and if necessary, the resonator is further depoled.

The plating-removal method of varying coupling avoids the time-lag problem and is therefore more compatible with full automation techniques. A method is to use laser shots to remove plating from the top electrode surface of the piezoelectric ceramic. The maximum effect on reducing coupling is to remove the material from the center of the ceramic (i.e., at a midpoint between the ends) which minimizes the change in capacitance (since less material is removed). Care must be taken to insure that the static capacitance is not reduced to the point where the resonating coil will not tune or the terminating resistance does not properly terminate the filter.

MAGNETOSTRICTIVE TRANSDUCERS

We limit our discussion in this section to the indirect attachment (wire coupling) of ferrite magnetostrictive transducers to metal alloy disks, bars, or rods. This "end-section" assembly therefore becomes a coupled two-resonator system; an example is the ferrite-disk combination of Figure 4.28.

Wire-Coupling Ferrite Transducer Resonators and Metal Resonators

The three basic attachment methods in the assembly of wire-coupled composite end sections are soldering, welding, and epoxy bonding. The first process step in all three methods is the careful cleaning of the ferrite rod. If solder bonding is to be used, the ferrite rod must be metal plated in the attachment area. A long used method of plating is to coat the ferrite with a high-temperature silver-glass paste, which is then baked at a high temperature. Because the ferrite is not polarized, there are no adverse effects on its magnetostrictive properties. Another method of plating is the use of RF-sputtering, where hundreds of parts can be plated at one time with good consistency. Following the plating step, the coupling wire is solder-attached to the ferrite by simple radiant heating from a Nichrome-wire coil or by heating with an RF-induction coil. If not already done, the coupling wire can be welded to the first metal resonator.

The metal plating step is avoided by the use of simple epoxy bonding. The coupling wire is pushed through a mound of epoxy and the pressure is maintained while the epoxy is cured. Epoxy can also be used to bond a small metal disk to the end of the ferrite rod, the disk diameter being that of the rod. The coupling wire is then welded to the disk and then to the first metal resonator. An advantage of this second method is that the ferrite/thin-disk assembly can be tuned before the wire attachment; the subsequent welding step is consistent enough to result in a predictable resonator frequency shift so no further tuning is necessary. A similar process can be used where a thin disk is soldered, rather than epoxy bonded, to the plated end of the ferrite.

In those processes where the wire attachment results in nonacceptable ferrite-resonator frequency variations, the ferrite rod must be tuned after its assembly to the coupling wire and metal resonator. The tuning process involves removing material from the end of the ferrite rod by either sanding or sandblasting with Al_2O_3 abrasive. The natural frequencies are monitored and the process is terminated when one or both natural frequencies fall within prescribed limits [6]. The process is complicated by the need to maintain a magnetic bias level equivalent to that seen in the filter. This is difficult because of the problems of measuring magnetic fields by direct or indirect means and because of the sensitivity of the ferrite rod's frequency to the position of the rod in the magnetic field.

Coils and Magnets

Unlike the piezoelectric ceramic transducer, which is a single-piece assembly, the magnetostrictive transducer requires a coil, one or more permanent magnets, and a support structure, in addition to the ferrite rod. A typical assembly was shown in Figure 2.11.

The transducer coil forms are normally made of a rexolite type of material that can be molded or screw-machined. The dimensions of the coil form are important because of the dependence of inductance and electromechanical coupling on the position of the windings with respect to the transducer rod. If the coil form holds the magnets, then the dimensional tolerances also affect the mechanical resonance frequency.

With regard to coil winding, consistency is the most important factor. The normal coil winding considerations, with respect to wire size and tension, number of turns, type of wind (random, layer), and starting and ending points, apply to the transducer coils as they do for other inductive components. When production volume is high, multiple-turret winders are more efficient than single-head designs, but they require longer set-up times and often more sophisticated maintenance. Regarding coil testing, Q-meter sample-plan measurements of Q and resonating capacitance, are usually sufficient.

Incoming inspection of magnets involves dimensional measurements, as well as sample-testing of saturation flux density. On the production line, the magnets can be magnetized either before or after assembly in the filter. If magnetized before assembly, they are usually saturated, stabilized, and stored for future use. A difficulty in this method is having a good means of storing without causing the magnets to touch and demagnetize. Magnetizing after assembly in the filter requires either a difficult measurement of Gauss level or a secondary measurement of inductance (or by electrical tuning, a measurement of the electrical circuit resonance frequency). If coil inductance, in air, and dimensional tolerances are held tight, the inductance or frequency measurement provides a good measure of the magnetic bias field in the transducer rod. Figure 2.13 showed how inductance varies with bias level. In all cases, if Alnico 5 magnets are used, they must be demagnetized at least 8 percent but not more than 50 percent, in order to be stable under operating-signal, temperature, and vibration conditions.

WIRE-TO-RESONATOR BONDING

The most popular method of attaching support wires and coupling wires to the mechanical filter resonator is through capacitor-discharge resistance welding [7]. Through this means, inter-resonator coupling variations can be held to less than 1 percent. Obtaining this high degree of consistency is not easy, but through good equipment design, tight process control, and uniform material characteristics, the low coupling variations can be achieved. Proper welding techniques will provide both consistent coupling and good mechanical strength.

Basic Welding Concepts

Before looking at specific techniques for welding mechanical filters, we will look at concepts that are basic to capacitor-discharge resistance welding. We first establish an equivalent circuit model for the weld set-up, then we look at the dynamic (time-response) characteristics of the weld process, and finally look at the weld itself.

The Equivalent Circuit

Figure 6.9 shows a pictorial diagram and a schematic diagram of a resistance welding system. The resistance welder is composed of a DC power supply used to charge a capacitor bank, a switch (usually a silicon-controlled rectifier) to discharge the capacitors, and a tapped pulse transformer for shaping the waveform and setting the pulse width of the secondary current. The secondary windings of the transformer are connected through a pair of cables to the weld head, electrodes, and the parts to be welded. Mechanically, a force is applied through one of the weld head elements, through the electrodes, the parts, and through the other half of the weld head. When the applied force reaches a prescribed level, the capacitor bank is discharged and the weld is formed.

In the equivalent circuit of Figure 6.9(b), we see the power-supply voltage V, capacitor-bank capacitance C, and the transformer secondary current I_s. The resistance R_s includes that of the transformer secondary windings, the cables, all connections, the weld head and electrodes, the resonator R_r and a portion of the wire-to-electrode resistance R_{ew}. The load resistance R_L is composed of the wire-to-resonator contact resistance R_{wr}, the wire resistance R_w, and the portion of R_{ew} contributing to the heating of the wire-to-resonator interface. Figure 6.10 shows some of the resistances in the secondary current I_s path. The resistances in the secondary circuit are on the order of tenths of milliohms; for example, the cable resistance may be 0.75 mΩ/meter, a connection is 0.1 mΩ, the wire-to-resonator junction 0.4 mΩ. The resistance R_L is typically 20 percent of the total secondary resistance. Knowing the equivalent circuit values helps in understanding the dynamics of the weld process.

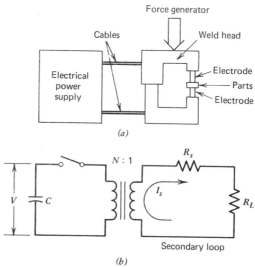

FIGURE 6.9.   Resistance-welding equipment, pictorial and schematic diagrams.

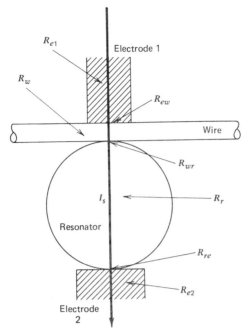

FIGURE 6.10.   Weld resistances.

Dynamic Characteristics

Involved in the dynamic (time-response) characteristics of the weld are mechanical, electrical, and thermal effects. These effects can best be described through an illustration.

Figure 6.11 shows the dynamic characteristics of a weld of the type shown in Figure 6.10. At time zero, a force is applied to the wire-resonator junction. This force is increased until at time T_1 the capacitor bank is discharged. Note that as the force is increased from time 0 to T_1, the contact resistance is decreased. Thus, the greater the force, the smaller the resistance R_L, and the more current that is needed to make the weld. Returning to the figure, as the capacitor bank is discharging, the current in the secondary loop is increasing, as is the power $I_s^2 R_L$ in the weld, and the temperature at the weld junction. Energy is being pumped into the weld junction much faster than it is being thermally conducted away from the junction into the resonator, wire, and electrodes. Therefore, the temperature continues to rise and finally melting takes place at time T_2. Depending on the hardness of the wire, the shape of the electrodes, and the force being applied, the wire or resonator may deform, causing the force to decrease slightly as the forging takes place. The temperature finally peaks at the point where the heat being conducted away from the junction is equal to the power dissipated in the junction. When the material at

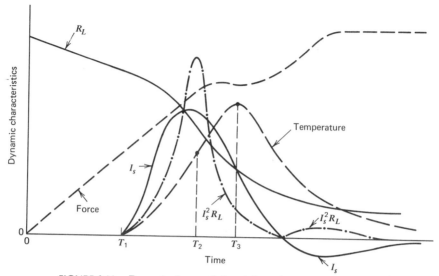

FIGURE 6.11.   Dynamic characteristics of the resistance-weld process.

the junction returns to a solid condition, the forging ends and the weld is complete.

The Weld

The result of the welding process is a wire/resonator bond, such as that shown in Figure 6.12. The wire may be deformed due to a forging action by the upper electrode. The deformation may be on the top and bottom sides of the wire. In addition, the resonator may be deformed by the wire. Beads usually form on the side of the wire or under the wire due to the molten material being forced out of the junction by the pressure of the electrode. The beads and the coupling wires also show discoloration. Depending on the welding process, there may be intermixing of the grains of the materials, or in the case of dissimilar metals, there may be an intermetallic compound formed at the wire-resonator junction. In other words, the weld is primarily the result of a metallurgical change. In cases where melting temperatures are not achieved, the weld is simply a mechanical bond.

Welding Mechanical Filter Coupling Wires

The challenge of mechanical filter welding is to realize consistent welds made at a high speed with automated equipment. Because of the large number of

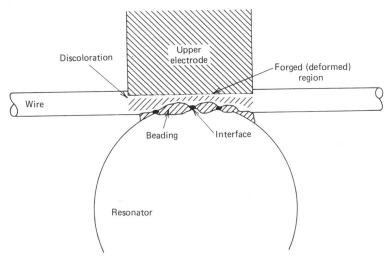

FIGURE 6.12.   The coupling-wire-to-resonator weld.

mechanical filter configurations, there are a wide variety of welding configurations used to accomplish the consistency, speed, and automation objectives.

Welding Configurations

Figure 6.13 shows three configurations used in coupling-wire/resonator welding. The first illustrates the use of a wide electrode in contact with the coupling wire. This scheme reduces wire deformation and provides a weld across the entire interface. A disadvantage in this type of weld is that the upper electrode surface and the flat portion of the resonator must be kept parallel or the weld will not be uniform.

Figure 6.13(*b*) shows a narrow electrode used to weld a broad resonator surface. This method provides a good weld and consistent coupling if the wire is allowed to deform and bow away from the resonator in the nonwelded area. Also, in this example the second electrode makes contact with the upper surface of the resonator, causing the current to flow through the top side of the resonator; this indirect weld provides a well-defined current path, but a slightly dissymmetrical weld from one side to the other.

The third configuration in Figure 6.13 is used for a point contact weld. A major advantage of this type of weld is that a narrow upper electrode can be used, therefore reducing the tolerance on parallelism. Also, the centering tolerance is less stringent than the previous case because the weld is automatically centered by the wire-resonator point of contact.

Various other welding configurations that are based on the previous three examples include inverting the wire and rod in Figure 6.13(*a*) and (*c*). Also,

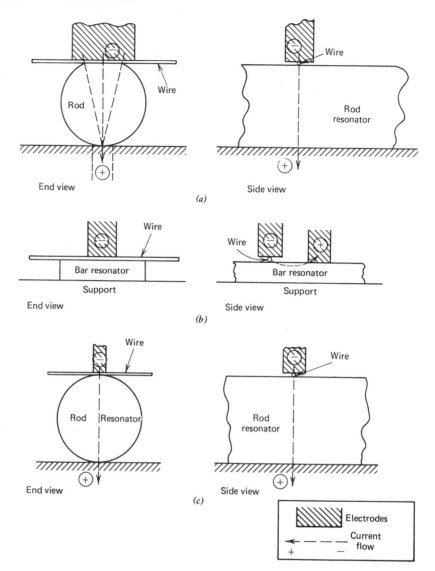

FIGURE 6.13. Coupling-wire, resonator-resistance welding configurations. (*a*) The electrode width exceeds the contact length, and current flow is through the thickness of the resonator. (*b*) Narrow electrode with two top-side electrodes. (*c*) Point-contact weld.

the indirect (series) weld of Figure 6.13(*b*) has been used with the rod resonators of Figure 6.13(*a*) and (*c*). Let's next look at methods of supporting and spacing the resonators.

Resonator Supporting and Positioning. Figure 6.14 illustrates three methods of holding in place and spacing mechanical filter resonators. The prism

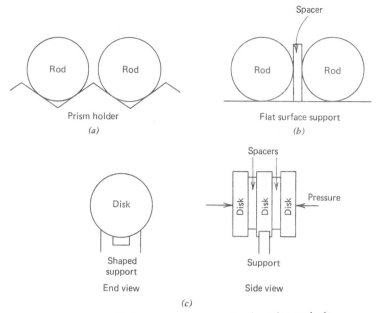

FIGURE 6.14.   Various resonator support and spacing methods.

holder provides spacing and support, and when used as the lower electrode it provides a two-point current path, which improves the consistency of the welds. If the prism is used as the lower electrode, heating, and therefore wear, occurs at the contact points. This can be avoided by using the indirect weld method shown in the side view in Figure 6.13(b).

A method of obtaining very exact spacing between the resonators is to use a precision spacer, which is attached to the flat support surface, as shown in Figure 6.14(b). A means of welding disk-resonator filters is to use dissolvable spacers and a support shaped to conform to the disk edge. This method is shown in Figure 6.14(c). When used as a lower electrode, the shaped support provides a consistent, low-resistance path for the secondary current. Tolerances on the resonator spacings, for the various methods shown in Figure 6.14, vary from ± 10 μm to ± 50 μm.

The Weld Sequence.   To this point, we have assumed a single power supply, one electrode-pair, and no movement of the parts. In most systems, the electrodes remain laterally stationary and the parts are moved. A major exception is where one electrode-pair is provided for each weld and the entire system remains stationary. In some cases, the wires and resonators are moved more than once to positions where multiple electrode-pairs are used. Each electrode-pair may have its own power supply, set to a fixed voltage and transformer turns ratio (which controls pulse width). The time and position sequence of each weld also varies with the manufacturer. One method is to start at one end of the filter and sequentially weld a single wire to adjacent resonators, whereas another is to weld out from the center. It is also possible to make two welds at one time by having a coupling wire between each electrode and the resonator. The last method works well because the power $I_s^2 R_L$ into each weld seeks a balance due to the weld resistance being lower in the more maturely formed of the two welds.

With regard to power supplies, one or many can be used. If one supply is used for all welds, and if more than one coupling wire size is used, then the voltage is varied to optimize each weld. The use of more than one power supply allows both the voltage and the pulse width to be varied to obtain an optimum weld.

Electrodes and Cleanliness.   The most commonly used electrode materials are the copper based RWMA Class alloys which combine hardness, a high annealing temperature, and wear resistance, with good electrical and thermal conductivity. Depending on the size and the shape of the electrode, the current required, the electrode material, and the applied force, the electrodes must be resurfaced from once every ten filters (in the case of one manufacturer) to once each day (in the case of another company). The resurfacing of the electrode is sometimes done without removing the electrode from the weld head. This saves time and maintains a consistent, secondary loop resistance. With regard to the importance of cleaning the resonators and coupling wires before welding, there is no consensus; some manufacturers clean the parts thoroughly, some do no cleaning.

Automated Welding

A fully automated welding station involves: (1) automatic feed of the resonators to the holder, (2) automatic coupling wire feed, (3) a belt or an X-Y table to locate the parts under the welding electrodes, (4) automatic wire cutting, and (5) computer control of the entire process.

Automatically placing the resonators in the holder is done with a gravity or vibratory feed from trays of parts, one tray for each frequency. The most common automatic wire feed involves the use of a wire spool. The wire is first straightened, then indexed into position, the welds take place, and the wire is cut.

Methods of positioning the parts under the welding electrodes involve combinations of belts, linkages to turn the holder over, X-Y tables, and stepping motors for rotation. The feeding, the positioning, and the welding is computer controlled. The computer may be a digital high-speed computer, a calculator, such as the HP 9825, a numerical control unit, or a microprocessor.

Welding Resonator Support Wires

In this section we will look at the use of butt welds for attaching support wires to mechanical filter resonators. Please note that this type of welding can also be used for coupling between resonators and for composite (indirectly attached) transducers, such as that of Figure 4.28.

Butt welding is simply another form of resistance welding, consequently the basic welding concepts previously discussed still apply. Therefore, we will concentrate only on specific butt welding techniques that differ from our previous coupling-wire-to-resonator welding discussions.

Butt Welding Methods

Figure 6.15 shows two methods of making a butt weld. The first method shows a headed wire supported by a hole in the electrode. The electrode forces the wire against the resonator, and then the capacitor bank discharges current, which passes from the electrode across to the head of the wire and then across the headed-wire/resonator interface where the weld takes place.

The second method shown in Figure 6.15 involves clamping the wire with the first electrode by the use of either a flat electrode and support, or as shown, by the use of a prism electrode and support. Next, the electrode and wire assembly (or the resonator) is moved so contact is made between the wire and the resonator. At this point, the capacitor bank is discharged and the weld is made. When used for a resonator support, it is essential that the wire contacts

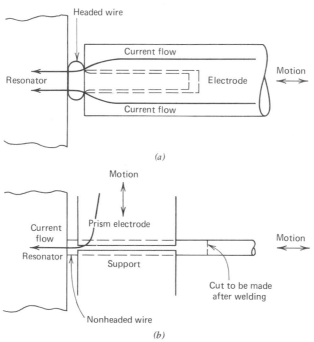

FIGURE 6.15.   Butt welding methods. (a) Electrical contact is made at the headed wire, and (b) contact is made along the length of the wire. In both cases the second electrode contacts the resonator.

the nodal point. Therefore, the tolerances on the position of the weld may be as tight as 20 μm.

The Support Wire

The support wires shown in Figure 6.15 are only two of a number of different styles used in mechanical filters. Some support wires have a small protrusion preceding the head that first makes contact with the resonator and then collapses under heat and pressure. In some automated equipment, the head is formed by the welding machine from a continuous spool of wire. The weld sequence involves forming the head, welding, cutting, forming the head, and so on. In some cases, where two support wires are needed on each resonator, two spools are used to form and weld two support wires at one time.

When butt welding is used for coupling wires, the wires are made from the iron-nickel alloys described in Chapter 3. These same alloys can be used for support wires, but their constant modulus of elasticity properties are not needed for this application. An example is the 52 kHz flexural-mode filter of Figure 7.13 that uses copper or iron wires, which are coated with nickel or tin for solderability to the solder-bath support.

Solder Bonding

An alternate method of coupling-wire attachment is solder bonding. Under certain conditions, a solder-bonded filter has greater structural strength than an equivalent welded filter. Also, a solder bond may be necessary when the spacing between resonators is too small to allow a butt-welding electrode between the resonators, and in the production of very low-cost filters using PZT ceramic resonators (see Figure 7.5) where solder bonding is both necessary and cost effective. The major disadvantage of solder bonding is the less consistent coupling between resonators. Also, when drilling is necessary, the cost of solder bonding is higher than the cost of welding.

Solder bonding usually requires that the coupling wire be either plated with an easily solderable material or pre-tinned with solder. With regard to the resonator, if a ceramic is used the plated electrode serves as a good bonding surface, but in the case of metal-alloy resonators, the bonding surface must be pre-tinned with solder. A common method of solder bonding that is used in place of butt welding, is as follows: (1) drill an oversized hole in the resonator, (2) heat and fill the hole with solder, (3) redrill to a smaller size, (4) insert the coupling wire and flux, and (5) using a controlled temperature oven, bond the parts.

FILTER ASSEMBLY

The assembly of electrical and mechanical components involves a number of hand operations. Resistors, capacitors, and cup cores are usually hand soldered (for high reliability) to metal-pin terminals. The enclosure cover is snapped, soldered, glued, or heat staked to the base assembly, again usually by hand. Other processes, such as the resonator lead-wire attachment and the resonator support assembly are often hand operations, but they require exceptional care because they directly influence the filter response characteristics.

Lead-wire attachment to ceramic transducer plating requires low-temperature soldering for a short period of time in order to prevent coupling-coefficient changes in the ceramic or deterioration of the plating. Insulated wire is usually pre-tinned, but this can create wire breakage problems if the tinned portion of the wire is not fully enclosed by the solder connection. Ideally, the lead connection process should be automated in order to eliminate the temperature, time, and position variations due to hand soldering.

As was shown in Figure 5.3, there are numerous methods of mounting the resonators and the mechanical filter structure as a whole. What all of the methods have in common is the necessity of providing a consistent mounting. The length of the support wire, from the resonator to the secondary support, must be the same from filter to filter. The wires, brackets, or silicone primary supports must be located at the resonator nodal points. The support material properties must remain the same and the support should be free from spurious modes of vibration. Deviations from the design values will result in resonator mistunings or variations in mechanical Q.

ADJUSTMENTS AFTER ASSEMBLY

When the filter specification requires tolerances on tuning and coupling that are tighter than the component or process tolerances, then it is necessary to make adjustments after the assembly of the filter. These adjustments involve the electrical-circuit tuning or termination, mechanical-resonator tuning, as well as changes in wire coupling and bridging. Let's first look at the electrical circuit.

Electrical-Circuit Tuning and Termination

Intermediate-band and wideband mechanical filters, as described in Chapter 4, require electrical tuning of the piezoelectric ceramic or magnetostrictive transducers. The input and output tuning components are either adjustable cup cores or fixed capacitors, which are used as shown later in Figures 7.1 and 7.2. The cup-core tuning simply involves rotating the tuning slug until the proper response is obtained. Capacitive tuning is first done with a calibrated variable capacitor, which is part of the tuning fixture. Having tuned the filter, the operator replaces the variable capacitors with fixed values.

There are a number of techniques used to tune mechanical filters. The simplest is to set the signal generator frequency to the center frequency and peak the output voltage with the tuning elements. This method does not result in the lowest passband ripple but can be quickly done with a signal generator and a voltmeter. A second method is to observe the detected filter response on an oscilloscope sweep, and then tune for the flattest passband response. A third method is to tune the electrical circuit to a prescribed resonance frequency by observing the input impedance. If the prescribed frequency is close to or inside the passband, the resonance effects of the other tuned circuits will interfere with the measurement. In this case, it is necessary to shift the frequency with a calibrated fixed capacitor or inductor and tune to a different specified frequency.

If the specified passband ripple is low, it is sometimes necessary to vary the terminating resistance observed by the filter. This can be done by adding series or shunt resistors, changing the transformer turns ratio, or by the use of a capacitance dividing network like that described in Chapter 9. This operation is usually performed on a test fixture where the operator switches calibrated resistors, transformers, and capacitors in and out of the circuit until the frequency response meets the specification. The test fixture elements are then replaced.

Mechanical-Resonator Tuning and Coupling Adjustments

In some designs, it is necessary to adjust the mechanical-resonator tuning after assembly because the necessary tolerances for the "no-tuning" condition cannot be met by the production process. These filters are usually tuned by trial-and-error or by pattern recognition. In other cases, the resonators are rough-tuned before assembly and then fine-tuned; these designs require a means of clamping resonators so that the resonator that is to be tuned is isolated from the others. Let's look at a narrowband two-resonator example of this method.

Figure 6.16 shows an electrical equivalent circuit of a piezoelectric ceramic transducer two-resonator narrowband filter. Across the filter output is a coil used to clamp the second resonator in order to tune the first resonator. The inductor L_c resonates at f_0 with the static capacitance of the transducer C_{02}.

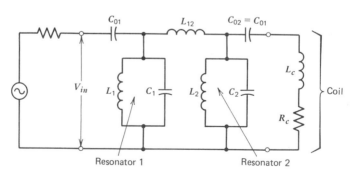

FIGURE 6.16. Two-resonator narrowband filter with output clamping coil.

Therefore, in the region of f_0 the electrical tuned circuit looks like the coil resistance R_c, which is in parallel with the second resonator. If the coil Q is within one-half to two times the ratio $r = C_2/C_1$, the second resonator acts like a short circuit, causing the right hand side of L_{12} to be grounded. Measuring the frequency at which V_{in} is a maximum, the first resonator is tuned until the peak of V_{in} is at a prescribed frequency. The peak, which is at the resonance frequency of the parallel combination of L_1, L_{12} and C_1, is unaffected by the frequency of the second resonator. The filter is then turned around and the output resonator is tuned. In an automated system, the tuning is done either with abrasive Al_2O_3 (sandblasting) or with laser shots.

An alternative to electrical clamping is to physically prevent the second resonator from vibrating. With the second resonator clamped, the operator observes the frequency of the V_{in} peak while tuning on the first resonator. Physical clamping has also been used on multiple-resonator filters that use one-quarter wavelength coupling.

In the quarter-wave coupling design, the second resonator is clamped while the first resonator is tuned until the peak of the input impedance (of V_{in}) is at the center frequency f_0. Next, the third resonator is clamped and the second resonator adjusted until, again, the peak falls at f_0. This process is repeated over the first half of the filter; the filter is then turned around and the output end of the filter is tuned.

We have mentioned laser and Al_2O_3 automatic tuning. Hand tuning techniques include filing, grinding, adding solder, or even changing the position of the clip mass shown later in Figure 7.40. Methods used in changing the coupling between resonators include filing, adding solder, and prying. In multiple-resonator filters, the coupling is varied in order to achieve a specified bandwidth. This is done by uniform filing of the wires. Adjusting the position of an attenuation pole is accomplished by filing on a bridging wire.

FILTER PRODUCTION AUTOMATION

At the beginning of this chapter, we looked at two production line examples (Figures 6.1 and 6.2) and discussed reasons for the degree of automation of each line. Let's next look at options regarding the kind of machines used in an automated line.

Some manufacturers have decentralized their production equipment into small self-contained machines controlled by microcomputers. This strategy reduces the machine complexity and prevents a machine failure from disabling an entire system. Also, because the machines are self-contained, process improvements can be made without affecting other parts of the system. Figure 6.17 shows a small machine designed to assemble transducers [8].

With regard to larger machines with centralized computing, these have the advantage of greater computational speed and memory and the advantage of less parts handling. Figure 6.18 shows a large automated machine that assembles

FIGURE 6.17.   Automatic transducer assembly machine. (Courtesy of NEC, Japan.)

composite transducers [5]. Figure 6.19 shows a central computer (right) which automatically controls disk resonator sorting and tuning equipment.

Most of the automated machines that are designed to perform a number of process steps use rotary indexing, where a set of parts is carried through four to eight stations. Figure 6.20 shows the basic elements of a hypothetical machine used to assemble composite transducers.

In the example of Figure 6.20, a resonator bar is fed from the vibratory bowl to Station 1 where flux is deposited on the bar. The resonator then moves to, and stops at, Station 2 where a solder preform is placed on the bar. The bar then moves to Station 3 where the ceramic transducer plate is placed over the preform, and then to Station 4 where the composite resonator is soldered. Because the soldering step takes the longest time, it is this step that limits the rate of producing the composite resonators. The parts are then cooled and cleaned. Other processes could have been added to this machine, such as the

FIGURE 6.18.   Automatic transducer assembly machine. (Courtesy of Fujitsu Ltd., Japan.)

FIGURE 6.19. Computer-controlled resonator sorting (left) and drill tuning (center) machine. [Courtesy of Rockwell International, USA]

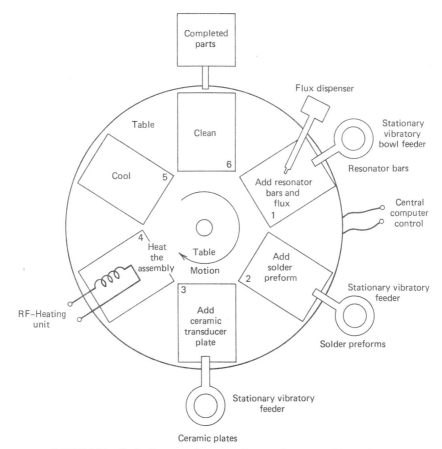

FIGURE 6.20. Basic elements of a composite transducer assembly machine.

269

lead-wire attachment, the support-wire attachment, and tuning. This type of machine requires no operator, only a person to load the feeders and a person to perform maintenance. A counter is normally used to shut down the machine after a specified number of parts have been built. Also, a check station can be added to automatically inspect for missing parts before soldering. The computer can be programmed to shut the machine down when inspection failures exceed a prescribed limit.

REFERENCES

1. A. Günther, H. Albsmeier, and K. Traub, "Mechanical filters meeting CCITT Specification," *Proc. IEEE*, **67**(1), 102–108 (Jan. 1979).

2. K. Yakuwa, T. Kojima, S. Okuda, K. Shirai, and Y. Kasai, "A 128 kHz mechanical channel filter with finite frequency attenuation poles," *Proc. IEEE*, **67**(1), 115–119 (Jan. 1979).

3. F. Crescenzi, S. Cucchi, and G. Volponi, "Mechanical filter for FDM channel modem," *Telettra Review*, **30**, 52–58 (1979).

4. J. Deckert, "Elastiche eigenshaften von loten für piezoelektrische verbundwandler," *Conference on Soldering and Welding in Electronics*, München, 47–50 (1976).

5. J. Tsuchida, M. Takado, K. Shirai, and Y. Kasai, "Automated mass production of mechanical channel filters," *Proc. 1979 IEEE ISCAS*, Tokyo, 1072–1075 (July 1979).

6. W. Domino and R. A. Johnson, "Microcomputer-controlled tuning of mechanical filter transducers," *IEEE Workshop on Automated Circuit Adjustment*, San Antonio (April 1980).

7. G. Straka, W. Amann, K. Böhmler, and H. Rudolf, "Präzisionsschweissen an mechanischen filtern," *Conference on Soldering and Welding in Electronics*, München, 75–78 (1976).

8. M. Watanabe, M. Kidokoro, T. Kouge, T. Kobessho, and T. Yano, "Large scale production of a pole electromechanical channel filter," *Proc. 1979 IEEE ISCAS*, Tokyo, 1076–1079 (July 1979).

APPLICATIONS OF
MECHANICAL FILTERS

Mechanical filters are used in systems that require narrowband selectivity. Beyond this requirement, mechanical filters can often satisfy the need for miniaturization, reliability, and low cost. In the case of fractional bandwidth requirements of less than 0.05 percent, or stability better than 1 ppm, crystal filters are most often used. For bandwidths greater than roughly 10 percent, *LC* or ceramic filters can be used; but in the frequency range of 200 Hz to 600 kHz and for bandwidths between 0.05 and 10 percent, the mechanical filter has gained wide acceptance and is used in voice and data communications, signaling, detection, control, and test equipment.

In this chapter each example illustrates not only a particular application, but a few characteristics unique to that filter. In this way you can get a broader understanding of uses and characteristics of mechanical filters than if we studied a few filters in detail. A more detailed understanding of a specific filter can be derived from the reference papers.

COMMUNICATION EQUIPMENT

The mechanical filter was first designed for use in communication equipment, specifically for high-performance HF radios. This work in the late 1940's was followed a decade later by telephone voice-channel filter designs, and then after another ten years, by mechanical filters for FSK (frequency shift keying) equipment.

In our study of communication equipment applications we first look at voice bandwidth radio and telephone filters, then narrowband signaling and pilot-tone filters for telephone systems, and finally narrowband (low data rate) FSK filters.

Mechanical Filters for Radios

In this section we look at three examples. The first two filters are very complex, with 12 and 14 resonators, and are used in high-performance single-sideband (SSB) radios. The third filter is very simple and inexpensive and is used in AM radios. Other designs having from seven to nine resonators are used in a variety of applications, such as amateur, citizen-band, and avionics radios operating in SSB, AM, CW, and FM modes. Although our example filters have voice bandwidths of 3 to 6 kHz, mechanical filters are manufactured that have bandwidths ranging from 200 Hz to more than 20 kHz.

A Disk-Wire SSB Filter

Disk-wire mechanical filters have been manufactured since the early 1950s. The filters at that time used magnetostrictive wire transducers, capacitor tuning, a maximum of nine disks, and were manufactured with somewhat more primitive equipment. The result was a filter with 20 dB insertion loss, 3 dB passband ripple, and moderate selectivity. In contrast, the modern SSB filter of Figure 7.1 used PZT transducers to reduce the insertion loss to less than 3 dB and the passband ripple to under 2 dB over a temperature range of $-55°C$ to $-85°C$ [1]. The 12 disks and the bridging wires make this filter very selective, as we can see in Figure 7.2.

The filter is physically symmetrical, so the attenuation poles $f_{\infty 1}$ and $f_{\infty 3}$, corresponding to the first and second bridging wires, are at the same frequency. The same bridging also produces identical-frequency poles at $f_{\infty 2}$ and $f_{\infty 4}$, as shown in Figure 7.2. The phase inversion necessary to produce attenuation

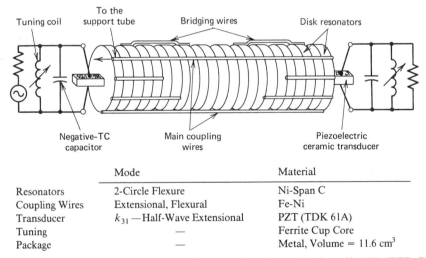

	Mode	Material
Resonators	2-Circle Flexure	Ni-Span C
Coupling Wires	Extensional, Flexural	Fe-Ni
Transducer	k_{31} —Half-Wave Extensional	PZT (TDK 61A)
Tuning	—	Ferrite Cup Core
Package	—	Metal, Volume = 11.6 cm^3

FIGURE 7.1. SSB disk-wire mechanical filter for HF-radio applications. [© 1979 IEEE. Reprinted, with permission, from *Proc. 1979 IEEE ISCAS*, Tokyo, 896–899 (July 1979).]

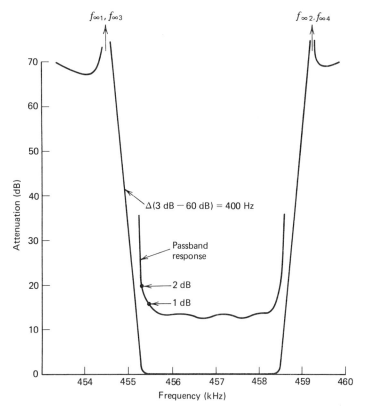

FIGURE 7.2.   Frequency response of the disk-wire mechanical filter of Figure 7.1. [© 1979 IEEE. Reprinted, with permission, from *Proc. 1979 IEEE ISCAS*, Tokyo, 896–899 (July 1979).]

poles is realized by the length of the bridging wires being greater than one-half wavelength.

The combination of a ferrite cup core and a negative temperature-coefficient capacitor balances out the positive temperature coefficient of the transducer capacitance. This allows operation to $-55°C$. Although similar filters are built with ferrite transducers, they have the disadvantage of not having temperature compensation of the electrical tuned circuits.

An SSB Filter with Torsional Resonators

A method of achieving low passband ripple and a stable response is to reduce the center frequency of the filter while maintaining the same bandwidth. An example is the SSB filter of Figure 7.3, which has a carrier frequency of 200 kHz [2].

The 200 kHz filter uses fundamental-mode torsional resonators and magnetostrictive ferrite transducers. The resonators are supported at their nodal

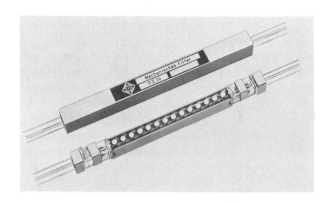

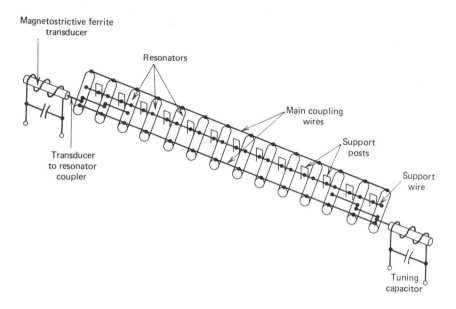

	Mode	Material
Resonators	Half-Wave Torsion	Thermelast 5409
Coupling Wires	Extensional	Thermelast 5409
Transducer	Half-Wave Extensional	Magnetostrictive Ferrite (S-3)
Tuning	—	Capacitor, 0-T.C.
Package	—	Metal, Volume = $14.4 \times 1.5 \times 1.2 \text{ cm}^3$

FIGURE 7.3. Sixteen-resonator SSB mechanical filter (Courtesy of Telefunken, F.R.G.)

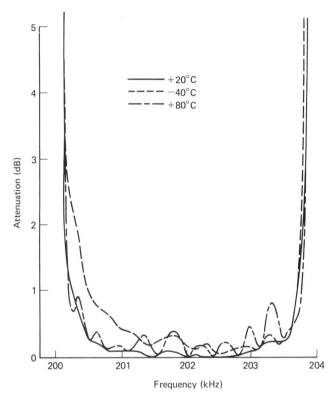

FIGURE 7.4.   Passband frequency response characteristics of the 16-resonator filter of Figure 7.3. [Reprinted, with permission, from *Telefunken-Zeitung*, **36**, 272–280 (May 1963).]

points by a single wire, which is attached to posts between the resonators. The resonators are spaced a quarter-wavelength apart; this reduces the sensitivity of the response to variations in coupling-wire length and provides space for the support posts. But this makes the filter very long. Use of the torsional mode of vibration increases the resonator Q from about 20,000 to about 30,000. The higher Q and the greater fractional bandwidth (B/f_0) results in less "rounding" at the passband edges, as compared to its 455 kHz disk-resonator counterpart. The frequency response of the 200 kHz filter is shown in Figure 7.4.

Also shown in Figure 7.4 is the frequency response of the 200 kHz filter at $-40°$C and $-80°$C. Note the deterioration of the response at $-40°$C due to electrical mistuning at that temperature.

A Filter for Low-Cost AM Radios

Millions of mechanical filters have been built for use in portable radios and car radios. Early filters included an H-shaped metal plate with two ceramic

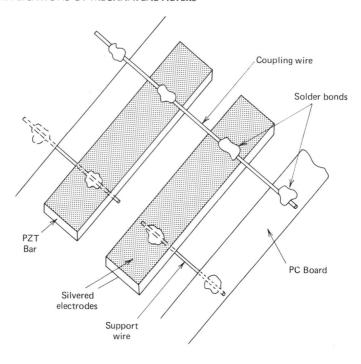

	Mode	Material
Resonators	Full-Wave Extension	PZT Ceramic
Coupling Wire	Flexural	Fe-Ni Alloy
Tuning	—	Optional Ferrite Core
Package	—	Plastic

FIGURE 7.5. Mechanical filter for low-cost radio applications. (Courtesy of Toko, Japan.)

transducers on the major surfaces of the legs. These filters suffered from low electromechanical coupling, which made it necessary to tune the input and output with coils. The low coupling problem was solved by simply using the ceramic transducers as the resonators, as is shown in Figure 7.5.

The filter of Figure 7.5 is designed for low-cost IF filtering in the 455 kHz and 262 kHz center frequency ranges. The resonator bars are a full-wavelength long and are supported primarily by the two support wires but also by the coupling wire. The coupling wire acts as a common ground and is soldered to the top of the PC-board mount. The support wires act as the input and output leads and are attached to the bottom side of the PC board. The result is a simple, easy-to-manufacture mechanical filter with temperature characteristics which are an order-of-magnitude better than a transformer coupled *LC* filter.

Figure 7.6 shows the frequency response of 4 kHz and 12 kHz bandwidth filters at 455 kHz. Additional selectivity can be obtained by using input and output transformers and external capacitor bridging between the input and

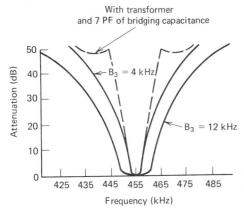

Figure 7.6. Frequency response of 4 kHz and 12 kHz mechanical filters of Figure 7.5 (Toko, Japan).

output terminals. The filters are manufactured to a ± 2 kHz nominal center frequency value, but the center frequency can be partially corrected in the radio by varying the terminating resistance (at the expense of additional insertion loss).

Mechanical Filters for Telephone Communications

Small size, low cost, and good performance have been the main reasons for the successful use of mechanical filters in frequency division multiplex (FDM) telephone equipment. Uses in FDM equipment include voice-channel and pilot-tone filtering and filtering in signaling circuits. With regard to these applications, major breakthroughs in manufacturing technology and automation allowed the mechanical filter to be the superstar of the 1970s. Small size was the major advantage over LC channel filters, although LC filters were able to maintain better delay performance for the same stopband selectivity. Although monolithic crystal filters were smaller, their passband and stopband performance was generally not as good as the mechanical filters. Active filter, switched capacitor, and CCD designs had reliability and dynamic-range problems that eliminated them in the 1970s, but we can be sure that both the new technologies and the old technologies will not stand still.

A major factor in the successful use of mechanical voice-channel filters in FDM systems was the decreasing cost of frequency generation and modulation. To reduce the amount of mixing, early systems used the direct modulation scheme shown in Figure 7.7 [3]. Note that each telephone user is assigned a different frequency range in the 60 to 108 kHz band. A problem with this method is that 12 discrete filters, ranging from 64 kHz to 108 kHz, are needed to separate the voice channels. Because of this wide frequency range, each filter had its own peculiar manufacturing problems and resultant high costs. Mechanical filters, for this type of system, have been manufactured in the USA [1], Japan [4], and Czechoslovakia [5].

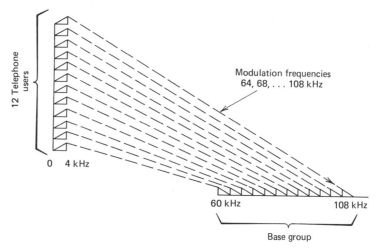

FIGURE 7.7. Direct modulation method. Twelve discrete lower sideband filters. [© 1974 IEEE. Reprinted, with permission, from *Circuits and Syst.*, 5–13, (Dec. 1974).]

Another system that is important in terms of mechanical filter applications is the multiple pregroup scheme shown in Figure 7.8. For applications outside of the USA, each *LC* channel filter requires a signaling filter centered at 3.850 kHz above the carrier frequency. For this purpose, 100 Hz bandwidth flexural-mode mechanical filters were designed and manufactured in Japan [6], [7] and Italy. These filters formed a technology bridge to new channel filter designs and the modulation methods shown in Figures 7.9–7.12.

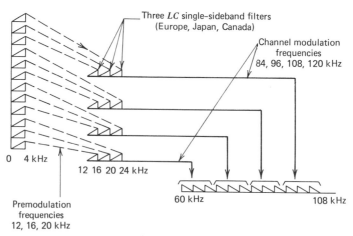

FIGURE 7.8. Multiple pregroup modulation method. [© 1974 IEEE. Reprinted, with permission, from *Circuits and Syst.*, 5–13 (Dec. 1974).]

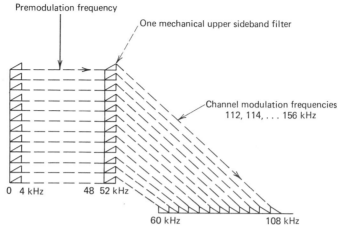

FIGURE 7.9.   Single-filter premodulation method I. [© 1974 IEEE. Reprinted, with permission, from *Circuits and Syst.*, 5–13 (Dec. 1974).]

A Twelve-Resonator Channel Filter

A major breakthrough came in the early 1970s with the introduction of the single-filter premodulation scheme shown in Figure 7.9 and Figure 7.12 [8]. The upper-sideband (USB) channel filters were at a single 48 kHz carrier frequency; twelve-signal modulation was used to form the 60 to 108 kHz base band. This concept was adopted by the West German telephone service, and all equipment suppliers were then required to use this general scheme. The channel filters used flexural-mode resonators and a single extensional-mode coupling wire and were of the form shown in Figure 7.13. Not shown are the input and output electrical tuning coils and temperature compensating capacitors.

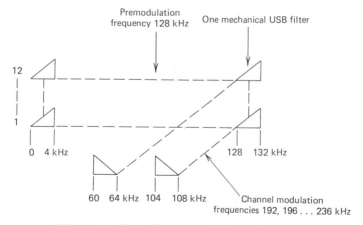

FIGURE 7.10.   Single-filter premodulation method II.

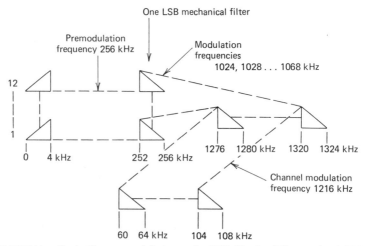

FIGURE 7.11. Single-filter premodulation method III. (Rockwell International, U.S.A.)

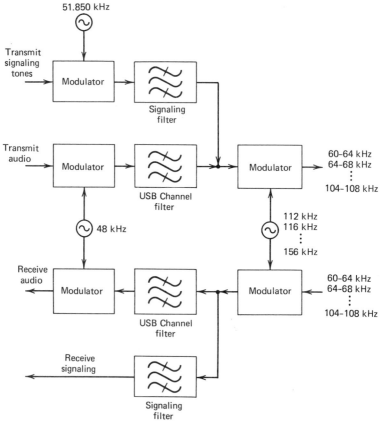

FIGURE 7.12. Simplified block diagram of an FDM channel modem using premodulation method I (Figure 7.9).(SEL, F.R.G.)

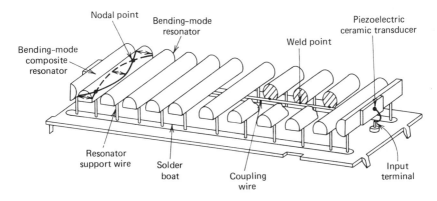

FIGURE 7.13.   Twelve-resonator channel filter; also see Figure 1.10. [Adapted, with permission, from *Proc. 1978 IEEE ISCAS*, New York, 330–335 (May 1978). © 1978 IEEE.]

	Mode	Material
Resonators	Fundamental Flexure	Thermelast 5409
Coupling Wires	Extensional	Thermelast 5409
Transducer	k_{31} —Extensional	Siemens Vibrit 391
Tuning	—	Ferrite Cup Core
Package	—	Metal Base, Plastic Cover, Volume $= 17\ cm^3$

Of particular importance is that the 48 kHz filter was designed with automatic production in mind. The production flow chart of Figure 6.1 showed that almost every process is automated. The filter requires no post-assembly resonator tuning. Even the electrical circuit tuning is automated, in spite of the fact that the filter meets the stringent 1/20 CCITT (± 0.11 dB) passband amplitude-variation specification shown in Figure 7.14.

As can be seen in Figure 7.12, the channel and signaling filters are connected in parallel at their output terminals on the transmit side and at their input terminals on the receive side. In order to minimize the effect of impedance variations of the channel filter on the signaling filter response, the zero-quads of the channel filter's characteristic function were adjusted so that its impedance peak was at the center frequency of the signaling filter. Amplifiers and isolation networks could have been used to prevent interaction between the two filters, but at the expense of degraded system behavior, lower reliability, and increased cost. The impedance-adjustment solution was the result of close cooperation between the system engineers and the channel and signaling filter designers.

A Ten-Resonator Filter with Finite Poles

The twelve-resonator monotonic filter was successfully implemented, but there remained further challenges for the filter and system designers. Though the stopband selectivity of the monotonic design was adequate, the fact that twelve

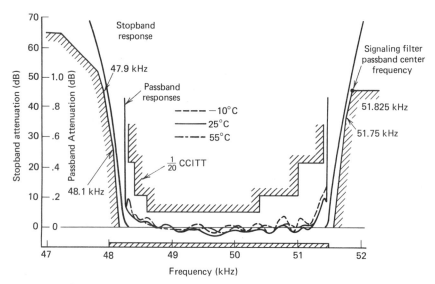

FIGURE 7.14. Frequency response of the 12-resonator channel filter of Figure 7.13. (Siemens, F.R.G.)

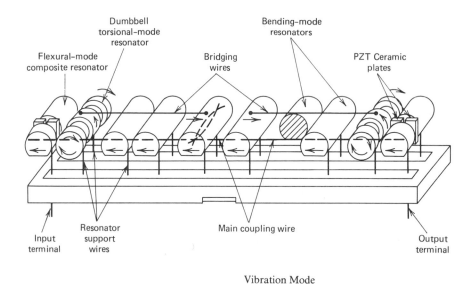

	Vibration Mode
Resonators	Fundamental Flexure and Full-Wave Torsion
Coupling Wires	Extensional
Transducer	Composite Flexure Resonator, k_{33}-Mode Ceramic

FIGURE 7.15. Ten-resonator channel filter with out-of-phase bridging; also see Figure 1.10. (Siemens, F.R.G.)

mechanical resonators and two selectivity-producing electrical resonators were used, caused a high absolute and differential delay. Also, if the tuning coils could be eliminated, both the size and cost of the filter could be reduced.

A solution to the coil-elimination problem was to increase the carrier frequency to 128 kHz, thereby reducing the fractional bandwidth of the new filter. This made it possible to use the modified Langevin k_{33}-mode composite resonator of Figure 7.15 to eliminate the tuning coils. The new transducer-resonator used two oppositely polarized ceramic plates sandwiched between two metal rods to achieve the needed electromechanical coupling [9].

At 128 kHz, the filter designers had a choice between torsional resonators and bending-mode resonators, both having approximately the same dimensions. The torsional resonator has the advantage of a greater mechanical Q, of 30,000, as opposed to 20,000 in the flexural-mode case. Against this advantage, the filter designer had to weigh an existing large capital investment in automated fabrication, tuning, and assembly equipment, as well as an investment in the theoretical and practical understanding of the bending-mode filter. It was decided that the basic flexural-mode resonator, having the same diameter but shorter length, would be used in conjunction with two dumbbell torsional-mode resonators used for bridging-wire phase inversion. Because of choosing a physically symmetrical design, the two-resonator bridgings produced confluent double attenuation poles in each stopband, as shown in Figure 7.16. The result of reducing the number of resonators was an improvement of the minimum group delay to less than 659 μsec.

As the result of going to a higher frequency, and the use of new, better-balanced modulators, the microphonic responses due to shock or vibration were maintained at a low value. Microphonics would have been a problem at 48 kHz, but their amplitude was reduced by the output electrical tuned circuit in much the same manner as described in Chapter 9.

Having mentioned the use of torsional resonators at 128 kHz, let's look at what was happening in Japan in the 1970s.

A Torsional-Mode Parallel-Ladder Filter at 128 kHz

In the mid 1970s, Japanese telephone equipment manufacturers introduced a variety of 128 kHz mechanical filters. Possibly the most spectacular was a parallel-ladder type that used acoustic wave-separating techniques to realize the combination channel filter and signaling filter shown in Figure 7.17. The choice of the 128 kHz carrier frequency was the result of studying longitudinal, flexural, and torsional modes in frequency ranges near 50 kHz, 120 kHz and 200 kHz [10]. Let's look at this study.

In order to meet a 1/20 CCITT specification and have substantial clearance at the corner points, the round-off at the edges of the filter passband must be small. Therefore, either the mechanical Q must be high or the fractional bandwidth must be large. For non-predistorted designs the following condition must hold, $2\pi f_0/Q < 31$. All modes fail this criteria at 200 kHz. Because a

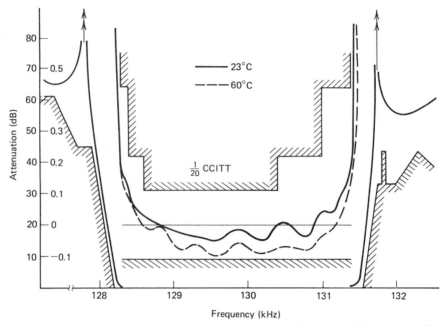

FIGURE 7.16. Frequency response of the double-resonator bridging filter of Figure 7.14. (Siemens, F.R.G.)

longitudinal-mode resonator is approximately 1.7 times the length of a torsional resonator, the longitudinal mode was eliminated because of length at 50 kHz and 120 kHz. The torsional-mode resonator at 50 kHz was too long, being greater than 2 times as long as a practical flexural resonator at that frequency. Because of the lower Q of flexural-mode resonators, due to acoustic losses in air, the flexural mode at 120 kHz was eliminated. The contest was finally between the 50 kHz bending-mode resonators and the 120 kHz flexural-mode resonators. On the basis of better amplitude sensitivity, smaller volume (for a diameter that was practical for manufacturing), and an ability to realize both attenuation poles and mechanical wave separation, the 120 kHz (actually 128 kHz) torsional resonator was chosen. Although some of the preceding argument may have been contested by other manufacturers, it shows the design process involved in choosing a resonator.

The parallel-ladder configuration was picked as a means of realizing finite attenuation poles. Because the parallel-ladder realization is not limited to frequency symmetry, the attenuation poles were placed as shown in Figure 7.18 in order to meet specific equipment specifications. The use of attenuation poles not only improved the delay response over the monotonic stopband-type filter, but, by eliminating resonators, reduced the size as well.

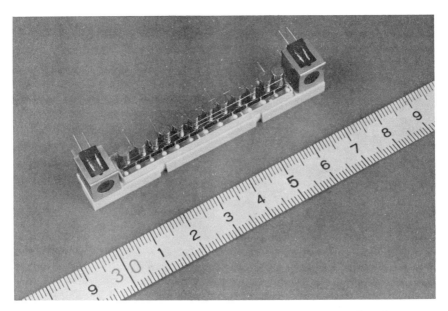

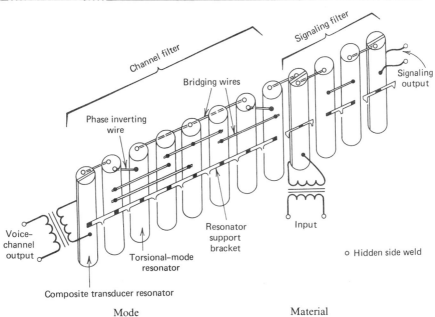

	Mode	Material
Resonators	Fundamental Torsion	Tokin TE-3
Coupling Wires	Extensional	Special NEC Alloy
Transducer	k_{15}—Shear → Torsion	PZT—Tokin NEPEC—11
Tuning	—	Ferrite Cup Cores (Channel Filter)
Package	—	Plastic, Volume = Transmit 10.6 cm^3, Receive 11.7 cm^3

FIGURE 7.17. Combined channel and signaling filters at 128 kHz. [© 1979 IEEE. Reprinted, with permission, from *Proc. 1979 IEEE ISCAS*, Tokyo, 1076–1079 (July 1979). Photograph, courtesy of NEC, Japan.]

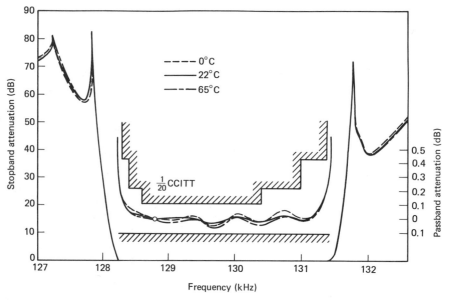

FIGURE 7.18. Frequency response of the telephone channel filter of Figure 7.17. (NEC, Japan.)

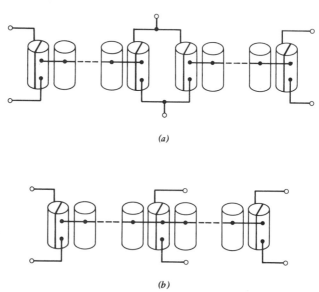

FIGURE 7.19. Channel and signaling filters using (*a*) an electrical parallel connection, and (*b*) a mechanical parallel connection. [© 1978 IEEE. Reprinted, with permission, from *Proc. 1978 IEEE ISCAS*, New York, 330–335 (May 1978).]

An additional means of reducing both size and cost is to use a common transducer for the channel and signaling filters. The 128 kHz filter uses the premodulation method II of Figure 7.10 and its channel modem is like that of Figure 7.12 except that the modulation frequency is 128 kHz rather than 48 kHz. This means that the channel and signaling filters have a common output (transmit side) or a common input (receive side). Rather than using a common electrical junction and two transducer resonators, as shown in Figure 7.19(a), a common transducer resonator [see Figures 2.18(e) and 7.19(b)] was used. Care was taken in the design to make the impedance of each filter a high value within the passband frequency range of the other filter, so as to reduce interaction. The input and output resonators attached to the channel filters are electrically tuned. The electromechanical coupling of the transducer is designed to be high, therefore providing wideband LC tuning circuits that merely have to be set to a frequency in order to obtain the required low passband ripple characteristics shown in Figure 7.18.

Other 128 kHz Channel Filters

In this section we describe two other channel filters designed for use at 128 kHz. The first is the eight-resonator filter shown in Figure 7.20 [11]. The beauty of this filter is its simplicity. The input and output transducers use the second flexural mode (first overtone) in order to realize phase inversion in the manner described in Chapter 4 (Figure 4.32). The resulting frequency response is shown in Figure 7.21. The temperature characteristics of this filter are very good, considering that the electrical circuit is tightly coupled and is used to provide stopband selectivity.

The eight-resonator channel filter also uses the input and output transformers to (*1*) match the impedance of the connected signaling filter and (*2*) provide a balanced circuit for operation of transistor single-balanced modulators and demodulators. By using the filter transformer, either an inductive component or a pair of active elements are saved. (See Chapter 9 for a discussion of SSB modulation techniques.) Most of the telephone channel modems using mechanical filters use hybrid circuits containing combinations of single-balanced, double-balanced, and passive modulators for the premodulation and channel modulation of the signals.

A second filter we will look at is the 10-resonator channel filter shown in Figure 7.22(a) [12]. Because of the high coupling coefficient of the transducer shown in Figure 7.22(b), this filter does not require electrical tuning. This results in both a size (about 6 cm^3) and a cost savings. The filter's frequency response is similar to that of Figure 7.21, in that four attenuation poles are realized by the two bridging wires. As in the previous filter, the end of one bridging wire and the start of the second are welded to different rod resonators. Without this separation, that is if the bridging wires ended and started on the same bar, only three attenuation poles could be realized instead of four. This is due to the very complex nature of mechanical filter resonators, as discussed in Chapter 3. The necessary phase inversion is realized by bridging at

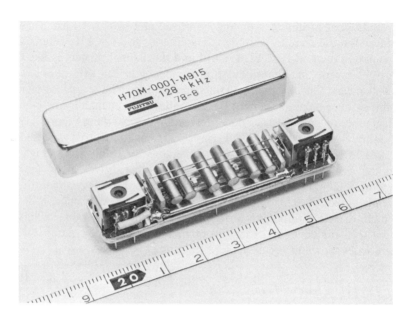

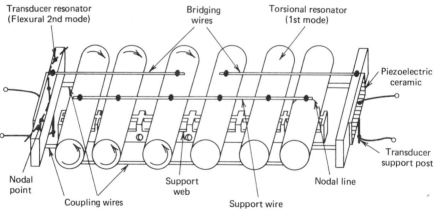

	Mode	Material
Resonators	Center—Torsional, End—Flexural	Elinvar
Coupling Wires	Extensional	Elinvar
Transducer Plate	k_{31}—Extensional	PZT Ceramic
Tuning	—	Ferrite Cup Core
Package	—	Metal, Volume = 10.7 cm^3

FIGURE 7.20. Eight-resonator 128 kHz channel filter that uses a second-mode flexural transducer for phase inversion. The input and output transformers are not shown in the drawing. [Adapted, with permission, from *Proc. 1979 IEEE ISCAS*, Tokyo, 1072–1075 (July 1979). © 1979 IEEE. Photograph, courtesy of Fujitsu, Japan.]

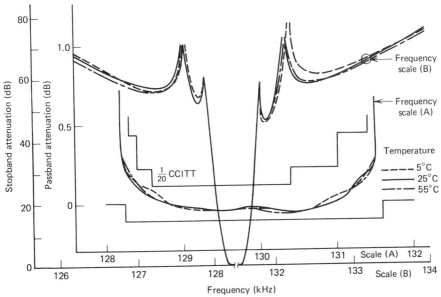

FIGURE 7.21.   Frequency response of the bridging-wire filter shown in Figure 7.20. [Adapted, with permission, from *Proc 1979 IEEE ISCAS*, Tokyo, 1072–1075 (July 1979). © 1979 IEEE.]

an angle. In this way, the out-of-phase regions of the starting and ending resonators are connected.

A Seven-Resonator Channel Filter at 256 kHz

Figure 7.23 shows a channel-modem block diagram that corresponds to premodulation method III of Figure 7.11. The unique features of this modem include: the use of three modulation steps, relatively high frequency (256 kHz) lower-sideband channel filters, selective *LC* filters for both off-carrier signaling and adjacent channel rejection, and an audio-frequency signaling filter [13].

The design of this modem was based on a study that showed that 256 kHz was an ideal carrier frequency for high performance disk-wire filters of moderate size (less than 25 cm³). In addition, it was shown that a single-disk bridging, lower-sideband filter was economical to build, and that the off-carrier selectivity could be obtained with audio lowpass filtering. The lowpass filter also solved the problem of having to build a selective signaling filter at 256 kHz. The lowpass filter is used in the following way. On the receive side of the modem, for example, the channel filter passes both the modulated voice message and the signaling tones. The signaling tone is then removed in the audio channel by the *LC* lowpass filter, and conversely, in the signaling channel the audio is removed by the 3.825 kHz mechanical signaling filter. Because of the better absolute frequency stability of the 3.825 kHz filter, as opposed to a 131.825 or a 252.175 kHz filter, the necessary selectivity is

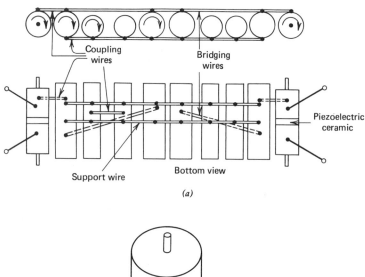

Coupling wires

Bridging wires

Piezoelectric ceramic

Support wire

Bottom view

(a)

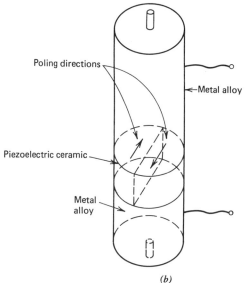

Poling directions

Metal alloy

Piezoelectric ceramic

Metal alloy

(b)

FIGURE 7.22. (a) A ten-resonator telephone channel filter with angled bridging, and (b) torsional-mode transducer. [Reprinted, with permission, from K. Sawamoto, T. Yano, K. Yakuwa, Y. Koh, and M. Konno, "Electromechanical filters developed in Japan," *L'Onde Electrique*, **58**(6–7), 482–487 (1978).]

achieved with a 50 Hz bandwidth (as compared to 100 Hz) two-resonator device rather than a four- or five-resonator filter.

The mechanical channel filter and its frequency response are shown in Figures 7.24 and 7.25. Figure 7.25 shows how the signaling tone at 3.825 kHz from the carrier frequency is "notched-out" by the lowpass filter attenuation pole at that same frequency. The attenuation poles on the carrier-frequency side of the passband are the result of the single-resonator bridging wires. At 256 kHz, the bridging-wire length is considerably less than one-half wave-

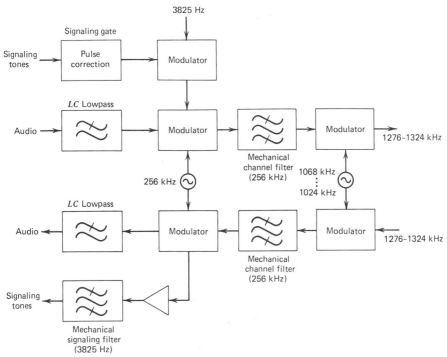

FIGURE 7.23. Simplified block diagram of a modern using premodulation method III (Figure 7.11). (Rockwell International, USA.)

length, so phase inversion is not possible and a lower-sideband filter response results.

As is often the case, the modem was designed around the channel filter. Because of unwanted modulation products and the use of a lower-sideband filter, it was necessary to use a double-modulation scheme. What was a disadvantage in terms of additional mixers, was an advantage in terms of system flexibility. Not only could each group of 12 channel filters be directly translated to a 60 to 108 kHz band, as shown in Figure 7.11, but to baseband frequencies anywhere in the range of 12 to 552 kHz, without the need of conventional group and supergroup equipment. Close cooperation and good communication between the system designers and the filter designers made their system practical.

Telephone Signaling Filters

One of the most common applications of mechanical filters is in telephone signaling circuits. The purpose of the signaling circuit is twofold: it communicates the on-hook or off-hook condition to indicate if the line is busy, and it transmits the number being called by means of dial pulses. In systems outside

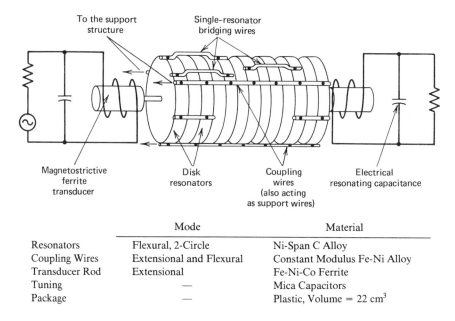

	Mode	Material
Resonators	Flexural, 2-Circle	Ni-Span C Alloy
Coupling Wires	Extensional and Flexural	Constant Modulus Fe-Ni Alloy
Transducer Rod	Extensional	Fe-Ni-Co Ferrite
Tuning	—	Mica Capacitors
Package	—	Plastic, Volume = 22 cm^3

FIGURE 7.24. Seven-disk channel filter with single-resonator bridging and magnetostrictive ferrite transducers. (Rockwell International, USA.)

of the USA, the signaling is done out-of-band, that is, the signaling frequency is higher than the voice band and is separated from the voice band by the channel filter (see Figure 7.14) or is separated by a lowpass filter (see Figure 7.25). The signaling filter must pass dial pulse data at a rate of up to 16 pulses per second, which means that its bandwidth must be at least 50 Hz but is more commonly about 100 Hz.

Depending on the modem frequency scheme, the signaling filter will be at an audio frequency of about 3.825 kHz [13] or at 3.825 kHz above the IF carrier frequency. The earliest mechanical signaling filters were used with *LC* channel filters at frequencies of 15.83 kHz, 19.83 kHz, and 23.83 kHz [6], [7]. These filters had two flexural-mode resonator bars and used an input inductor and output transformer for impedance matching, selectivity improvement, and microphonic signal rejection. Data taken over a period of 10 years showed that these filters had excellent reliability. By the fourth year of operation, the failure rate had dropped from an initial value of 110 Fit to about 20 Fit, and by the end of the sixth year, there was no indication that an increase in failure rate was approaching [14].

Other signaling filter designs include Telettra's three-resonator filter at 27.825 kHz, two- and three-resonator filters designed by West German companies at 51.830 kHz, and a wide variety of types at 131.830 kHz. These filters are all designed without input and output inductors or transformers.

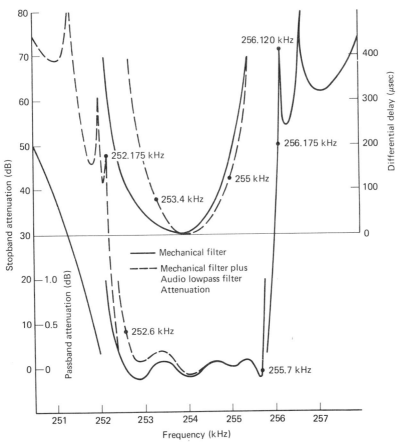

FIGURE 7.25. Frequency response of the seven-resonator channel filter of Figure 7.24 plus a three-coil lowpass filter. [Adapted, with permission, from *Proc. 1974 IEEE ISCAS*, San Francisco, 127–131 (Apr. 1974). © 1974 IEEE.]

An example of two of the 131.830 kHz filters is shown in Figure 7.26 [15]. The transducers are like those of the torsional-mode NTT design of Figure 7.22, except that they are positioned at the end of the metal rod rather than being sandwiched between two rods. As in the case of Langevin transducers discussed in Chapter 2, the end position decreases the electromechanical coupling but improves the resonator stability. A highly stable transducer resonator is necessary because of the small fractional bandwidth of the filter (0.10 kHz/131.8 kHz). The narrow bandwidth of a signaling filter, as compared to a channel filter, makes it necessary to have small values of coupling between the resonators. This is accomplished by placing the coupling wire near the resonator nodal points. For the purpose of cost saving and size reduction, the two filters of Figure 7.26 are enclosed in a single package.

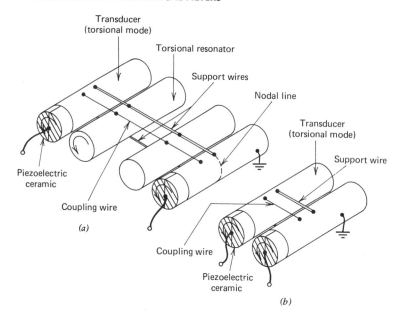

	Mode	Material
Resonators	Fundamental Torsion	Elinvar
Coupling Wires	Extensional	Elinvar
Transducer	k_{15}—Shear	PZT Ceramic
Package	—	Metal Hermetic Seal, Volume = 5.1 cm^3

FIGURE 7.26. (*a*) Four-resonator transmit filter, and (*b*) two-resonator receive filter for telephone signaling applications. (Fujitsu, Japan.)

Telephone Pilot-Tone Filters

In telephone multiplex systems, pilot tones both control received signal levels and sound alarms in case of abnormal variations of the incoming tone. For each group of messages to be transmitted, a pilot tone is generated at frequencies that do not interfere with the voice channel. A typical frequency is 84.080 kHz. These constant-level sinusoidal signals are picked off in the receive channel by a narrowband pilot-tone filter. The signals are then amplified, detected, and finally compared with a fixed reference; the difference (error) voltage is used to change the gain of the system. It is important that the insertion loss not change with either time or temperature, because the signal levels of the entire group are a function of the filters' loss.

It is possible to build mechanical pilot-tone filters that have less than a 0.2 dB variation of insertion loss over the temperature range of +10°C to +50°C. Figure 7.28 shows typical frequency-response variations of the two-resonator filters of Figure 7.27 [16]. This filter was designed with a higher electromechanical coupling coefficient than is necessary to terminate the 28 Hz bandwidth.

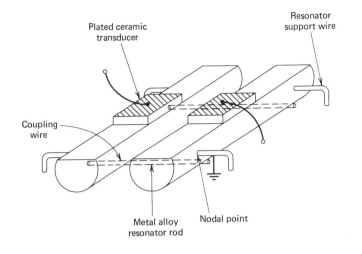

Plated ceramic
transducer

Resonator
support wire

Coupling
wire

Metal alloy
resonator rod

Nodal point

	Mode	Material
Resonators	Fundamental Flexure	Thermelast 5409
Coupling Wires	Torsional	Thermelast 5409
Transducer	k_{31}—Extensional	SEL PPK 32
Package	—	Metal Base, Plastic Cover, Volume = 9.70 cm^3

FIGURE 7.27. Two-pole pilot-tone filter. (SEL, F.R.G.)

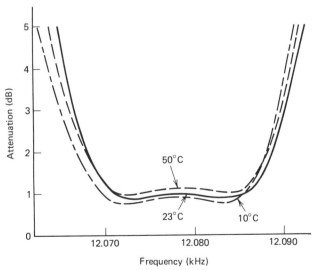

FIGURE 7.28. Temperature characteristics of the pilot-tone filter of Figure 7.27. [Adapted, with permission, from *Proc. 1976 IEEE ISCAS*, München, 743–745 (Apr. 1976). © 1976 IEEE.]

This allowed both matching into a specified terminating resistance of 17 kΩ and loss compensation resulting from the positive variation of the piezoelectric transducer's static capacitance.

The 12.080 kHz filter type is used for all pilot tones through the use of additional modulators. By choosing an optimum center frequency and needing only one type of filter, both size and cost savings result. This is an example of the need to weigh the trade-offs between a nonoptimum set of filters at various frequencies and a single filter/modulator design with various mixer frequencies.

Mechanical Filters for FSK Systems

Frequency shift keying (FSK) modulation schemes have evolved in a similar way to telephone FDM modems. Early designs involved using a group of filters at the lowest frequency band; in the FSK case, this was the audio band of 300 Hz to 3.4 kHz. Second generation systems used groups of six, and of ten filters, some groups being located in the audio band, others slightly above the audio band. Later systems used a single mechanical filter design which was located at a frequency high enough to eliminate the use of tuning coils. In this latter case, each mechanical filter in the basic group has its own modulation frequency, as in the case of the premodulation FDM telephone filters described in previous sections. Let us first look at a six-resonator monotonic design used as one of six filters in the frequency band of 3.960 to 4.560 kHz.

A Six-Resonator Monotonic Design

Filters designed for use in FSK systems have requirements on both amplitude and differential delay. The amplitude and delay specifications are between that of a maximally flat amplitude filter (Butterworth) and a maximally flat delay (Bessel or Thompson) design. The six-resonator mechanical filter shown in Figure 7.29 is a design of this type; specifically, it is a TBT (Transitional Butterworth-Thompson) filter having the rounded amplitude and nearly linear phase characteristics shown in Figure 7.30. The filter uses tuning-fork resonators and identical low-stiffness U-shaped coupling wires. The mechanical coupling varies according to the position of the coupling wire on adjacent forks [4].

The six-resonator filter is one of six filters spaced 120 Hz apart starting at 3.960 kHz and is used to transmit data at a 50 baud rate. A set of filters with increased stopband selectivity, but with greater differential delay, is used in the receive channels.

Although these filters have excellent frequency response characteristics, the tuning coils make them somewhat large. The tuning coils are needed because of the low electromechanical coupling of tuning-fork resonators. If the filters were designed at higher frequencies, the coils could be eliminated, but the frequency stability of the filter would not be adequate.

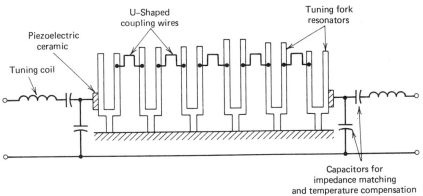

FIGURE 7.29. A six-resonator FSK filter based on the TBT design method. (Courtesy of Kokusai Electric Co., Japan.)

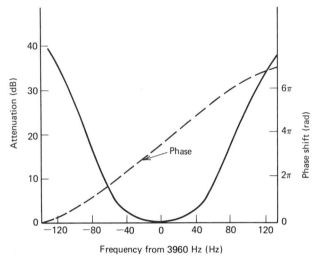

FIGURE 7.30. Amplitude and phase responses of the six-fork filter of Figure 7.29. (Courtesy of Kokusai Electric Co., Japan.)

Bar Flexural-Mode for FSK

By using a high coupling-coefficient ceramic transducer and a bar-flexural mode of vibration, it is possible to design stable, inductorless FSK filters. The filters shown in Figure 7.31 were designed at 4050, 8100, and 16,200 Hz for data rates of 50, 100, and 200 baud [17]. The two-resonator sections employ one-piece metal-alloy H-elements to which are bonded piezoelectric ceramic transducers. Because of reduced system requirements, only a two-resonator filter is needed in the transmit channel and a four-resonator filter in the receive channel. The elimination of coils and the low number of resonators results in a two-resonator package size of only 11 cm^3. In addition, only a single transmit and a single receive filter type is designed for each data rate, resulting in lower manufacturing costs.

Figure 7.32 shows the frequency response of a 50-baud two-resonator transmit filter. Note the flattened delay response and the rounded passband amplitude response, in addition to the attenuation poles. The attenuation poles result from bridging the input and output terminals with a capacitor as shown in Figure 7.31(b). The receive filter shown in Figure 7.31(c) is a cascade of two, two-resonator sections, isolated by 6 dB pad.

The use of resistive pads or isolation amplifiers between mechanical filter sections improves the consistency of each filter's termination but results in less stopband selectivity. In other words, a cascade of two decoupled two-resonator sections of the same bandwidth will have less selectivity than a four-resonator filter composed of direct or capacitively coupled two-resonator sections. The design problem then becomes one of weighing ease of manufacturing (i.e., less cost and capital equipment) versus improved stopband selectivity.

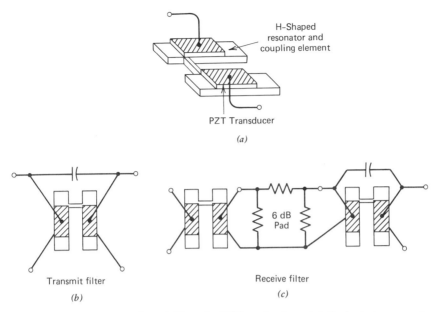

FIGURE 7.31. H-shaped bar-flexural filters for FSK applications: (a) the basic two-resonator section, (b) the transmit filter with capacitor bridging for generating attenuation poles, and (c) the receive-channel filter with an isolation pad. [Adapted, with permission, from *Proc. 1979 Ultrason. Symp.*, New Orleans, 119–122(Sept. 1979). © 1979 IEEE.]

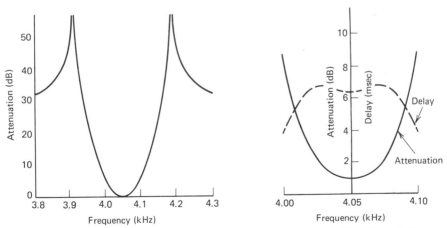

FIGURE 7.32. Attenuation and delay responses of the 50 baud transmit filter shown in Figure 7.31(b). [Adapted, with permission, from *Proc. 1979 Ultrason. Symp.*, New Orleans, 119–122 (Sept. 1979). © 1979 IEEE.]

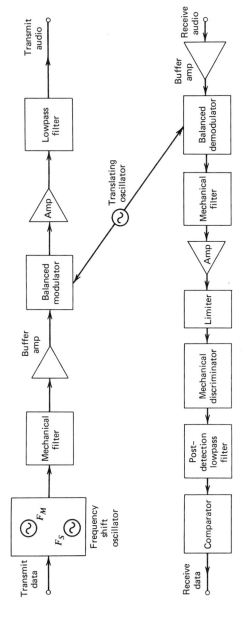

FIGURE 7.33. FSK modem using a premodulation scheme and mechanical resonators. (Courtesy of Rockwell International, USA)

Reference [18] describes a set of premodulation FSK filters that use a shunt capacitor to electrically couple two, two-resonator sections. Although these filters have the excellent selectivity of four-resonator designs, the inconsistency of each filter's impedance to the other filter makes it difficult, when cascading, to obtain consistent results in a nonautomated manufacturing environment. This would not be the case, if in addition to the amplitude response the static capacitance and electromechanical coupling were finely adjusted. Figure 7.33 shows a block diagram of the FSK modem in which the filters are used. A version of this modem was made where both the mechanical discriminator and the frequency shift oscillator used mechanical-filter resonators.

An FSK filter that makes use of three parallel-ladder resonators is described in Reference [4]. This is a five-resonator audio-band filter consisting of input and output coil-and-static-capacitance electrical resonators and flexural-mode bar resonators with unequal-size PZT transducers.

SIGNALING, DETECTION, AND CONTROL

There are many applications for mechanical filters similar to that of telephone signaling. Most of these applications require narrow-bandwidth stable filters. The bandwidths are as narrow as possible to reduce noise and adjacent signals but are wide enough to pass modulated signals with little distortion. The frequency-shift over temperature and with age should be low in order to keep the bandwidth as narrow as possible. Mechanical filters are well suited to meet these dual requirements of narrow bandwidth and good stability.

Mechanical Filters for Radio Navigation

Mechanical filters have been designed for various types of navigation equipment, such as Omega, Loran-C, and Decca receivers. In these applications, the mechanical filter is able to realize the necessary system selectivity near the front end of the receiver, the filter center frequency being the same as the incoming signal frequency.

Filters designed for Decca Navigation are two-resonator devices with bandwidths ranging from 16 Hz to 23 Hz and center frequencies of 5 kHz, 6 kHz, 8kHz, and 9 kHz. These are ideal center frequencies and bandwidths for mechanical filters. Crystal filters in this frequency range tend to be bulky and expensive, and ceramic filters often lack the necessary frequency stability. These limitations also apply to filters for Omega navigation receivers.

Filters for Omega Navigation Receivers

Omega is a long-range, very low frequency (VLF) radio navigation system providing global maritime and aircraft coverage. A worldwide network of eight

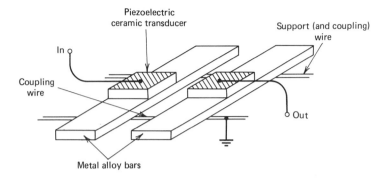

	Mode	Material
Resonators	Fundamental Flexure	Ni-Span C
Coupling Wires	Torsional	Fe-Ni
Transducer	k_{31}—Extensional	PZT (TDK 61A)
Package	—	Plastic or Metal

FIGURE 7.34. Flexural-mode mechanical filter for omega navigation. [Reprinted, with permission, from R. A. Johnson, "Mechanical filters using disk and bar flexure-mode resonators," *L'Onde Electrique*, **58**(2) 141–148 (1978).]

stations transmitting at 10.2 kHz, 11.05 kHz, 11.333 kHz, and 13.6 kHz provides a hyperbolic lines-of-position grid that is based on phase difference measurements from pairs of transmitters. As a plane or ship moves from one half-wavelength "lane" into another, a computer updates its position. By this process, position can be established within a one-to-two-mile radius anywhere in the world. To accomplish this, the phase shift of an Omega signal through the receiver must be held stable to within 2 degrees, regardless of the level of the incoming Omega and off-channel signals. Mechanical filters exceed this requirement over a very wide dynamic range. A typical variation is 0.5 degrees from an input level of − 10 dBm to the equipment noise floor [19].

Most Omega receivers operate with an antenna pre-amplifier driving a narrowband mechanical filter. Filter bandwidths range from 10 Hz to 120 Hz, with 20 Hz to 30 Hz being most typical. Figure 7.34 shows a two-resonator design where the coupling wires are welded to the underside of the bar; Figure 7.35 is a typical frequency response of this type of filter. If more stopband selectivity is needed, the filter can be designed with a less rounded passband shape or with an additional resonator. A cascade of two filters has also been used. The center frequency shift of the two-resonator filters, over a $\pm 30°C$ temperature range, is less than ± 0.1 times the bandwidth. The stability, as a function of time, is governed by the equation $\Delta f_0 = 0.02\ B \log(t/t_0)$, where the center frequency shift Δf_0 and the 3 dB bandwidth B are in Hz, and t is in days. The initial time t_0 is 10 (days), which is the time to manufacture the

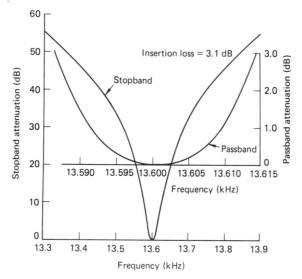

FIGURE 7.35. Frequency response of the 30 Hz bandwidth Omega filter of Figure 7.34. [Reprinted, with permission, from R. A. Johnson, "Mechanical filters using disk and bar flexure-mode resonators," *L'Onde Electrique*, **58**(2) 141–148 (1978).]

filter. A mechanical filter with a 50 Hz bandwidth will therefore shift less than 3 Hz in ten years.

A Wideband Filter for Loran-C

Loran-C is a 100 kHz medium-range navigation system primarily used for ships. Loran-C receivers are designed to detect only surface transmission and to reject reflected waves from the ionosphere. This is done by detecting only the first few cycles of a transmission, but to do this the filter must have a bandwidth on the order of 20 to 30 kHz. Because of numerous interfering signals near 100 kHz, the Loran-C filter must also have good stopband selectivity.

Figure 7.36 shows an extensional-mode mechanical filter designed for Loran-C receivers [20]. In order to achieve the fractional bandwidth, of about 25 percent, it was necessary to use high-coupling coefficient Langevin trans-ducers (Figure 2.21). The wide bandwidth made it possible to use the electrical elements to obtain four-resonator selectivity. In addition, the capacitive bridg-ing produces the two attenuation poles shown in Figure 7.37.

Spurious mode responses are often a problem in the manufacture of wideband mechanical filters but in this case they were kept out of the passband and are almost equally spaced on either side of the center frequency. There is a range of 40 percent of the center frequency that is spurious free. Note that B/f_0 exceeds the limit in Figure 1.11.

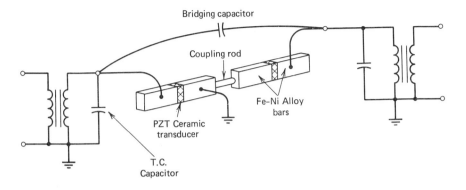

	Mode	Material
Resonators	Half-Wave Extensional	Seiko Spron 300
Coupling Rod	Extensional	Seiko Spron 300
Transducer	k_{33}—Thickness	PZT Spem 5C (Sumitomo)
Tuning	—	Ferrite Cup Core
Package	—	Metal, Volume = 51 cm^3

FIGURE 7.36. Extensional-mode wideband filter for Loran-C applications. [Courtesy of Daini Seikosha (Seiko), Japan.]

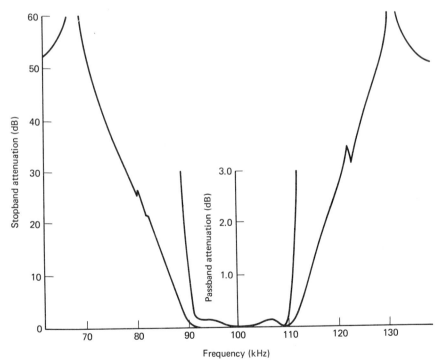

FIGURE 7.37. Frequency response of the Loran-C filter of Figure 7.36. (Courtesy of Seiko, Japan.)

Mechanical Filters for Train Control

Mechanical filters have been used in train control systems since the mid-1960s. The first filters were electrically coupled tuning forks for Japan's high-speed railroad system, where they were used in overspeed control. In the early 1970s, in the United States, control systems utilizing mechanical filters were designed for both train and people movers. One use was in automatic train protection (ATP) where the filters were used for overspeed control, block occupancy, and door control safety. Another use was in automatic train operation (ATO) applications, such as door control, train identification, and graphical displays.

Flexural-Bar Fail-Safe Filters

Figure 7.38 shows a mechanical filter designed for fail-safe operation [19]. This is one of 15 filters, having center frequencies from approximately 5 kHz to 10 kHz, that are used for automatic train control. With regard to safety, it is important that there be no signal output from the filter in case of a filter failure. In other words, if the ground is removed or if the coupling wires break off of the bars, it is important that the filter does not pass other frequencies or distorted signals. This is guaranteed in the case of the fail-safe filter in Figure 7.38 in the following ways.

The input and output of the fail-safe filter are isolated by a mechanical barrier which prevents input-to-output shorts. Because this is a simple ladder

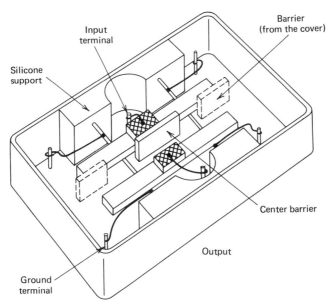

FIGURE 7.38.   Fail-safe mechanical filter for automatic train control applications. (Courtesy of Rockwell International, USA)

network, all other short circuits are to ground and therefore result in no output. Likewise, any open circuits due to breakage of lead wires or the coupling wires will result in zero output. A no-output condition shuts the automatic system down until a repair can be made. The train is operated in a manual mode in the interim. The ladder topology of the mechanical filter can be contrasted with either parallel-ladder or lattice configurations which are not fail-safe to either open or short circuits. Protection against losing the filter ground is through redundancy (four ground leads) or by detecting DC current running through two of the ground leads and the filter.

A second advantage of the mechanical filter technology for train control is the reliability of the filters. Because of the simple construction of narrowband two-resonator filters, typical mean time between failures is on the order of 3×10^7 hours, or about one failure in 400 parts over a time-span of 10 years.

Paging Systems and Long Distance Monitoring and Control Equipment

This section is primarily devoted to applications of tuning-fork filters. The operating frequency range of tuning-fork filters is between 100 Hz and 30 kHz. Because the transducers are mounted outside of the nodal points, as opposed to inbetween the nodes in the flat-bar resonator configuration, the electro-mechanical coupling is low, which means that more ceramic must be used to obtain a specified filter bandwidth. The larger ceramics, as well as the problem of realizing high-Q mounts, often leads to lower resonator Q's and reduced stability. Therefore, tuning-fork filters are usually used in applications where a single tone is to be detected, and the center frequency only needs to remain within its own nominal passband.

Remote Control and Monitoring Systems

Mechanical filters are widely used in remote control systems where switches or relays are excited by signals passed through a frequency-comb set of mechanical filters. A typical example is shown in Figure 7.39. A set of mechanical-filter oscillators is used to generate the transmit tones. If Relay No. 2 is to be excited at the remote site, the switch of Oscillator No. 2 is thrown, which sends a signal to the remote terminal. All of the remote-terminal mechanical filters are excited, but only Filter No. 2 passes the tone, which causes Relay No. 2 to be excited.

The system of Figure 7.39 can also be used as a monitoring system where the condition of the system is a function of the oscillator switch positions. The "remote terminal" now acts as a receiver.

Tuning-Fork Oscillators

At low frequencies, mechanical filters provide an inexpensive means of frequency generation. The best oscillator performance is obtained when the

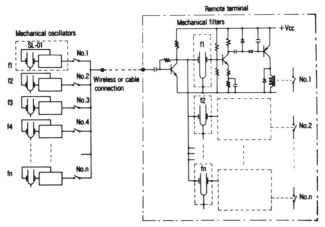

FIGURE 7.39. A remote control system using mechanical filters. (Courtesy of Seiko, Japan.)

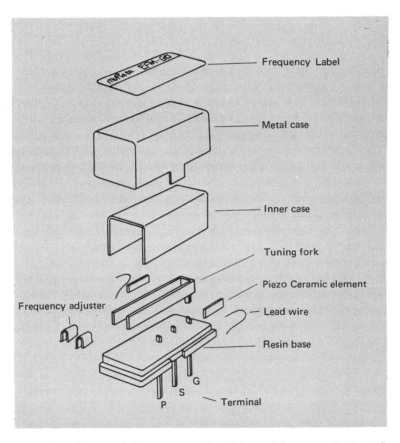

FIGURE 7.40. Low-frequency fork resonator used in oscillator and detector applications. (Courtesy of Murata Corp., Japan.)

307

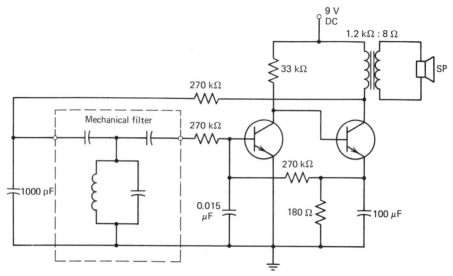

FIGURE 7.41. Simple filter-resonator oscillator circuit. (Courtesy of Murata Corp., Japan.)

filter resonator is used as a four-terminal network, that is, as a filter rather than a simple two-terminal impedance element. Figure 7.40 shows a one-pole tuning-fork filter, which can be used as either an oscillator or as a detector.

When the filter is properly terminated, the phase shift from input to output is approximately 90° at the resonance frequency (minimum attenuation point). Since the phase shift, from the output of the amplifier through the feedback network to the input of the amplifier, must be 180 degrees, it is necessary to use RC elements to provide the remaining phase shift $\phi = -\tan^{-1}(R/X_c)$. Figure 7.41 shows an oscillator circuit with RC elements at the input and output of the mechanical filter. Not having the proper phase shift or the proper termination can cause frequency deviations, spurious modes of oscillation, combinations of spurious and desired tones, and instability.

The insertion loss of the tuning-fork filter at its minimum loss point may vary from 6 dB to 13 dB, depending on the specific filter type and manufacturer. To be safe, the amplifier gain should be greater than 30 dB. Some tuning-fork filter manufacturers provide the associated phase shifting and amplifier circuits in a hybrid IC, or the filter and associated circuits as a complete unit.

REFERENCES

1. R. A. Johnson and F. L. Fanthorpe, "Mechanical filters for single-sideband applications," *Proc. 1979 IEEE ISCAS*, Tokyo 896–899 (July 1979).

2. M. Börner, E. Dürre, and H. Schüssler, "Mechanische Einseitenbandfilter," *Telefunken-Z.*, **36**, 272–280 (May 1963).

3. C. F. Kurth, "Channel bank filtering in frequency division multiplex communications: an international review," *Circuits and Syst.*, 5–13 (Dec. 1974).

4. M. Onoe, "Crystal, ceramic, and mechanical filters in Japan," *Proc. IEEE*, **67**(1), 75–102 (Jan. 1979).

5. "The electromechanical filters of Tesla Strasnice," *Tesla Electron.*, **5**(2), 57–59 (1972).

6. K. Endo et al., "New channel translating equipment," *Fujitsu Sci. Tech. J.*, **3**(1) (Mar. 1967).

7. M. Kogo et al., "Mechanical filters and oscillators for control equipment," *NEC Tech. J.*, **(82)** (Nov. 1969), in Japanese.

8. H. Albsmeier, A. E. Günther, and W. Volejnik, "Some special design considerations for a mechanical filter channel bank," *IEEE Trans. Circuits Syst.*, **CAS-21**, 511–516 (July 1974).

9. A. Günther, H. Albsmeier, and K. Traub, "Mechanical channel filters meeting CCITT specification," *Proc. IEEE*, **67**(1), 102–108 (Jan. 1979).

10. T. Yano, T. Futami, and S. Kanazawa, "New torsional mode electromechanical channel filter," in *Modern Crystal and Mechanical Filters*, D. F. Sheahan and R. A. Johnson, Eds. New York: IEEE Press, 1977.

11. K. Yakuwa, T. Kojima, S. Okuda, K. Shirai, and Y. Kasai, "A 128-kHz mechanical channel filter with finite-frequency attenuation poles," *Proc. IEEE*, **67**(1) (Jan. 1979).

12. K. Sawamoto, T. Yano, K. Yakuwa, Y. Koh, and M. Konno, "Electromechanical filters developed in Japan, Part 2: channel EM filters, "*L'Onde Elect.*, **58**(6–7) (1978).

13. R. A. Johnson and W. A. Winget, "FDM equipment using mechanical filters," in *Proc. 1974 IEEE ISCAS,* San Francisco, 127–131 (Apr. 1974).

14. K. Yakuwa, S. Okuda, Y. Kasai, and Y. Katsuba, "Reliability of electromechanical filters (in Japanese)," *Fujitsu*, **24**(1), 172–179 (1973).

15. K. Yakuwa, M. Yanagi, and K. Shirai, "DA-series channel translating equipment using pole-type mechanical filters. Part 2: components and devices," *Fujitsu Sci. Tech. J.*, **15**(3), 23–45 (Sept. 1979).

16. W. Borowski and P. Wollmershäuser, "Modern FDM pilot filters," *Electrical Comm.*, **53**(2), 148–152 (1978).

17. T. Kawana and H. Kawahata, "Preshift mechanical filter for voice frequency telegraph transmission system," *Proc. 1979 Ultrason. Symp.*, New Orleans, 119–122 (Sept. 1979).

18. R. A. Johnson and W. D. Peterson, "Build stable compact narrow-band circuits," *Electron. Design*, 60–64 (Feb. 1, 1973).

19. R. A. Johnson, "Mechanical filters take on selective jobs," *Electronics*, 81–85 (Oct. 13, 1977).

20. Y. Kawamura, "Mechanical filters find promising applications," *JEE*, 52–55 (Dec. 1978).

SPECIFYING
AND TESTING
MECHANICAL FILTERS

This chapter is written primarily for the user of mechanical filters. Understanding the principles outlined in this chapter will help the user obtain the filter he needs at the lowest cost. Emphasis is placed on subjects that relate mainly to mechanical filters, though all important specifications and test circuits, relating to filters in general, are discussed.

SPECIFYING MECHANICAL FILTERS

A mechanical filter is usually considered a component, but because of its complexity, it is best viewed by the filter user as a subsystem. This view emphasizes the fact that repeated interaction between the filter user and filter designer is usually necessary and should be expected. The following is an example of effective communication:

USER: The user studies the requirements and contacts the designer to get a "feeling" for the design problem. A preliminary specification is then written by the user, which specifies ranges for each important parameter. Environmental conditions are also included. Care must be taken not to over-specify at this point and jeopardize realizability, performance, or cost.

DESIGNER: The designer studies the preliminary specification and sends the user the following: costs, time for designing, prices, and package size. Exceptions to the specification are also listed. Contact is made if the user's requirements are not clear.

USER: The user studies the trade-offs and writes a final specification. Prototype filters are ordered according to this specification.

DESIGNER: Prototype filters are designed, and a vendor specification is written in a format that is best for manufacturing testing. This specification may be sent to the user if there is some question about test methods or if the user's specification is not complete enough for his own inspection and testing. Whether the final specification be the customer's or the supplier's, it should be a clear description of the performance requirements and the environmental conditions. The specification should also include a complete description of the test methods and a detailed drawing of the package shape, markings, and pin locations.

The following is a check list of performance and environmental specifications.

Must Be Specified	Sometimes Specified
Passband attenuation and frequency points (or center frequency and bandwidth)	Phase or delay
	Carrier frequency rejection
	Time response
Insertion loss	Intermodulation distortion
Passband ripple	Microphonic responses
Stopband attenuation	Dielectric strength
Spurious response rejection levels	Insulation resistance
	Moisture resistance
Terminating resistance range	Aging
Signal levels	Weight
Operating and storage temperature	Marking information
	Other specifications
Shock and vibration	(CCITT, MIL, EIA, etc.)
Maximum package dimensions	
Hermetic or nonhermetic seal	
Pin locations	

Some of the preceding items are specified at room temperature and some over the operating temperature range. Although an operating temperature range (note the word *range*) is specified, most factory measurements are made at room temperature and the temperature extremes. This is a good approximation, because most of the resonance modes, in mechanical filters, have similar temperature coefficients. Also, a test voltage level is specified. It is assumed that at voltages below the test level the filter will meet the specifications. If high voltage levels are required, an additional set of measurements at the maximum voltage level can be specified. As often as possible, combined testing should be done on the prototype-filter level so as to reduce the amount of factory testing. The type of testing that is undesirable can be illustrated by the extreme case of having to make combined measurements of phase over a range of signal frequencies and amplitudes, vibration frequencies and amplitudes, and temperatures.

Loss and Selectivity

In this section we look at the amplitude versus frequency response characteristics of mechanical filters and how they are specified. By the term *amplitude versus frequency*, we are referring to the variation of the filter output voltage (relative to a fixed reference V_R), as a function of the applied frequency. The output voltage is the voltage V_L generated across the load resistance R_L, shown in Figure 8.1(a). The fixed reference voltage V_R can be any one of the following: the maximum output voltage in the passband, the output voltage at a specified frequency, the source voltage V_S in Figure 8.1, or the output voltage V_0 in Figure 8.1(b).

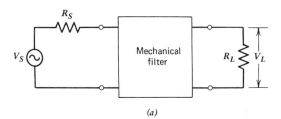

(a)

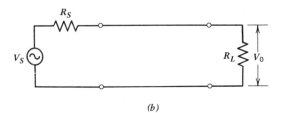

(b)

FIGURE 8.1.   Basic circuits for measuring frequency response and insertion loss.

In equation form, the filter amplitude response is expressed as an attenuation (loss),

$$A = 20 \log_{10}\left(\frac{|V_R|}{|V_L(f)|}\right), \tag{8.1}$$

where $|V_R|$ and $|V_L(f)|$ are the magnitudes of the reference voltage and the frequency-dependent output voltage.

Insertion Loss

Equation (8.1) is used to calculate insertion loss by setting the reference voltage V_R equal to the voltage V_0 of Figure 8.1(b). V_0 is the voltage across the load resistance when the filter is removed and replaced by short circuits, as shown in the figure. It is important to include the reactive part of the termination as part of the filter; in other words, the mechanical filter is considered to be the network between the source resistance R_S and the load resistance R_L. Therefore, when the filter is removed from the circuit and replaced by short circuits, as shown in Figure 8.1(b), all series or shunt capacitance and inductance must be removed also. In cases where not enough of the reactance can be removed to obtain valid reference voltage measurements, it is necessary to use alternate calculations or methods of measurement. One of the easiest methods is to simply measure the source voltage V_S in Figure 8.1 and calculate V_0 from

$$V_0 = V_S\left(\frac{R_L}{R_L + R_S}\right). \tag{8.2}$$

Equation (8.2) can be substituted into Equation (8.1) to give us the insertion loss

$$IL = 20 \log_{10}\left(\frac{R_L}{R_L + R_S} \times \frac{|V_S|}{|V_L|}\right). \tag{8.3}$$

In discussing insertion loss, we have not mentioned the frequency at which it is measured. Because insertion loss is a special case of the attenuation Equation (8.1), it tracks the amplitude versus frequency curve of the filter. Most often though, the insertion loss is defined as a single value of attenuation measured at the minimum attenuation frequency, or is specified as the loss at a specific frequency point, such as the center frequency f_0. This is illustrated in Figure 8.2.

Using the insertion loss Equation (8.3), it is possible to have an insertion gain, even under the condition where $R_S \gg R_L$. This can be shown by an example where the filter is replaced by an ideal transformer.

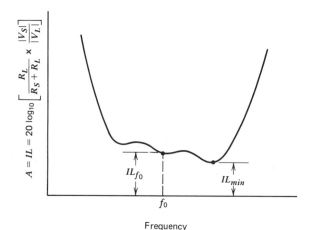

FIGURE 8.2.   Illustration of insertion loss as a function of frequency and at single frequency points.

Example 8.1.   Various transducer tuning configurations can lead to the source and load resistances being unequal. For example, if the input end of a magnetostrictive transducer filter is tuned with a shunt capacitor and the output end with a series capacitor, then $R_S \gg R_L$. Let's assume that $R_S = 20$ kΩ, $R_L = 2$ kΩ, and that the internal losses are so low that we can set them equal to zero. Under these conditions, we can replace the filter with an ideal transformer of turns ratio $(10)^{1/2}:1$, which makes the input impedance look like 20 kΩ. The output voltage V_L under these conditions is

$$V_L = \frac{V_S}{2} \times \frac{1}{\sqrt{10}},$$

and from (8.3) after canceling the V_S terms,

$$IL = 20 \log_{10} \left[\frac{2}{2 + 20} \times \frac{2\sqrt{10}}{1} \right] = -4.81 \text{ dB}.$$

In other words, the filter matched the source and load impedances to provide an insertion gain of 4.81 dB. In practice, an insertion gain is rare, but a surprisingly low loss is common.

Example 8.2.   We are required to measure the insertion loss of a 455 kHz filter that is terminated with 20 kΩ ($R_S = R_L = 20$ kΩ). We would like to use Figure 8.1 to measure the loss, but the test fixture has 20 pF across both the input and output terminals. Shorting between the input and output, as in Figure 8.1(b), places 2×20 pF $= 40$ pF ($X_C = 8.7$ kΩ) across the load resistance R_L making the measurement of V_0 invalid. A solution to this problem is to place a 500 Ω resistor across R_L, which eliminates the effect of the shunt capacitance. The voltage across R_L, which we will call V_0' is,

$$V_0' = V_S \frac{500 R_L/(500 + R_L)}{R_S + [500 R_L/(500 + R_L)]} = \frac{V_S}{42}.$$

Therefore

$$V_S = 42 V_0'.$$

We can now calculate the insertion loss from Equation (8.3).

$$IL = 20 \log_{10}\left(\frac{R_L}{R_L + R_S} \times \frac{|V_S|}{|V_L|} \right) = 20 \log_{10}\left(0.5 \times \frac{42|V_0'|}{|V_L|} \right)$$

$$= 20 \log_{10}(21) + 20 \log_{10}\frac{|V_0'|}{|V_L|} = 26.44 \text{ dB} + 20 \log_{10}\frac{|V_0'|}{|V_L|}.$$

Passband Specifications

Because insertion loss is usually specified at a single point, most often the passband response curve is also referenced to that same point. Figure 8.3 shows two typical curves, which illustrate methods of specifying a mechanical filter passband.

The curve in Figure 8.3(a) shows increasing attenuation in the direction of the top of the page, whereas the curve in Figure 8.3(b) shows decreasing amplitude in a downward direction. Note that in Figure 8.3(a) the attenuation is with reference to the filter response at 800 Hz from the carrier frequency, that is, at 128.800 kHz. The amplitude reference in Figure 8.3(b) is the minimum attenuation (maximum amplitude) point rather than a fixed frequency.

The two response curves in Figure 8.3 satisfy very different demands. The curve in Figure 8.3(a) is that of a 12-resonator filter designed to pass voice information. The specifications require low passband ripple and sharp corners. The curve in Figure 8.3(b) represents the normal rounded response of a two-resonator filter; the passband specification includes only a minimum passband ripple value A_p and two frequencies corresponding to the 3 dB attenuation points.

Two-resonator and three-resonator filters are sometimes specified with regard to center frequency and bandwidth tolerances rather than points at the band edges. This is often less desirable, in a factory environment, because of the calculations involved. For instance, using the 3 dB attenuation points, the center frequency and bandwidth are calculated from

$$f_0 = \frac{f_{3H} + f_{3L}}{2} \text{ (algebraic value)} \tag{8.4}$$

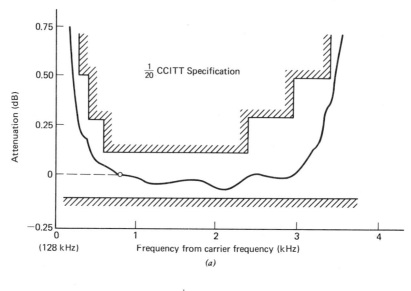

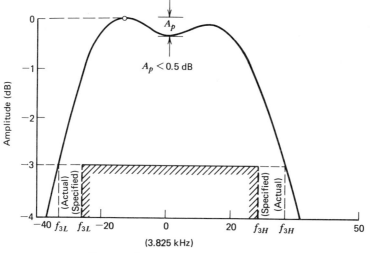

FIGURE 8.3. Passband response specifications. The curves must fall outside of the shaded regions.

and

$$B = f_{3H} - f_{3L}, \tag{8.5}$$

where f_{3H} and f_{3L} are the respective frequencies corresponding to the actual filter attenuation at the high- and low-frequency 3 dB points.

In Figure 8.3(b) we see that the variation between the maximum amplitude point and the minimum amplitude zero-slope point, is the passband ripple A_p. An alternative to simply specifying passband ripple is to specify passband response boundaries as was done in Figure 8.3(a). Passband ripple in mechanical filters should not be confused with ripple based on adjacent peaks and valleys, as is sometimes specified with low-Q filters.

A question that has not been discussed is how the user estimates the frequency response curve of a mechanical filter in order to write a realistic specification. Let us defer this question to the following section.

Stopband and Transition-Band Specifications

It is helpful for the filter user to have a way of estimating the frequency response of a mechanical filter before contacting the designer. Knowing that a mechanical filter can be modeled with R's, L's, and C's in the ladder or bridged-ladder form, use can be made of computer-aided analysis programs and generalized tables and curves. Hand-held calculator programs are available for calculating monotonic amplitude response curves. These curves can also be found in Handbooks [1] and [2]. The estimation problem is more difficult when attenuation poles are needed, because most tables assume equal stopband attenuation minimums (so-called elliptic-function responses) and $N - 1$ or $N - 2$ attenuation poles, where N is the number of resonators [3]. As was discussed in Chapter 4, these conditions are rare in the design of mechanical filters; therefore, the mechanical filter user must depend on, and interact closely with, the designer in order to satisfy both performance and cost objectives when attenuation poles are needed.

The stopband and transition-band of a filter having a monotonic response can be specified with two points or straight lines, whereas, a filter with attenuation poles often requires a broken-line specification. Examples of these types of specifications are shown in Figure 8.4; also shown are nominal-value frequency response curves.

It is important to note, in Figures 8.3 and 8.4, that a margin is always allowed between the specification and the nominal response curve. In Figure 8.4(a), the margin is expressed as an amount y dB or x Hz. The passband and stopband margins are a function of the following:

Component and process tolerances

Yield and tuning-time tradeoffs

Shifts in component values due to temperature and aging

Measurement accuracy

An idea of necessary room temperature tolerances can be derived from histograms similar to that shown in Figure 8.5 of a 10,000 unit sample of filters, which were built for a low-cost radio application. In this case, the specification was written loose enough that the production yield would be high.

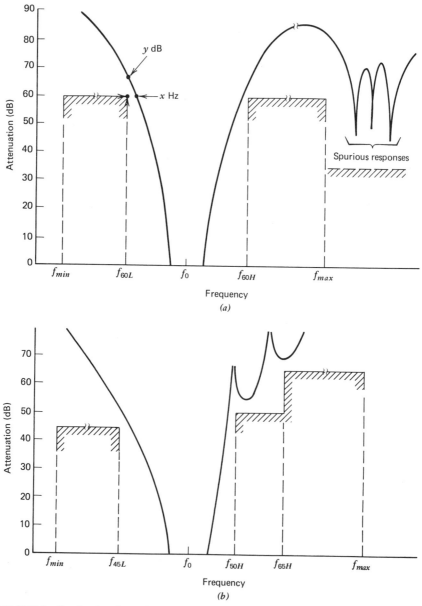

FIGURE 8.4. Stopband and transition-band specifications. (*a*) A monotonic response filter, and (*b*) a filter with finite-frequency attenuation poles.

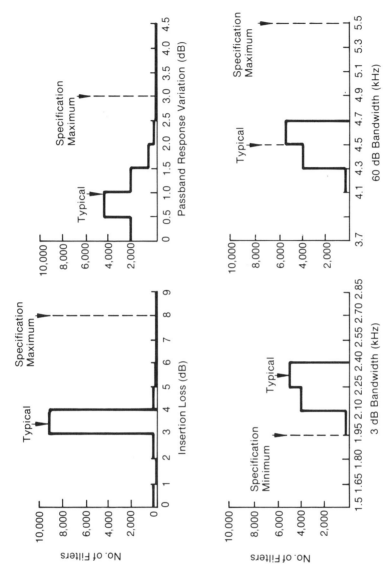

FIGURE 8.5. Typical production results on a 455 kHz disk-wire upper-sideband filter. (Courtesy of Rockwell International, USA.)

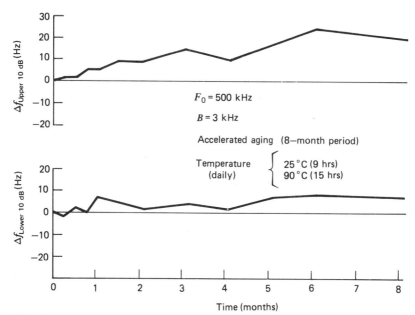

$F_0 = 500$ kHz

$B = 3$ kHz

Accelerated aging (8–month period)

Temperature (daily) { 25 °C (9 hrs), 90 °C (15 hrs)

FIGURE 8.6. Disk-wire mechanical filter aging. (Courtesy of Rockwell International, USA.)

Estimates of frequency shift with temperature change can be made from the curves in Figures 2.17(b), 3.27, and 3.28, and the data of Table 3.5. Estimates of aging can be derived from curves like those in Figure 2.17(a) and Figure 8.6. Let's next consider the subject of spurious responses.

In Figure 8.4, lower-frequency f_{min} and upper-frequency f_{max} specification limits are shown. It is assumed that there is either additional selectivity provided in the system to attenuate spurious responses, or that outside of these limits, no attenuation is necessary. In cases where attenuation is necessary, but only a moderate amount of additional selectivity is available, it may be necessary to add a reduced stopband limit to the specification like that of the dashed line above f_{max} in Figure 8.4(a).

Amplitude, Delay, and Time Response Curves

Although each filter design is unique, the question is often asked as to what is the response of a typical mechanical filter. Understanding the risks in answering a question like this, I will draw on my experience and state that the typical mechanical filter is a 0.1 dB equal passband-ripple design. To be more specific, the stopband, group delay, and time responses are close to that of a 0.1 dB filter, although the passband amplitude ripple and ripples in the delay curve are usually greater than those of the 0.1 dB theoretical design. To give the designer or the filter user a starting point, amplitude, group or time delay T_G, and time t response curves, from the set in Reference [1], have been reproduced as Figures 8.7, 8.8, 8.9, and 8.10. Referring to the curves, n equals the number

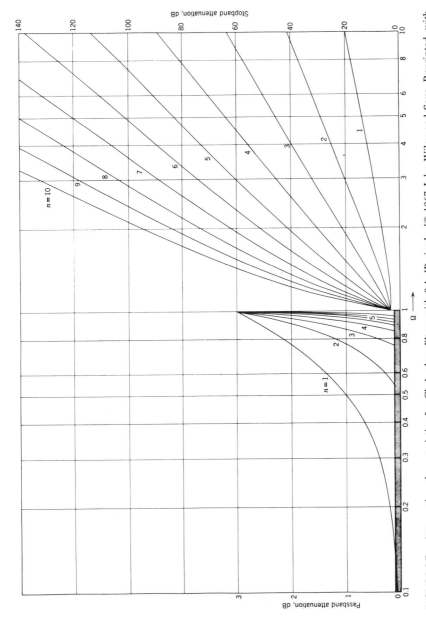

FIGURE 8.7. Attenuation characteristics for Chebyshev filters with 0.1 dB ripple. [© 1967 John Wiley and Sons. Reprinted, with permission, from *Handbook of Filter Synthesis* by A. I. Zverev, Wiley, New York (1967).]

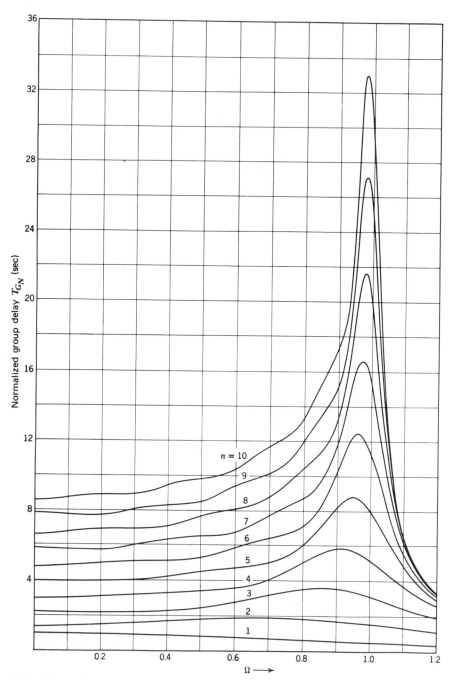

FIGURE 8.8. Group delay characteristics for Chebyshev filters with 0.1 dB ripple. [© 1967 John Wiley and Sons. Reprinted, with permission, from *Handbook of Filter Synthesis* by A. I. Zverev, Wiley, New York (1967).]

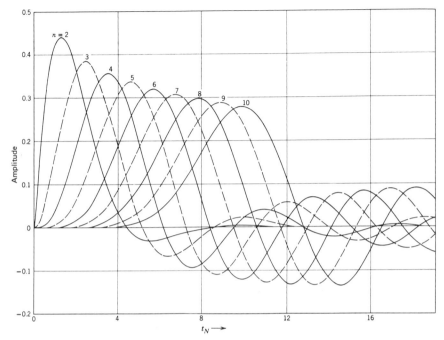

FIGURE 8.9. Impulse response for Chebyshev filters with 0.1 dB ripple. [© 1967 John Wiley and Sons. Reprinted, with permission, from *Handbook of Filter Synthesis* by A. I. Zverev, Wiley, New York (1967).]

of resonators. Equations for denormalizing the time and frequency axis are:

$$f = \Omega\left(\frac{B_{3dB}}{2}\right) + f_0 \tag{8.6}$$

$$T_G = \frac{T_{G_N}}{\pi B_{3dB}} \tag{8.7}$$

$$t = \frac{t_N}{\pi B_{3dB}}, \tag{8.8}$$

where frequency and bandwidth are in Hz and time and group delay are in seconds.

Phase, Delay, and Time Response

By the terms *phase response* and *delay response,* we mean the phase shift and group delay of the mechanical filter as frequency is varied. By *time response* we

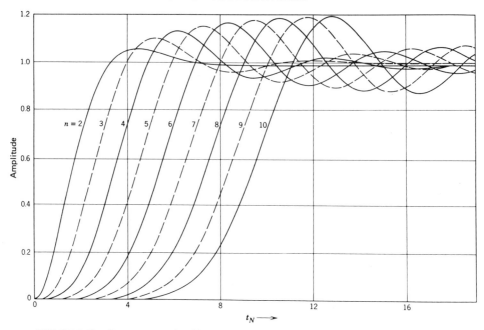

FIGURE 8.10. Step response for Chebyshev filters with 0.1 dB ripple. [© 1967 John Wiley and Sons. Reprinted, with permission, from *Handbook of Filter Synthesis* by A. I. Zverev, Wiley, New York (1967).]

mean the amplitude of the filter output as a function of time, as opposed to frequency. Because they are so closely related, let us look at phase shift and group delay together.

Phase Shift and Group Delay

There is often confusion between the filter user and the designer regarding the words *phase shift* and *group delay*. Therefore, we will carefully define these two terms in the manner most common to filter manufacturers. By phase shift through the filter, we mean the phase difference between the output sinusoidal signal V_L (across the load resistor R_L) and the source sinusoid V_S which acts as a reference. The reference signal phase ϕ_S is not affected by the filter characteristics, whereas the phase across the input terminals is a function of the filter characteristics. Therefore, the phase of the signal at the input terminals is not used as a reference.

Phase shift has a further and more important meaning, which is the change in the phase shift through the filter as frequency is varied. For instance, the phase shift (as a function of frequency) between frequency f_1 and frequency f_2 is

$$\Delta\phi = \left(\phi_S - \phi_L\right)_{at\,f_1} - \left(\phi_S - \phi_L\right)_{at\,f_2}.$$

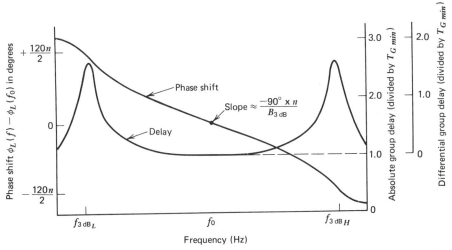

FIGURE 8.11. Typical phase shift and delay response curves based on 0.1 dB ripple data. n is the number of resonators.

If we measure phase with reference to $\phi_S = 0$,

$$\Delta\phi \doteq \left[(\phi_L)_{f_2} - (\phi_L)_{f_1}\right]_{\phi_S=0}. \tag{8.9}$$

The term *phase shift*, when used in this book, is $\Delta\phi$ as defined in Equation (8.9). Now we are ready to define the term *group delay*.

Group delay, which is sometimes called time delay, envelope delay, or simply delay, is defined as:

$$T_G \doteq \Delta\phi/\Delta\omega|_{as\ \Delta\omega\to0}, \tag{8.10}$$

where $\Delta\omega = 2\pi(f_2 - f_1)$. In other words, group delay is simply the slope of the phase (in radians) versus frequency (in radians per second) response curve. Figure 8.11 shows typical phase-shift and group-delay curves. Note that group delay is shown as absolute (or total) delay, which is the value of Equation (8.10), and as differential delay, which is simply the difference between the absolute delay and its minimum value within the filter passband. Generally, the maximum allowable differential delay variation across a band of frequencies is specified, rather than the absolute delay.

Some common requirements regarding delay and phase are as follows:

Delay

Differential delay over a frequency band (radio or telephone)

Absolute delay at a single frequency or at the minimum (telephone)

Delay difference at two frequencies (FSK modems)

Phase

Phase shift variation at a specified frequency due to temperature, time, amplitude, or vibration.

Phase shift difference between two filters over a band of frequencies (phase tracking).

Example 8.2. Questions regarding phase shift variations are often asked by users of mechanical filters. As an example, what is the phase shift variation of a 455 kHz signal applied to a 455 kHz center frequency, 500 Hz bandwidth, six-resonator filter, as the operating temperature is varied from $+23°C$ to $+60°C$?

Because the signal is applied at the center frequency, we can estimate the change in phase from the approximation equation in Figure 8.11:

$$\frac{\Delta\phi}{\Delta f} = \frac{-90° \times n}{B_{3dB}}, \tag{8.11}$$

where n is the number of resonators. In this example, Δf is the frequency shift of the filter [the sign in Equation (8.11) should be changed], rather than the frequency shift of the applied signal. Assuming that the maximum bandwidth change is -50 Hz and the maximum frequency change is $+50$ Hz,

$$\Delta\phi_{max} = \frac{90° \times n \times \Delta f}{B_{3dB}} = \frac{90 \times 6 \times 50}{(500 - 50)} = 60 \text{ degrees.}$$

This solution can be viewed in two parts: (1) the bandwidth narrows symmetrically, increasing the slope $\Delta\phi/\Delta f$, but the phase shift at center frequency does not change; and (2) the filter then shifts Δf in center frequency, the phase shift being proportional to the new slope.

Time-Domain Response

Figures 8.9 and 8.10 show time responses to impulse and step functions applied to 0.1 dB ripple filters having two to ten resonators. The amount of ringing or overshoot can be decreased by reducing the passband ripple, the ringing and overshoot being equal to zero in the round-top Gaussian case. Rough estimates of delay and rise times are shown in Figure 8.12. T_G can be found from normalized curves and Equation (8.7). The rise time T_r, measured between the 10 percent and 90 percent points, shows a small increase with the number of resonators but can be approximated by $T_r = 1/B$, where T_r is in seconds and the filter bandwidth B is in Hz.

Input and Output Parameters

It is important that both the mechanical-filter user and manufacturer clearly specify both signal levels and source and load impedances. How these parameters are specified is illustrated in Figure 8.13, which shows a mechanical filter terminated in a parallel combination of resistors and capacitors and driven by a signal voltage V_S.

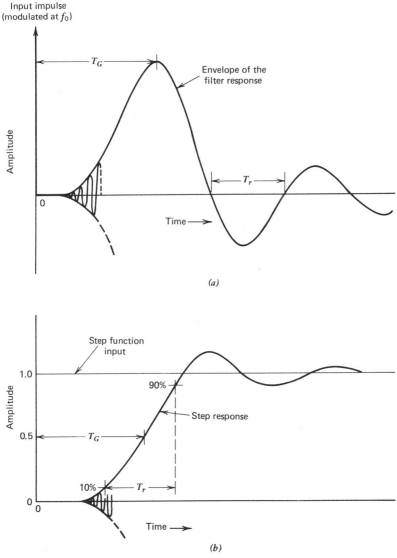

FIGURE 8.12. (*a*) Impulse response, and (*b*) step response curves, showing approximate delay and rise times.

The question often arises as to whether the filter response measurements should be based on a specified value of V_S or a specified value of V_L. The filter user sometimes prefers specifying V_S based on knowing what the voltage will be at that point in the circuit. The filter manufacturer normally prefers specifying V_L because the production people normally use a voltmeter across the output terminals; the output voltage is set at the specified value of V_L and

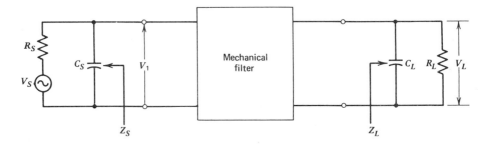

R_S, R_L = Parallel source and load resistance

C_S, C_L = Parallel source and load capacitance

Z_S, Z_L = Source and load impedances (terminating impedances)

FIGURE 8.13. Terminating circuit parameters. R_S, R_L = parallel source and load resistance, C_S, C_L = parallel source and load capacitance, Z_S, Z_L = source and load impedances (terminating impedances).

the frequency response is measured. Usually a number such as -10 dB [0.775 V(RMS)] or -20 dB [0.245 V(RMS)] is used, in order to simplify the measurements. The subject of voltage levels is discussed further in Chapter 9.

It is also necessary to specify terminating circuit parameters, such as resistance, and capacitance or inductance. Figure 8.13 shows a filter terminated in a parallel combination of resistance and capacitance. These element values represent the termination that must be supplied by both the application circuit and the test circuit. Looking back into the driving circuit, the impedance Z_S is composed of the parallel combination of R_S and X_{C_s},

$$ Z_S = - \left. \frac{jR_S X_{C_s}}{R_S - jX_{C_s}} \right|_{\text{at } f_0} . \tag{8.12}$$

The resistance and capacitance values are composed of active-circuit real and imaginary impedances, series or parallel resistors, resonating capacitance or inductance, and stray capacitance. The circuit could have been represented by its series equivalent circuit; transformations from parallel to series are shown in Figure 4.37.

The filter user and the filter designer should work closely to determine the best source and load impedances. In the case of narrowband (no tuning coils) mechanical filters, the source and load resistance values are determined by the dimensions and the dielectric constant of the transducer material. This limits the terminating resistance to values roughly between 10 kΩ and 100 kΩ. Intermediate-band and wideband filters can use transformers or capacitor impedance-matching circuits, and therefore, can achieve a wider range of terminating resistance values. This subject also is discussed in more detail in Chapter 9.

Whether the source and load impedances are required to be capacitive or inductive depends on the type of transducer and the internal transducer tuning. Magnetostrictive transducer filters will always require a capacitive source and load for tuning or taking into account stray capacitance. Piezoelectric-transducer filters may require external coils or transformers, or if the tuning coils are inside the filter, or if no tuning coils are used, capacitance values will be specified.

Environmental Specifications

The environmental specifications should always include the operating and storage temperatures and the maximum shock and vibration levels that the filter must withstand. In addition, the type of enclosure (hermetic or non-hermetic seal) must be specified. Let's first look at the temperature specifications.

Operating and Storage Temperature Range

Filters are normally tested at room temperature and at the limits of the specified operating temperature range. Military specifications are typically -55 to $+85°C$, and commercial product specifications are typically 0 to $+60°C$. Cost savings and performance improvement can sometimes be achieved by proper positioning of the filter in the equipment to take advantage of hot or cool zones, in order to reduce the operating temperature range.

Regarding the storage temperature limits, $-55°C$ and $+95°C$ are values commonly specified. The limit on the high temperature end is determined by the transformer or transducer lead wire limit, which, for polyurethane insulation, may be as low as $105°C$ or as high as $130°C$. Also, some bonding materials tend to deteriorate at temperatures above $100°C$. If polystyrene capacitors are used, the storage temperature must be kept below $85°C$.

Turning to the subject of shock and vibration, a typical shock specification is 30 g's for an 11 msec duration, whereas vibration is commonly specified at a 0.152 cm (0.06 in) peak-to-peak amplitude from 10 Hz to 55 Hz and 10 g's from 55 Hz to 2000 Hz. There is considerable variation in the shock withstanding capability of different filter types. This variation ranges from 15 g's to more than 1000 g's. As an example, Figure 8.14 shows how the shock level that disk-wire mechanical filters can withstand varies, according to bandwidth and the type of design. The curves have a positive slope, because as the bandwidth is increased, the coupling wires and support wires increase in diameter, therefore making the filter more resistant to shock damage. Often a mechanical-filter manufacturer will specify all filters at a low shock level (like 15 g's), because they know that all of the filters will withstand that level, and therefore it is not necessary to run shock tests on each filter design. In reality, some of

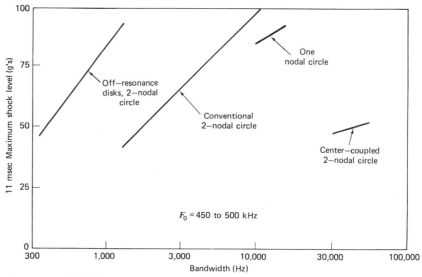

FIGURE 8.14. Variation of maximum shock level tolerance as a function of filter bandwidth, for various types of disk-wire filters. (Courtesy of Rockwell International, USA.)

those filters will meet a 100 g specification. The filter user should therefore specify what he needs and then interact with the vendor to see if these needs can be met. In Chapter 5, means for increasing ruggedness were discussed.

TESTING MECHANICAL FILTERS

Mechanical filter testing is important from the conception of the device to the use of the filter in equipment. Testing is done by the designer, the factory assembly operator, the quality-control final-test operator, the user's incoming inspector, the system designer, the modem test operator, and sometimes by a field maintenance person. Each person must perform the tests correctly if correlation is to be achieved; this is not a simple task. For instance, in making amplitude response measurements the following conditions may be experienced: different test equipment may be used, the calibration procedure may be different, and there may be differences in the cabling, the test fixtures, and the amount of RF-noise in the test area.

It is the purpose of this section to discuss the basic testing of mechanical filters, in particular in the area of measuring frequency response characteristics. In addition, special tests are covered such as measuring time response, intermodulation distortion, and the measurement of nonlinear element jumps in the frequency response. Measurements related to shock, vibration, humidity, insulation resistance, and so on, are common to most electronic components and therefore are not covered in this book.

Frequency Response Testing

In this section we examine test circuits and test equipment for measuring amplitude, delay, and phase as a function of frequency.

Amplitude-Response Test Circuit

Figure 8.15 shows a test circuit for measuring the amplitude response of a mechanical filter as a function of frequency. The circuit can represent specific components and pieces of test equipment or it can represent elements within a complex filter test set.

Let's look first at the frequency generator, that is, the signal generator or synthesizer. First, the generator must have the frequency and amplitude stability to be able to make measurements compatible with the filter specifications. Also, either the generator must provide an adequate voltage level or an amplifier must be added between the generator and the test fixture. If an oscilloscope trace of the response is required, the generator must provide a DC ramp voltage that is proportional to frequency. Ideally, the output impedance of the generator is low; if not, the output should be shunted with a low value of resistance or cascaded with a low output-impedance amplifier. The generator output signal should have a harmonic content at least 20 dB below the fundamental when measuring passband characteristics. When measuring stopband behavior over a broad frequency range, care should be taken not to mistake a harmonic signal for a spurious mode of vibration. For instance, when the fundamental signal is at $f_0/2$, the second harmonic is at f_0 and passes through the filter. This is not a problem in the test sets we will study in the next section—they use frequency tracking and narrowband filtering.

The output impedance of the frequency-generator circuit should be low compared to the specified source resistance, less than 1 percent if possible. This allows us to treat the generator circuit output as an ideal voltage source, which acts as a reference for insertion loss measurements and also acts as a short circuit across the coax cable and connector capacitance (to ground) between the generator and the test fixture.

The series resistance $R_S - R_G$ is located in the test fixture in order to reduce stray capacitance effects between the filter side of the resistor and the filter terminals. Between the series resistor and the filter is the tuning circuit, which is composed of the specified values of shunt or series capacitance, or inductance, or sometimes transformers or capacitive impedance matching networks. Shown as an example is the parallel combination of a fixed capacitance C_F plus the circuit stray capacitance C_S, totaling the specified generator shunt capacitance C_G.

The resistance and capacitance values should generally be within 1 percent of their specified value although in high performance telephone channel filters, for example, the capacitor tolerances should be tighter than one-half percent. Figures 9.12 and 9.13 show the effects of mistermination on passband ripple.

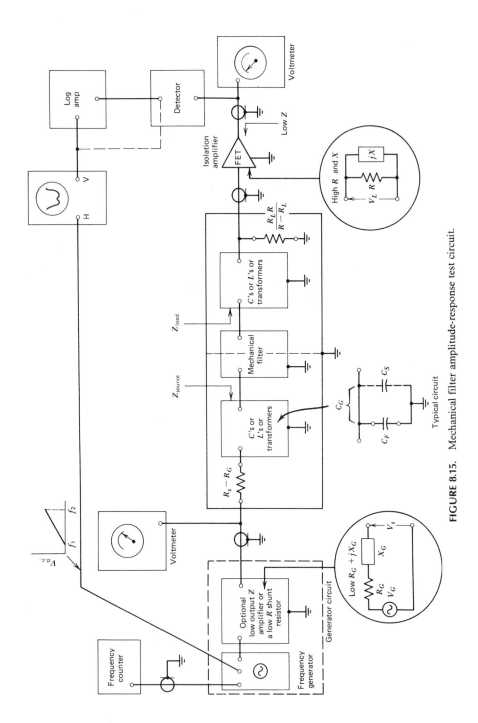

FIGURE 8.15. Mechanical filter amplitude-response test circuit.

In test fixtures that can be used for more than one type of mechanical filter, the various terminating resistors and capacitors can be mounted on crystal bases. The proper values are then plugged into crystal sockets that are part of the test fixture. If these components are to be used over an extended period of time their stability must be considered; metal-film resistors and mica capacitors are a good choice.

In laying out the test fixture, shielding is an important consideration. If stopband levels of greater than 80 dB (referenced from the passband) are not achieved, it may be necessary to shield the input circuit from the output circuit by use of a grounded conductive plate. In addition, there must be adequate grounding and connections between the filter ground terminals, shields, the test fixture, and the test equipment. This topic is discussed further in Chapter 9.

At the output side of the filter, the test fixture should look into a high impedance, such as the input of a unity gain FET isolation amplifier or the high input impedance provided by most voltmeters. Unlike the input side of the filter, where the low-impedance source shorts out stray capacitance, the output strays, from the filter terminals to the input of the amplifier, are additive. This stray capacitance must be kept constant and included in the total specified terminating reactance.

If the output impedance of the isolation amplifier is low, and the amplifier has unity gain, then the voltmeter will read the voltage across the specified load resistance R_L. By measuring the voltage V_S across the output of the generator circuit shown in Figure 8.15, the insertion loss can be calculated to within a 0.1 dB error from Equation (8.3) as long as the condition $R_S > 50$ $|R_G + jX_G| \simeq 50R_G$ is satisfied. Any amplifier gain or loss can be directly subtracted from or added to the insertion loss.

With regard to a visual display of the filter amplitude response, it is necessary to first detect the signal (i.e., convert the AC signal to a proportional DC signal), and then, if a logarithmic display is desired, feed the signal through a log amp to the vertical input of the oscilloscope.

The Relationship between V(dB) and V(RMS)

In regard to specifications and amplitude response measurements, there is often confusion regarding the voltage reference value when the voltage is expressed in dB. Historically, a 0 dB voltage level has meant that the voltage is equal to 0.775 V(RMS). This voltage reference is derived from power level measurements where 0 dBm = 1 mW. In a 600 system, 1 mW of power P dissipated in a 600 Ω resistor R requires $V = (RP)^{1/2} = 0.775$ Volts (RMS). If a piece of test equipment has 50 Ω input and output resistances, the voltage reference value is usually based on the voltage required to dissipate 1 mW across 50 Ω, which is 0.224 V(RMS). Another reference, which is finding wide use, is simply 0 dB = 1.0 V(RMS). Because of this potential source of confusion, it is best to specify voltage along with the dB level, for instance -10 dB, 0.245 V(RMS).

Measuring Phase and Delay

In this section, we look at methods of measuring phase, phase shift, and group delay. By *phase* we mean the phase of the output voltage V_L with respect to the phase of the source voltage V_S (see Figure 8.1). If we take the source voltage phase ϕ_S as our zero reference, then the output phase is

$$\phi_L = 360(t_S - t_L)f \text{ (in degrees)}, \qquad (8.13)$$

where $(t_S - t_L)$ is the difference in time (in seconds) between the zero crossings of the source and load sine waves, and f is the frequency of the signal.

Although the time and frequency quantities in Equation (8.13) can be measured on a dual trace oscilloscope, it is more convenient to make the measurements with a phase meter or with a network analyzer. In the latter two cases, the source and output signals are simply fed into the test instrument, which displays the phase. Phase shift $\Delta\phi$ (as a function of frequency) was defined in Equation (8.9) and can be measured by use of the above equipment.

Group-delay measurements can be made directly with the help of modulated signals or can be calculated from phase measurements and Equation (8.10). By modulating the source signal with a frequency much smaller than the filter bandwidth (typically 1 percent of the bandwidth) and comparing the detected source and output audio signals, we can measure the group delay by measuring the time difference between the zero crossings of the two detected wave forms. This delay, which is literally the envelope delay, is equal to the group delay when the modulation frequency goes to zero.

In making delay measurements, using the phase shift divided by the frequency method of Equation (8.10), a balance must be found between: (1) decreasing the frequency shift (in computer terms, the frequency step size) in order to more closely measure the delay at a specific frequency, and (2) increasing the frequency shift to reduce the effect of jitter in the phase measurements. Fortunately, most mechanical-filter delay-response curves have a parabolic shape in the steep, critical region just inside the passband edges (as an example, see Figure 8.8). It can be shown that the delay measurements in these steep regions are not greatly influenced by the frequency step size because of the parabolic shape of the curve. A method of determining the step size is to start with a very narrow step and increase the step until the jitter is removed.

Test Equipment

Test equipment for measuring the frequency response of a mechanical filter can vary from the basic components shown in Figure 8.15 to network analyzers that can be computer controlled and which measure amplitude, phase, and group delay. In this section, we discuss and compare the individual component test circuit (like that in Figure 8.15), the manual test set, the computer-controlled spectrum analyzer, and the computer-controlled network analyzer.

The major advantage of the individual-component test circuit is that it is inexpensive. One of its drawbacks is that all of the measurements are taken

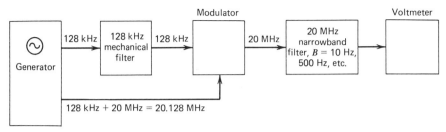

FIGURE 8.16.   Modulation scheme for eliminating noise through use of narrowband filtering.

manually. Another drawback is that a broadband voltmeter must be used. The broadband voltmeter picks up both the output signal harmonics and noise, making it difficult to obtain good stopband measurements.

Most integrated test equipments make use of narrowband (10 Hz, for example) filters to reduce noise. A method of doing this is shown in Figure 8.16. The generator (synthesizer), modulator, narrowband filter, and voltmeter are all part of an integrated test set, spectrum analyzer, or network analyzer.

Test sets, like the Wandel Goltermann WM-50, are used for analog testing of mechanical filters. The frequency sweep can be manual or automatic but is not computer controlled. The manual sweep is used in some factories and engineering labs in filter tuning. This option is sometimes not available in digital equipments. Most integrated test-sets have crystal-controlled frequency sources, so measurements can be made to 0.1 Hz accuracy. Also a broad range of frequency sweep options are necessary, ideally from 10 Hz to 1 MHz. If the filters to be measured have bandwidths greater than 10 Hz, a 35 Hz sweep range is adequate. The amplitude stability and measurement accuracy needed is a function of the specified passband ripple; 3 dB passband ripple filters can be measured with a 0.1 dB test set, whereas a ± 0.11 dB ripple filter needs a test set with ± 0.02 dB, or better, accuracy and precision. A test set like the WM-50, using an expanded-scale voltmeter, has a resolution of 0.02 dB with better than ± 0.02 dB accuracy. Stopband selectivity preferably is measured with an accuracy better than 1 dB from the 0 dB reference, but in most test equipment this degree of accuracy is barely achievable at -60 dB levels.

Spectrum analyzers can be adapted to work as network analyzers, as in the case of the HP 3585A or the modified Tektronix FL5. In these examples, the measurements can be computer controlled or the filter can be manually swept, making its use applicable to both lab work and high-volume automatic testing. In addition, its spectrum analyzer capability makes it useful for intermodulation distortion measurements. A major drawback of the spectrum analyzer is that it does not measure phase or delay.

Network analyzers are designed to make digital frequency response measurements of amplitude, phase, and delay, of two-port devices such as mechanical filters. A network analyzer like the HP 3040A is computer controlled but does not have a manual sweep capability. In digital test equipment, as in analog equipment, the sweep rate (wait-time between measurements) must be set low enough to prevent distortion of the response. The sweep rate must be further lowered when using a very narrow-bandwidth tracking filter (see Figure 8.16) to make low-noise measurements in the filter stopband frequency range. For each filter specification, a balance must be found between the number of measurement points, the wait-time, the bandwidth of the tracking filter, and the accuracy of the measurements.

Making reliable frequency response measurements is a difficult job. But filter manufacturers have everyday experience in measuring filter characteristics, and therefore the filter user should view them as resources for setting up a test system.

Special Testing

In this section we look at three additional tests. These are for intermodulation distortion, step and impulse response, and for nonlinear responses.

Intermodulation Distortion

Intermodulation (IM) involves the mixing of two or more signals in a nonlinear device to produce new frequencies at sums and differences of integer multiples of the original frequencies. The amplitude of the products that fall in the filter passband range are a measure of the IM distortion of the original input driving signals. If the input signals are outside of the filter passband frequencies, the resultant intermodulation products within the passband are called out-of-band intermodulation distortion. If the input signals are inside the passband, the products are called in-band intermodulation distortion.

Intermodulation can be viewed as a combination of harmonic generation and mixing. For instance, the third-order IM products, in which we are interested primarily, are:

$$f_{\text{IM}_{1,2}} = 2f_1 \pm f_2, 2f_2 \pm f_1, \qquad (8.14)$$

where f_1 and f_2 are the input signals. From Equation (8.14), we see that f_{IM_1} is the result of generating the harmonic $2f_1$ and the mixed product $2f_1 \pm f_2$.

In making filter IM measurements, the plus sign in Equation (8.14) is usually ignored and we look for the products

$$f_{\text{IM}}|_{f_1 < f_2 < f_L} = 2f_2 - f_1 \ (\text{Out - of - band})$$

$$f_{\text{IM}}|_{f_L < f_1 < f_2 < f_H} = 2f_1 - f_2, 2f_2 - f_1 \ (\text{In - band}),$$

where f_H and f_L are the high- and low-frequency passband limits.

Figure 8.17 shows a test circuit for measuring both in-band and out-of-band IM distortion. The signal generators should have low second-harmonic distor-

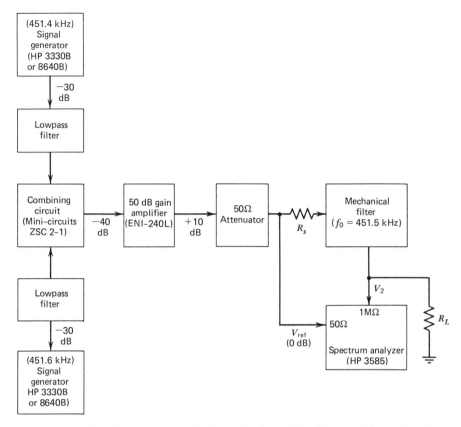

FIGURE 8.17. Test circuit for measuring in-band and out-of-band intermodulation distortion of mechanical filters. Bracketed levels correspond to typical in-band IM values for V_{ref} = 0 dB(0 dB = 0.224V).

tion, but if not, the lowpass filters can be used to pass the fundamental and attenuate the harmonics. The three-port combining circuit has low IM distortion only at low signal levels, so it must be operated at low levels and followed with an ultra-linear amplifier. The IM measurements are made with the minimum spectrum-analyzer filter bandwidth in order to reduce the noise level to at least 6 dB below the IM products. Because this is a 50 Ω system, 0 dB is equal to 0.224 V(RMS).

Figures 8.18 and 8.19 show IM distortion measurements of piezoelectric-ceramic and magnetostrictive-ferrite transducer filters. The solid lines are based on taking averages of several intermediate-band 450 kHz and 455 kHz single-sideband filter types; the dashed lines are simply extensions of the solid lines to show the slope intercept points. Note that 0 dB = 0.774 V(RMS). The first-order response is the filters' output voltage at the frequency of the generator signal. IM distortion is often defined as the difference, in dB,

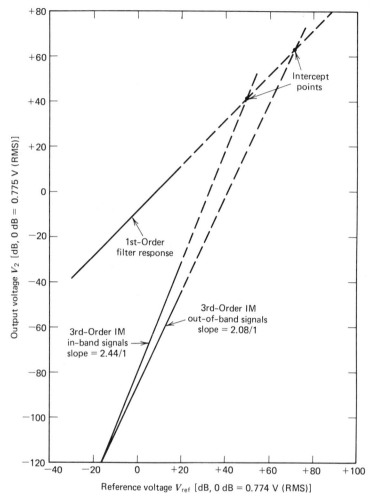

FIGURE 8.18. Typical third-order intermodulation distortion of piezoelectric-ceramic transducer mechanical filters.

between the first-order response and the third-order IM, at a chosen reference level. In Figure 8.18 at a reference voltage of 0 dB, the in-band IM distortion is −80.0 dB − (−8.5 dB) = −71.5 dB. For out-of-band distortion, the generator frequencies are shifted into the passband to find the first-order response.

We note in Figures 8.18 and 8.19 that the slope of the third-order products vary from 1.85/1 to 2.44/1. Highest and lowest slope values of individual filters ranged from 1.7/1 to 3.1/1. The reasons for this wide variation are not clear nor are the reasons why the slopes are not 3.0/1, as in the case of ideal amplifier IM distortion. Even though both tuning coils and PZT transducers are used in the ceramic-transducer filters, these filters have better IM distortion

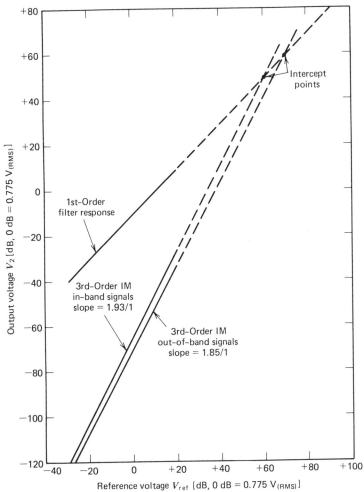

FIGURE 8.19. Typical third-order intermodulation distortion of magnetostrictive-ferrite mechanical filters.

than the ferrite-transducer designs tuned with mica capacitors. At a reference voltage level of 0.774 V, the difference between the third-order and the first-order responses is almost 20 dB greater in the ceramic transducer case. The ferrite-transducer filter IM distortion shown can be improved by as much as 12 dB by increasing the number of coil turns. Increasing the coil turns means that the filter input impedance is higher and, for a fixed input voltage, the input power is lower, and the ampere turns NI are reduced. The reduction of NI and therefore the flux density through the ferrite rod may explain the improved performance. The in-band ferrite-transducer filter IM distortion can also be reduced by increasing the magnetic bias level, but this does not help the out-of-band distortion.

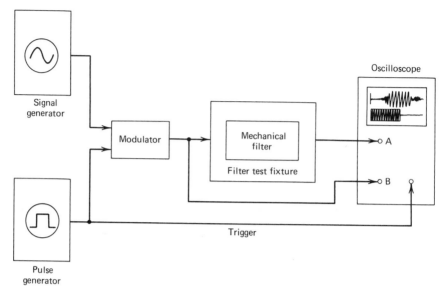

FIGURE 8.20. Test circuit for time-response measurements.

From Figures 8.18 and 8.19, we can observe that the out-of-band IM distortion is generally less than the in-band distortion. This is probably related to the fact that the in-band driving signals are present at the output transducer, whereas they have been filtered in the out-of-band case, and therefore are not available to produce additional IM products.

Time Response

Figure 8.20 shows a basic circuit used for making time-response measurements. The sine-wave signal generator is set at a specified frequency, which is usually the filter's center frequency. The output of the pulse generator is normally a step function (square wave), which, when narrowed, simulates an impulse. The two signals are fed into a modulator; one output of the modulator goes directly into the B-channel of a dual trace oscilloscope, the other output goes into the mechanical-filter test fixture. The filter (fixture) output is fed into the vertical A-channel of the oscilloscope. The leading edge of the step output from the pulse generator is also used to trigger the oscilloscope sweep.

Nonlinear Responses

In this section we consider nonlinear effects which result in sharp increases or decreases in the amplitude response of a filter as the frequency is swept. The solid curve of Figure 8.21 shows a sudden drop in amplitude from point A, as the applied frequency is increased; the dashed curve shows a sudden increase in amplitude from point B as the frequency is decreased. These jumps are the result of a nonlinearity in one of the mechanical filter resonators.

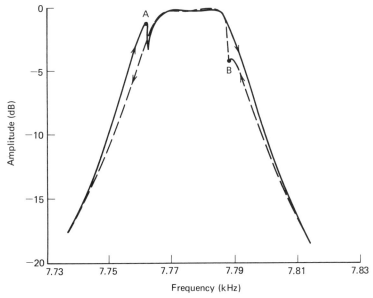

FIGURE 8.21. Frequency response jumps resulting from a nonlinearity in a two-resonator flexural-mode filter.

Only a small percentage of mechanical filters exhibit the sudden-jump phenomena shown in Figure 8.21. Factory personnel can be trained to recognize the difference between this phenomena and spurious responses, such as those shown in Figure 5.11, and take corrective action to remove the nonlinearity. But before we discuss the means of identifying this problem let us look at some basic concepts relating to nonlinear resonators.

Figure 8.22 shows the frequency response of a spring-mass resonator where the spring's stiffness (spring "constant") increases with increasing amplitude [4]. Note that the response looks much like a universal response curve that has been bent to the right. If the spring's stiffness decreased with amplitude, the curve would bend to the left. Let's follow the amplitude response starting from point A. When the increasing frequency reaches f_2 the continuous curve

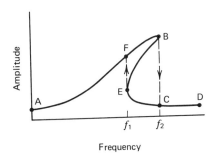

Frequency

FIGURE 8.22. Amplitude jumps in the frequency response of a nonlinear resonator that has a spring whose stiffness increases with amplitude.

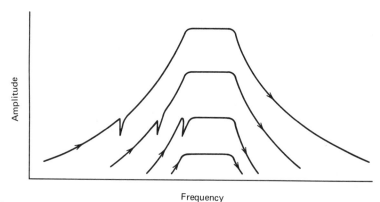

FIGURE 8.23. Effects of increasing drive signal amplitude on the frequency position of the amplitude jumps caused by nonlinearities.

reverses its direction, but since we are increasing the frequency, this is not possible. What does occur is a drop in amplitude from point B to point C. The response then continues to point D. If at D we decide to sweep down in frequency, then when we reach f_1 we find that we must jump from point E to point F, and then we are able to continue to point A.

Applying the concepts in Figure 8.22 to the frequency response curve of Figure 8.21, we can state that the response jumps act like they are caused by a nonlinear spring that has a stiffness which increases with amplitude.

Also observed is that the jumps occur at a constant amplitude. This effect is shown in Figure 8.23. As the driving amplitude to the filter is decreased, the jump moves into the filter passband and finally doesn't appear at all. The fact that the jump frequency varies with the filter input voltage helps the factory operator to identify the discontinuity as a nonlinearity rather than as a spurious mode. The operator can also identify the problem by noting that the

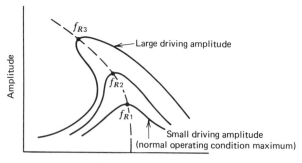

FIGURE 8.24. Decreasing resonance frequency effect resulting from driving a mechanical resonator at increasing signal levels.

jump occurs on the opposite side of the passband when reversing the direction of the frequency sweep. Since the nonlinearities are usually caused by particles or lead wires that are tangled against the resonator, the operator recleans the filter and repositions the lead wires to eliminate the problem.

The nonlinear responses described to this point are atypical, as compared to the normal nonlinear behavior of mechanical resonators operating at high amplitude levels. The frequency shift as a function of energy density that was shown in Figure 5.4, and the frequency reduction as a function of signal level shown in Figure 9.6, are caused by the decreasing-spring-stiffness type of behavior shown in Figure 8.24. Under normal operating conditions, a metal, ferrite, ceramic, or composite resonator will have a slightly lower resonance frequency as it is driven harder, but it will not exhibit the jump behavior previously discussed.

REFERENCES

1. A. I. Zverev, *Handbook of Filter Synthesis.* New York: Wiley, 1967.
2. ITT Staff, *Reference Data for Radio Engineers.* Indianapolis: Sams, 1975.
3. R. Saal, *Handbook of Filter Design.* Berlin: AEG-Telefunken, 1979.
4. W. T. Thomson, *Vibration Theory and Applications.* Englewood Cliffs: Prentice-Hall, 1965.

USING MECHANICAL FILTERS

This chapter was written with the user of mechanical filters in mind. We can view this material as providing answers to the question, "I have a mechanical filter and am ready to put it in my circuit; what do I do next?" Specifically we look at the mechanical filter in its environment, as illustrated in Figure 9.1, and show how to avoid a reduction in its performance. We consider questions such as: where should the filter be placed in the circuit; what are the allowable input signal levels; how do we avoid feedthrough and ground loops; how do we minimize or reduce the effects of temperature, magnetic fields, vibration, DC voltages; and how do we best terminate the filter? There is no single answer to each of these questions because the answers are dependent, for example, on the resonator vibration modes, the transducer types, such as PZT or ferrite, and whether the filters are electrically tuned or untuned.

CIRCUIT LAYOUT

Characteristics approaching the ideal can only be obtained by proper application of the mechanical filter in the user's circuit. Excellent stopband rejection can be destroyed by improper shielding or poor grounding. Amplitude and phase characteristics can be ruined by operation at too high a voltage level. The mechanical filter is a very complex device and plays a major role in determining system performance and therefore the circuit layout should be well thought out.

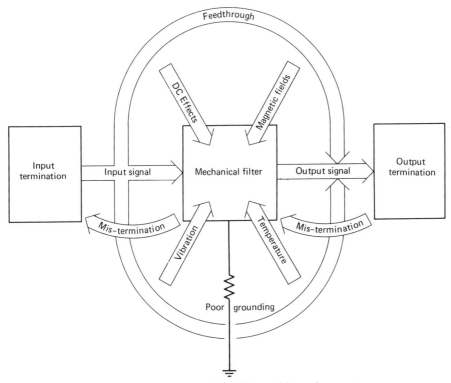

FIGURE 9.1.   The mechanical filter and its environment.

Location in the System

In receiver design, the essential rule-of-thumb is to position the filter as close to the front end as possible. For instance, when used in an IF circuit, the filter usually follows the modulator or the first IF amplifier. The reason for placing the filter near the front is to eliminate unwanted signals and noise before further amplification or gain control takes place. The function of the amplifiers and AGC is to control the desired signal level rather than have the desired signal be controlled by unwanted signals or noise. Let's first discuss the use of a mechanical filter between two linear amplifiers.

Using a Mechanical Filter in an Amplifier Stage

Through the use of a radio receiver IF-circuit example, basic concepts involved in using a mechanical filter between two amplifiers are illustrated. Figure 9.2 shows the first two stages of a 500 kHz IF circuit. The input signal comes from a balanced mixer and is applied to two channels. The channel shown in Figure 9.2 can be disabled by a −9V DC signal. With the channel turned on, the 500

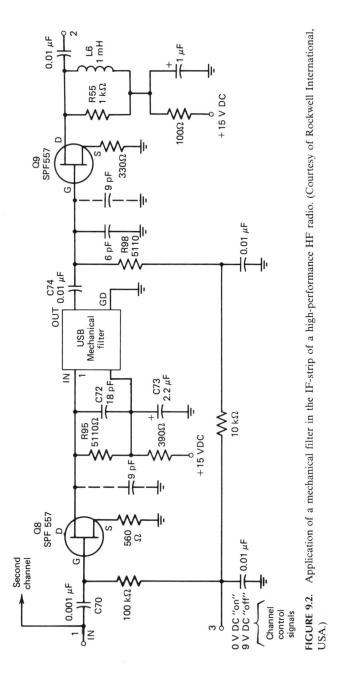

FIGURE 9.2. Application of a mechanical filter in the IF-strip of a high-performance HF radio. (Courtesy of Rockwell International, USA.)

kHz sideband signals at the input terminal 1 are applied, through the blocking capacitor C70, to the FET Q8, and are amplified.

The FET was chosen over a bipolar transistor or an op amp for the following reasons:

The input impedance is high (100 kΩ) which prevents loading of the balanced mixer.

The output impedance is high (> 40 kΩ). This provides a stable source impedance for the upper sideband (USB) mechanical filter.

The amplifier can be disabled by application of a negative voltage to the gate.

The circuit provides moderate gain of about 14 dB with low intermodulation distortion.

The mechanical filter is designed to be terminated with 5 kΩ in parallel with 30 pF. Looking back toward Q8, from the filter's input terminals, we see the parallel combination of 18 pF, 5.11 kΩ, the output impedance of Q8, and circuit board strays of 9 pF. The output capacitance of Q8 is about 3 pF, making the total shunt capacitance equal to the required value of 30 pF. If the output impedance of Q8 is 232 kΩ, the input of the filter would be terminated into 5 kΩ. Should the output impedance be at the minimum value of 40 kΩ, then the source impedance will be 9 percent low. Parenthetically, C73 is used for RC filtering of any ripple voltages which may be present on the supply voltage.

The input transducer coil of the mechanical filter provides a DC path from the +15 V DC supply to the drain of Q8. This allows a higher bias voltage to be applied at the drain of Q8. The DC current is approximately 5 ma, which is close to the allowable limit in magnetostrictive transducer filters. The DC path from the gate is through the 100 kΩ resistor and through the channel control circuit.

The mechanical filter's output terminating circuit is composed of the following shunt element values: 5.11 kΩ fixed resistance, 6 pF fixed capacitance, 9 pF stray board capacitance, a nominal value of 15 pF equivalent input capacitance of the Q9, and a 100 kΩ minimum shunt input resistance of the Q9 FET circuit. Combining these shunt elements, the filter termination becomes 30 pF across 4.86 kΩ, for the minimum FET Z_{in} case.

The capacitor C74, in Figure 9.2, prevents the mechanical filter transducer coil from causing a DC short. The 1 mH choke L6 provides a DC path for the +15 V supply to the drain of Q9, which improves the linearity of the output Q9 FET amplifier by establishing a higher drain-to-source bias voltage.

Looking at the control signal path, we should note that it provides a path around the filter, which, if we are not careful, will reduce the stopband attenuation of the amplifier/filter stage. This problem is solved by adding the 0.01 μF capacitor, which, in combination with the 100 kΩ, 10 kΩ, and 5.11 kΩ resistors, theoretically provides more than 120 dB of attenuation of the 500 kHz sideband signals.

A final consideration is that the filter be mounted in the circuit properly. This particular design has a plastic case and four terminals. The two input terminals are internally connected to the transducer coil leads only. The output terminals are internally attached to the coil leads, and one of the two terminals is attached to the metal resonator assembly. This terminal must be grounded. If the filter had been soldered backward in the circuit, the ground terminal would be connected to the input terminal 1, which would have placed the entire filter structure at the signal potential. The stopband response would be attenuated less than 30 dB. Also, if the printed circuit (PC) board had been laid out wrong, and the ground terminal was connected to ground through a 0.01 μF capacitor (31.8 Ω at 500 kHz), the stopband attenuation would not exceed 70 dB from the input signal reference. The subjects of terminal configurations and partial grounds will be discussed in subsequent sections.

Impedance Considerations in Amplifier Circuits

When using an amplifier to drive a mechanical filter, we would like the output impedance of the amplifier to be either very high or very low with respect to the specified filter source resistance. This is because we want a stable source resistance. If the amplifier impedance falls in a middle range, it must be designed to be very stable because its impedance value strongly influences the actual source resistance value seen by the filter. If the output impedance of the

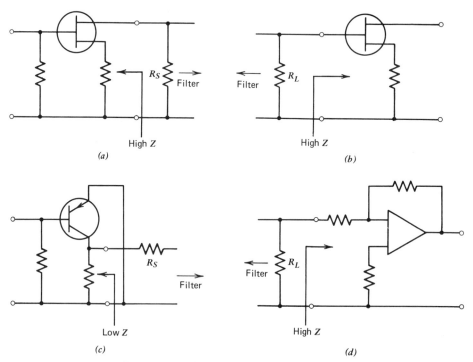

FIGURE 9.3. Active amplifier termination circuits.

amplifier is very high, a shunt resistor can be used as the filter load, or if the output impedance is very low, then a series resistor can be used. These conditions are shown in Figure 9.3(a) and (c) for a high-impedance FET circuit and a low-impedance bipolar emitter-follower.

With regard to using an amplifier in the filter's load circuit, it is necessary to either use a high input-impedance amplifier or a stable mid-impedance amplifier. An amplifier with a very low input impedance will cause a high insertion loss. Examples of amplifiers with high input impedances are shown in Figure 9.3(b) and (d).

Connecting Filters in Parallel

Often, system requirements dictate that an array of mechanical filters must be connected in parallel. An example is the remote control circuit that was shown in Figure 7.39. Some of the factors that enter into the design of parallel circuits are:

The frequency spacing between filters.

Terminal connections; whether the input or the output terminals are connected.

The filter impedances and surrounding circuit impedances.

The allowable insertion loss, both the absolute value and the variation.

The simplest type of connection to realize is that shown in Figure 9.4(a). The figure shows an array of three mechanical filters driven by a low-impedance generator, for example an emitter-follower. By making the generator impedance very low, each filter can be terminated in its own specified source resistance R_{S1}, R_{S2}, and R_{S3}. This is possible because the generator shorts all of the impedances other than the series resistor connected to the nongrounded input terminal. This configuration allows filters with overlapping passbands and widely varying impedances to be connected in parallel. The disadvantage of this arrangement is that the emitter follower has no gain and there is an approximate 6 dB loss across the source resistor. If we reduce the loss (or realize gain) and drive from a higher impedance generator, we must deal with impedance matching problems like those of Figure 9.4(b).

Figure 9.4(b) shows a parallel connection of the outputs of three mechanical filters. The array of filters is connected to a high input-impedance amplifier in order to obtain a low insertion loss and stable terminating conditions. Each filter must work into its specified terminating resistance R_{L1}, and so on where R_{L1} is composed of series and parallel combinations of R_{11}, R_{21}, and Z_{31}. Parenthetically, the shunt resistors R_{11}, R_{12}, and R_{13} can be removed if it is not necessary to match the loss of each channel. If the filters are widely separated in frequency, each filter output impedance Z_{out1}, and so on, will present a nearly constant impedance to the other filters. As an example, a narrowband ceramic-transducer filter will have an output impedance $Z_{out} = 1/(j2\pi f C_0)$ where C_0 is the output transducer static capacitance and f is a frequency

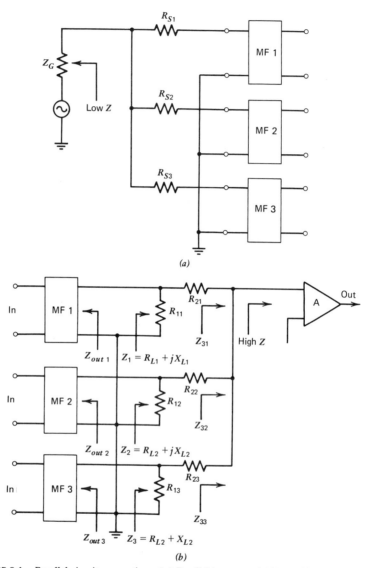

FIGURE 9.4. Parallel circuit connections. (*a*) Parallel inputs, and (*b*) parallel output circuits.

outside of the filter's own passband. The proper value of R_{21} (and R_{11}, if the loss is to be adjusted) is found by assuming a value for R_{21}, R_{22}, and R_{23} and calculating the impedance $Z_1 = R_{L1} + jX_{L1}$ at the center frequency of filter number 1. R_{21} is then adjusted to a value that makes R_{L1} equal to the specified terminating resistance. Next, Z_2 is calculated and R_{22} is adjusted. This process is repeated until all of the filters' impedances are matched (and if the shunt values R_{11}, etc., are used, the insertion losses will be matched also). In this

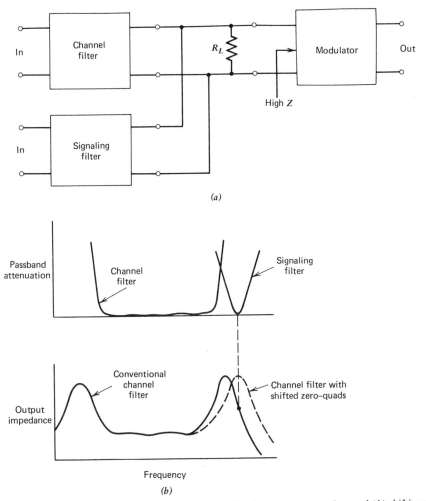

(a)

(b)

FIGURE 9.5.   Paralleling closely spaced filters. (a) Parallel output connection, and (b) shifting the channel filter output impedance to coincide with the passband of the signaling filter.

example we assumed that the reactance X_{L1} is either higher than some minimum specified value or is part of the output tuning circuit.

Our third example helps us understand the problems involved in paralleling closely spaced filters. Figure 9.5(a) shows a parallel connection of a narrow-band telephone signaling filter and an intermediate-band voice-channel filter. The filters are paralleled without series resistors in order to have a low insertion loss. Figure 9.5(b) shows sketches of the amplitude responses of the two filters and the output impedance response of the channel filter. The output impedance of the signaling filter looks like a capacitive reactance outside of its passband. Impedance variations within the signaling-filter passband cause only small changes in the channel-filter stopband selectivity. Of greater importance is that the conventional channel-filter design has an output impedance peak at a frequency below the signaling-filter passband. Because the modulator shown in Figure 9.5(a) has a high input impedance, the signaling filter sees R_L in parallel with the channel-filter impedance. In the signaling-filter passband, the channel-filter impedance is neither constant nor at its highest level. These conditions can cause variations in the signaling filter's amplitude response. By varying the position of the zero-quads of the characteristic function (see Figure 4.3 for an illustration of zero-quads) the peak of the channel-filter impedance can be aligned with the passband of the signaling filter [1], [2].

If, in the above example, insertion loss and minimizing the number of components were not critical, then series resistors, attenuator pads, or amplifiers could have been used to isolate the two filters and prevent the impedance variations of the channel filter from affecting the signaling-filter response. This was avoided in our example by the filter users and the filter designers working closely together to find compatible system and filter solutions.

The Use of Modulators with Mechanical Filters

Mechanical filters are used with a variety of modulator circuits. For single-sideband operation, these include single- and double-balanced circuits with active and passive components. The modulators are designed with both thick and thin film components in integrated and hybrid form [1], [3]. An example of a double-sideband suppressed-carrier integrated circuit modulator is the 1496, which contains eight transistors, a diode, and three resistors. Carrier suppression is obtained through the device's differential amplifiers [4] as opposed to the balanced transformers often used in telephone voice-channel modems [5]. Some telephone equipment manufacturers use the mechanical filter input transformer as the modulator's balanced transformer in order to reduce the number of components in the modem [3], [6]. With regard to the operation of the filter, the most important modulator characteristic is that its impedance be precise and stable with both temperature and time.

Signal and DC Voltage and Current Levels

In designing the circuit around a mechanical filter, attention should be given to both signal and DC levels. High signal levels are usually not large enough to cause permanent damage to the filter but result in nonlinearities and modified performance. As an example, if too large a signal voltage is applied across the input of the filter, the passband frequency response will be flat. In other words, a 2 dB ripple filter will appear to have almost 0 dB ripple. This is not only a potential testing problem but a system problem as well. Maximum signal voltage across the input terminals varies with the filter design, but for single-sideband filters, this voltage is usually in the range of 1 to 10 V(RMS).

In the case of piezoelectric-ceramic transducer filters that have only two or three resonators, large input signals cause changes in both the passband

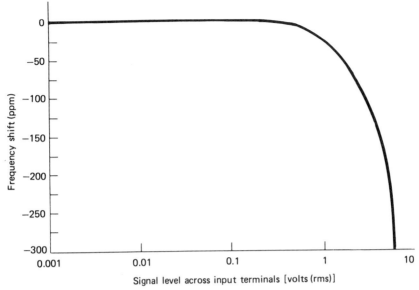

FIGURE 9.6. Variation in center frequency with input voltage level (10.2 kHz center frequency, two-resonator narrowband design).

amplitude response and the center frequency. Figure 9.6 shows the effect of the signal level on center-frequency shift. This data was taken on a two-resonator flexural-mode filter designed for a center frequency of 10.2 kHz. The high voltage levels cause a drop in the elastic modulus of the input transducer ceramic, which results in a lower resonance frequency. This effect is similar to that of the spring-mass resonators of Figure 8.24. In filters of this type, the specified maximum input voltage levels range from 0.5 to 3 V(RMS). Now let us look at DC effects.

Excessive levels of fixed or transient DC voltage or current will result in permanent damage to a mechanical filter. Of primary concern when using a ceramic transducer filter is the effect of voltage on the polarization of the transducer material. Therefore, before voltage breakdown occurs, in most cases the polarization and the filter's performance characteristics will have changed. As was discussed in Chapter 2, the depoling effect is both voltage and time dependent so very narrow transients may not cause excessive permanent damage. The maximum value of a fixed DC voltage across the terminals is also dependent on the transducer's thickness and its material formation. In most cases, the user is safe if the DC voltage is kept under 50 volts.

When using magnetostrictive-transducer filters, the circuit designer should try to avoid if possible, but at least be cautious in, passing DC current through the input or output filter terminals. Large levels of DC current through the transducer coils will change the magnetic bias level and therefore change the filter's frequency response. Another danger is the heating effect of DC current on the coil itself, which results in reduced reliability. For reasons of changed bias or heating, the current through the transducer coil should not exceed 3 to 10 mA (depending on the filter design).

The Physical Layout

Before laying out a printed circuit board or even testing a filter, it is necessary to understand the function of each filter terminal and how the terminal configuration relates to proper grounding and shielding. It is also important to know the effects of transformers or mechanical filters that are in close proximity.

Filter Terminal Configurations

Mechanical filters are designed with as few as three terminals and as many as eleven. The number of terminals depends on the number of functions the filter performs in the circuit and the number of options the filter manufacturer wants to provide for the filter user. The number of terminals is kept to a minimum for the purpose of reducing the manufacturing cost.

Figure 9.7 shows various mechanical-filter terminal configurations. The outermost dashed lines show the boundaries of the filter enclosure. Also shown are the capacitors and inductors that correspond to the capacitance and inductance of the PZT-ceramic and ferrite transducers. Note that the piezoelectric ceramic transducer is always attached to the metal wire/resonator structure at one end. In all cases, at least one external terminal is also attached to the wire/resonator structure; this is the ground terminal and must be connected to the user's circuit ground. Sometimes the stud on a metal case, or the metal case itself, acts as the ground terminal, but again it must be tied to the circuit. Grounding the wrong terminal results in either no output or very poor stopband rejection. The filter user can assume that if a metal enclosure is used, the enclosure is shorted to the wire/resonator structure.

If the filter is to be used in a balanced circuit, five or more terminals must be available. An exception is the configuration shown in Figure 9.7(d) where one end of the filter is balanced and the other end is unbalanced (i.e., one end of the transducer coil is grounded).

Figure 9.7 shows two six-terminal filters that have four floating terminals and two ground terminals. If the input or output of one of these filters is to be used as a balanced transformer for a modulator circuit, the coil can be center tapped and tied to ground [3]; this connection is shown by dashed lines. Figure 9.7(g) and (h) are even more general in that the transformer tap can also be used as a parallel output at a lower impedance [2] or simply a low-impedance output where one of the coil terminals on each end is not attached. In the magnetostrictive circuit, two of the three terminals can be used for the signal and two can be used as terminals for the resonating capacitor.

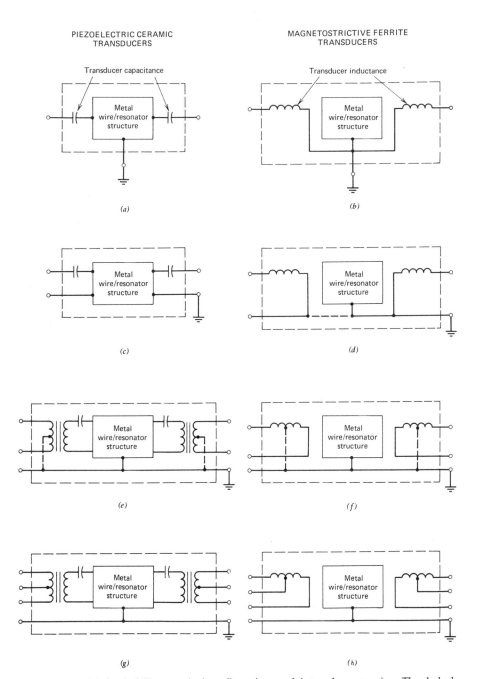

FIGURE 9.7. Mechanical-filter terminal configurations and internal construction. The dashed lines correspond to the package outline and optional center-tap grounds.

Circuit Mounting

Most new mechanical filters are designed for PC mounting. Subsequent wave soldering subjects the filter terminals to a controlled temperature-time environment. If a plastic enclosure is used, the controlled temperature reduces the chances of a terminal-to-plastic loose bond. Temperature control also prevents damage to internal components and lessens the possibility of loosening the coil or transducer leads. When a soldering iron is used, an effort should be made to work quickly with a low-temperature iron. A maximum temperature of 300°C applied for 3 seconds should not cause damage.

The mechanical filter is basically a device that obeys the law of reciprocity. Therefore, if it is used between equal source and load impedances, either end can be used as the input. If the filter is reversed and the response changes, one of the following may be the cause: the source and load resistances are not the same, the filter is over-driven, or the same terminals are not grounded. The last case can be explained by making reference to Figure 9.8. When operating the filter in an unbalanced circuit, one of the coil leads must be shorted to the ground terminal. If terminal 1 is connected to equipment ground, we are shorting the stray capacitance C_{s1} (which is from terminal 1 to a neutral point on the coil) to the grounded elements of the filter, which are the enclosure, the shield, the filter elements, and the ground strap. If we short terminal 2 to ground, we short out C_{s2}. If C_{s2} is not equal to C_{s1}, we have changed our operating conditions and the resultant frequency-response characteristics. This problem not only occurs when the filter is turned around on a PC board but is typical of the type of problem that occurs between the filter manufacturer's test and the user's incoming inspection.

In mounting the filter, another consideration is the effect of external magnetic fields on magnetostrictive filters. If the filter is enclosed in a nonmagnetic case, it is probably susceptible to stray magnetic fields or the

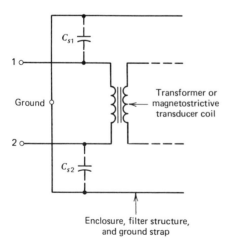

Enclosure, filter structure,
and ground strap

Figure 9.8. Balanced transformer stray capacitance to ground.

effect of close-by magnetic materials. Both of these will affect the transducer bias field and the coil inductance. For this reason, two magnetostrictive transducer filters should not touch but rather should be separated by at least one-half the width of the filter enclosure. Magnetic effects are best determined by experiment, where the filter response is observed as the magnetic materials, coils, transformers, or other filters are moved into close proximity to the filter under test. Although this is not a very common problem, it is a difficult problem to fix after the PC board has been laid out.

Grounding and Shielding

Obtaining good stopband signal rejection is dependent on having good grounds and sufficient input-to-output shielding. From a circuit analysis standpoint, these are similar problems, as can be seen from the dual circuits of Figure 9.9. In this simple example, the filter on the left side is not properly grounded; a small series resistance R_s separates the filter from the source and load common ground. The dual network also contains a filter but the filter is bridged by a large resistance R_b. From a knowledge of network theory, we can state that these two networks will have the same frequency response. Therefore, an understanding of one of the conditions is helpful in understanding the other.

Figure 9.10 shows the effect of a 5 Ω resistance (separating the filter from ground) on the frequency response of a single-sideband filter used in a transceiver. The figure also shows the effect of bridging the filter with a 0.25 pF capacitor and then reversing the lead-to-terminal connections and measuring the frequency response again with the same 0.25 pF bridging capacitor. Reversing the leads has the same first-order effect as bridging with a large inductor or the effect of magnetic feedthrough. In some cases, a controlled amount of reactive bridging or feedthrough is desirable to obtain better selectivity near the passband, but both the manufacturer and the user must understand that the polarity of the ceramic transducer and the phase of the coil must be correct in order to provide more, rather than less, selectivity.

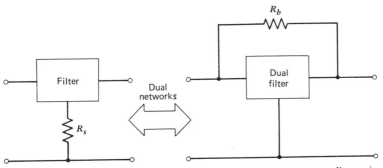

FIGURE 9.9. Dual networks showing the identical nature of insufficient grounding or input-to-output feedthrough.

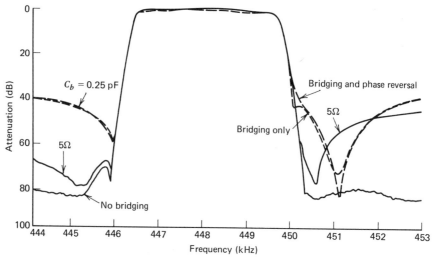

FIGURE 9.10. The effect of capacitive bridging and a resistance from the filter ground to test-circuit ground on the frequency response of a 12-resonator SSB filter ($C_b = 0.25$ pF, $R_s = 5\Omega$).

With regard to grounding, there should be a common ground between the input circuit, the shielding, the filter ground terminal (or metal case), and the output circuit. To reduce electromagnetic feedthrough, the input and output leads should be short and positioned as far away from each other as possible. Coaxial-cable leads and metal shields or ground planes are also helpful. The circuit should also be tested with the filter removed to determine if there are other paths which bypass the filter.

TERMINATION

A problem often faced by the equipment circuit designer is that of providing the correct filter-terminating impedances. This may happen because the filter specification is not available or is not clearly written. Also, the filter user may neglect parasitic effects or may have a preconceived idea regarding termination that may apply to ideal cases but not to an actual filter. For example, the circuit designer thinks that the filter should be terminated by the value of its input impedance. Impedance measurements are made, the filter is terminated, and the frequency response test shows excessive passband ripple. What went wrong?

What the filter user did not understand is that all filters have internal losses which cause the required termination to be different than the lossless termination value. In a mechanical filter, these losses are usually highest in the transducer resonators and the coils at the end of the filter and will often

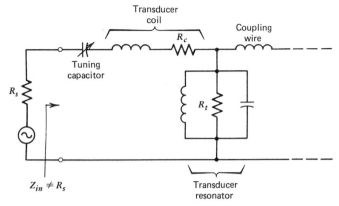

FIGURE 9.11. Input terminating section of a magnetostrictive-transducer filter.

provide a large percentage of the required termination. Figure 9.11 illustrates the effect of internal losses on the value of the required terminating impedance.

Figure 9.11 shows the input end of a magnetostrictive-transducer filter. Coil losses are proportional to the resistance R_c and transducer losses are proportional to $1/R_t$. If the filter was lossless, the input impedance Z_{in} would be approximately equal to an ideal resistance value R_i and the filter would be terminated by that resistance. If the transducer losses were equal to zero (R_t = infinity), the terminating resistance R_s would need to be the difference value ($R_i - R_c$). In other words, the coil losses and the terminating resistance provide the required termination and

$$Z_{in} = R_c + R_i \neq R_s.$$

A similar condition takes place if the transducer resonator Q is low. Therefore, when losses are present we can, in general, say that the terminating resistance is not equal to the input impedance. This means that the filter user should follow the rule that a mechanical filter must be terminated in its specified source and load impedance.

If the users instructions or the specification calls for a resistance of 5 kΩ in parallel with 100 pF, the circuit designer must provide an impedance, at the filter center-frequency, that is equivalent to 5 kΩ in parallel with 100 pF. Included in this impedance should be all strays, discrete R, L, and C values, and active element impedances as measured from the filter terminals. If there is a question regarding the impedance seen by the filter, a measurement should be made with a reactance bridge. The measurement should be made at the filter center-frequency with power applied to the active elements. If the reactive part of the impedance varies from circuit to circuit, enough to degrade the response, it may be necessary to use a trimming capacitor and a network analyzer test set. As the frequency of the analyzer output signal is varied across the passband, the filter output is observed; simultaneously the capacitor is adjusted to obtain the best frequency response. Having obtained this response, the variable capacitor can be left in the circuit or its value can be measured and a fixed value substituted.

If the proper resonating capacitance is not known (or if it is specified to be within a range of values) the user can simply vary the input and output capacitance values to obtain the maximum output at the specified center frequency of the filter. To obtain the minimum passband ripple response, the user will then have to sweep the passband frequency range to find the best value of capacitance.

Effects of Varying the Termination

We will first look at the effects of improper termination on filters that have four or more resonators. In these types of filters, variations in terminating impedance have the greatest effect on passband ripple, as opposed to center frequency shift.

Passband Ripple Variations in Multiresonator Filters

The sensitivity of a mechanical filter's passband ripple to changes in terminating impedance is dependent on a number of factors such as, internal losses, the nominal value of passband ripple, the number and location of attenuation poles, the amount of predistortion, and the degree of coupling of the electrical circuit and the transducer resonator to the main body of the filter. Each filter design is unique, and therefore the user must be careful in applying data from one kind of filter to another. Keeping this warning in mind, we will look at the effects of varying electrical tuning and resistance on the passband ripple of a nine-disk magnetostrictive-transducer mechanical filter.

Figure 9.12 shows how passband ripple is affected by ± 10 percent changes in resonating capacitance. The nominal (solid) curve is ideal, in that the filter

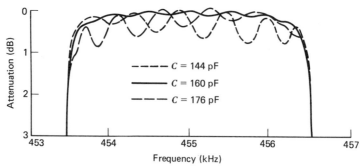

FIGURE 9.12. Variation of passband ripple with changes in resonating capacitance. Nine-disk mechanical filter with magnetostrictive ferrite transducers.

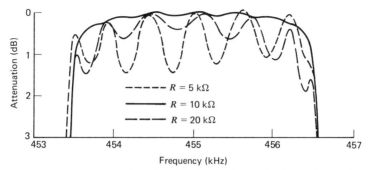

FIGURE 9.13. Variation of passband ripple with changes in terminating resistance. Nine-disk mechanical filter with ferrite transducers.

has no mistunings or miscouplings, although small tuning or coupling variations will generally not make the filter more sensitive to electrical tuning. If the transducer coupling coefficient had been greater, or if some of the resonating capacitance had been placed inside of the filter, then the sensitivity would have been less. Keeping the capacitance within ± 2 percent of the specified value is usually safe.

The curves in Figure 9.13 show the variation of passband ripple with changes in terminating resistance. As with tuning variations, the ripple sensitivity is dependent on a number of factors. Examples are transducer coil losses, acoustic losses, and internal resistor padding.

Mechanical filters designed with piezoelectric ceramic transducers exhibit mistermination effects similar to magnetostrictive-transducer filters. Usually, the ceramic-transducer filter is tuned by the inductance of a ferrite cup core (pot core), rather than a capacitor. The Q of the cup core is considerably greater than a ferrite transducer coil, which means lower electrical losses and lower insertion loss, but often greater sensitivity to terminating resistance changes than that shown in Figure 9.13. With the change of C's to L's, Figure 9.12 gives us an indication of the sensitivity of a piezoelectric-ceramic transducer filter to changes in inductive tuning.

Termination of Narrowband Filters

In this section we examine the effect of improper termination on narrow-bandwidth low-number-of resonator filters. Specifically, we use a two-resonator, piezoelectric-ceramic transducer filter to show the changes in ripple and center frequency resulting from variations in terminating resistance.

Example 9.1. Figure 9.14(a) shows an electrical equivalent circuit of a two-resonator ceramic-transducer filter. This filter has a typical symmetrical network topology where the source and load resistances are equal and the resonator frequencies are equal. Figure 9.14(b) shows the transformation of

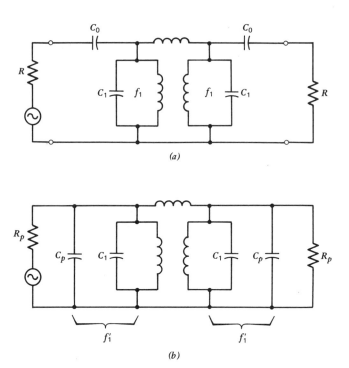

(a)

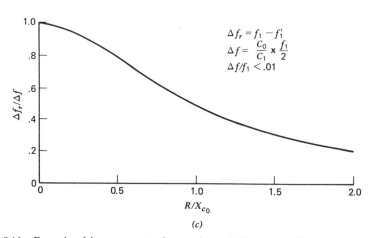

f_1' f_1'

(b)

$$\Delta f_r = f_1 - f_1'$$
$$\Delta f = \frac{C_0}{C_1} \times \frac{f_1}{2}$$
$$\Delta f/f_1 < .01$$

(c)

FIGURE 9.14. Example of how resonator frequencies and filter center frequency varies with changes in terminating resistance.

the series connected R and C_0 into a parallel equivalent circuit (equations describing these transformations were shown in Figure 4.37). The shunt capacitance C_p can be combined with the resonator capacitance C_1 to form a tuned circuit that is resonant at frequency f_1'. As the terminating resistance R is varied, both the parallel resistance R_p and the frequency f_1' are changed. The frequency variation is shown in normalized form in Figure 9.14(c).

As was discussed in Chapter 4, the circuit of Figure 9.14(b) is the narrow-band prototype filter from which the design in Figure 9.14(a) is derived. This means, that as the resistance R varies, resonator frequencies of the prototype vary and the center frequency shifts. If only one resistance deviates from the value R, then one resonator's frequency shifts and the passband ripple increases. The amount of frequency shift or passband ripple increase is dependent on both the coupling coefficient of the transducer, which is proportional to the pole-zero spacing Δf, and the design R/X_{C_0} ratio which determines the slope of the frequency shift curve. In addition to the variations in passband ripple due to resonator frequency shifts, the ripple is also a function of the value of R_p.

Choosing Termination Components

Unless otherwise stated, the filter user should terminate the filter with "zero" temperature coefficient (TC) R's, L's, and C's. The components should also have high Q values and good aging characteristics. With regard to inductive components, ferrite cores are used exclusively. Resistors and capacitors that can give satisfactory performance are shown in Table 9.1. The thick and thin film resistors and the NPO ceramic capacitors are ideal for hybrid circuits; the thin film tantalum resistors can be precision trimmed to obtain an exact termination for telephone channel filters.

TABLE 9.1.   Resistor and Capacitor Termination Components for Mechanical Filters

	Temperature Coefficient (ppm/°C)	Aging (%/20 Years)	Q
Resistors			
Standard carbon composition	± 1000	2.0	—
Carbon film	-250	5.0	—
Metal film	$+50$	0.01	—
Thin film Ta	-100	0.1	—
Thick film Cermet	± 150	1.0	—
Capacitors			
Mica	0 to 70	0.1	1000
NPO ceramic	0 ± 30	0.05	1000
Thin film Ta	$+200$	-0.5	400
Polystyrene	-120 ± 30	0.2	5000

In some cases, low temperature coefficients are not desired. For example, PZT-ceramic transducer filters are often tuned with a combination of a negative-TC capacitor and a low-TC coil to balance the positive-TC of the ceramic (see Figure 2.16 for the ceramic transducer characteristics). Also, low-ripple ferrite-transducer filters may require positive-TC capacitors to balance the negative-TC of the coil inductance (see Figure 2.14). Because it is difficult to obtain positive-TC capacitors, ferrite-transducer filters are usually designed with high electromechanical coupling and therefore low electrical tuning sensitivity. In the case of the PZT-transducer filters, the temperature compensation is usually internal and does not need to concern the user.

In most ceramic-transducer high-performance filters, the tuning coils are inside the filter enclosure. Inductors or transformers for inexpensive radio filters are often external to the filter but usually can be purchased with the filter.

Varying the Termination of Externally Tuned Intermediate-Bandwidth Filters

Often a circuit designer wants to use a mechanical filter in a network that presents a terminating resistance that is different from the filter's specified value. If the tuning is external, there are solutions available to the user that do not require a redesigning of the filter. For example, if the transducers are piezoelectric, a transformer can be used in the place of the tuning coil. The inductance of the transformer as seen from the filter terminals must be equal to the value required to tune the filter. The turns ratio is N_t/N_f, where N_t represents the turns on the electrical termination side of the circuit and N_f corresponds to the turns on the filter side of the circuit. In relation to the specified termination R_s and the user's circuit termination R_u, we can write

$$\frac{N_t}{N_f} = \sqrt{\frac{R_u}{R_s}} \, .$$

If the transformer is used on the input of the filter and the turns ratio N_t/N_f is less than unity, then care should be taken to insure that the voltage step-up does not cause nonlinear effects in the filter performance. If the turns ratio is greater than unity, then parasitic capacitance across the source or load will have a greater effect on performance than in the case of a low terminating resistance R_u. Let us next look at a method of varying the terminating resistance of a magnetostrictive transducer filter.

In the place of a transformer, magnetostrictive transducer filters can use a capacitance dividing network for impedance matching. We will assume in the following discussion that our starting point for impedance matching the filter to the source or load resistance R_t is the parallel circuit composed of C_p and R_p is shown in Figure 9.15. C_p is the specified parallel resonating capacitance and R_p is the specified parallel resistance. In Figure 4.37 we showed parallel-to-series and series-to-parallel transformations like that shown in Figure 9.15(a). If the

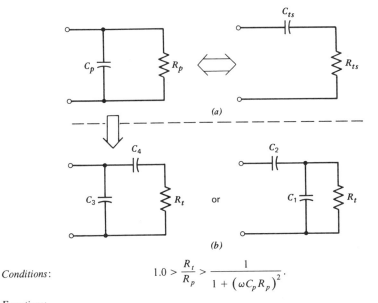

Conditions:
$$1.0 > \frac{R_t}{R_p} > \frac{1}{1 + \left(\omega C_p R_p \right)^2}.$$

Equations:

$$C_1 = \left[\frac{1 + \left(\omega C_p R_p \right)^2}{\omega R_p \left(\frac{R_t \left[1 + \left(\omega C_p R_p \right)^2 \right] - R_p}{R_p} \right)^{1/2}} \right] \times \left[\frac{1}{\left(\frac{R_p}{R_t \left[1 - \left(\omega C_p R_p \right)^2 \right] - R_p} \right) + 1} \right] ;$$

$$C_2 = C_p \left[1 + \frac{1}{\left(\omega C_p R_p \right)^2} \right] \times \left[1 + \frac{1}{\left(\frac{\left(\omega C_p R_p \right)^2 R_p}{R_t \left[1 + \left(\omega C_p R_p \right)^2 \right] - R_p} \right)^{1/2} - 1} \right]$$

$$C_3 = C_p - \frac{1}{\omega R_p} \left(\frac{R_p}{R_t} - 1 \right)^{1/2} ; \quad C_4 = \frac{1}{\omega R_p} \left[\left(\frac{R_p}{R_t} - 1 \right)^{1/2} + \left(\frac{1}{\frac{R_p}{R_t} - 1} \right)^{1/2} \right]$$

FIGURE 9.15.   Terminating resistance transformations.

filter specification calls for series capacitance tuning, we can first convert, by use of the equations of Figure 4.37, the series circuit of C_{ts} and R_{ts} to C_p and R_p and then transform to the desired value of resistance. Our impedance matching problem is therefore as follows: Given a specified R_p and C_p, find a capacitive matching network that has a termination resistance R_t and that resonates with the coil inductance. Two circuits that perform this task are shown in Figure 9.15(b). Conditions, as well as equations for the transformations shown in Figure 9.15, are also shown in the figure. The conditions implicitly state that we cannot terminate the filter in a resistance higher than R_p or lower than R_{ts}. Also, the conditions apply to both circuits of Figure 9.15(b). Our choice of circuit is, therefore, based on criteria such as the resultant capacitor values or the necessary position of the capacitors in order to include stray capacitance with the shunt values. If there is appreciable stray capacitance across both the filter terminals and the load R_t, a pi network of capacitors can be formed by removing capacitance from C_p and then transforming the remaining parallel circuit to the C_2, C_1 circuit of Figure 9.15(b). Let's illustrate these concepts with an example.

Example 9.2. A mechanical filter, with a center frequency of 455 kHz, has a tuning capacitance and termination resistance specification of $C_p = 110$ pF and $R_p = 20$ kΩ. Stray capacitance across the filter terminals is 10 pF, stray across the load is 15 pF, and the terminating transistor input resistance is 5 kΩ.

The first step in the transformation is to remove the 10 pF stray capacitance from the 110 pF specified capacitance, as is done in Figure 9.16(b). Next, the 100 pF and the 20 kΩ are transformed to the network of Figure 9.16(c) through use of the equations for C_1 and C_2 shown in Figure 9.15. Finally, 15 pF of stray capacitance is removed from the value of C_1. If we had very little stray capacitance across the load, we could have terminated the filter in the smaller capacitance values shown in the alternate solution of Figure 9.16(d). Again, these element values were calculated from the equations of Figure 9.15.

Looking for Causes of Excessive Passband Ripple

One of the most common problems in using any bandpass filter is improper termination. The filter is soldered into your circuit, the passband frequency

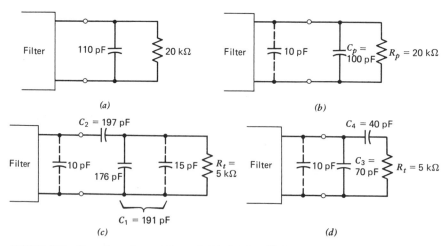

FIGURE 9.16. Transformations of Example 9.1. (a) Specified termination, (b) removal of 10 pF stray capacitance, (c) transformation of the terminating resistance and removal of 15 pF of stray capacitance, and (d) alternate solution assuming that there is no stray capacitance across the termination.

region is swept, and lo ... there are large ripples or no output. Is the filter the problem, or is the circuit the problem? If you think that the circuit is at fault, consider the following steps:

1. Check the specification for the terminating resistance and the tuning component values (if the filter is externally tuned).

2. Is the tuning component a capacitor, an inductor, or a transformer? Is the tuning component in series or in parallel with the source or load resistance?

3. Are the components connected to the proper filter terminals? Are the proper terminals grounded?

4. Is stray capacitance taken into account?

5. Are the input and output impedances of the active circuits the proper values? Have they been checked on a reactance bridge with power applied to the circuits? Are the series or parallel equivalent-circuit values being measured?

6. If there is uncertainty regarding the filter tuning, add variable capacitors and sweep the response. Adjust for best response.

If you think that the problem is in the filter, you can do the following:

1. Determine the internal configuration and its relationship with the external terminals. The configuration may be one shown in Figure 9.7, plus there may be series or shunt tuning and resistance components.

2. First check for opens or shorts between terminals and to ground (the enclosure, if it is metal). Coil resistances are a few Ohms or a few tens of Ohms.

3. If the filter has piezoelectric transducers, check the transducers capacitance with a capacitance meter. Input and output capacitance values usually have the same nominal value.

4. X-ray and look for possible damage due to handling.

Component applications engineers state that filter mishandling is the major cause of failures. There is a misconception that because the mechanical filter is composed of metal alloy elements, it can be dropped and no damage will occur. Therefore, mechanical filters are sometimes poured from one bin to another, and damage does occur. Mechanical filters should be carefully packaged and handled like vacuum tubes or crystal filters or crystal oscillators. As with vacuum tubes, mechanical filters should be left in their packaging until the time they are to be mounted in the equipment.

SPECIAL CIRCUIT CONFIGURATIONS

In this section we look at circuits used to minimize the effects of microphonic and spurious responses, as well as methods for enhancing the stopband selectivity of a mechanical filter.

Microphonic and Spurious Responses

Microphonic responses are signals generated at the output of the mechanical filter due to the filter being subjected to external vibration. A reduction of the amplitude of these responses can be achieved by mounting the filter in a position on the circuit board or chassis that will be subjected to the lowest g-level. If this is not sufficient, the filter can be mounted on a soft foam or a high-damping silicone pad. An alternate solution is the use of additional filtering at the output of the mechanical filter.

Figure 9.17 shows microphonic response curves corresponding to three different circuit conditions. The highest level curve is for the condition where no additional filtering is used, the next level corresponds to the use of an RC highpass filter following the mechanical filter, and the third curve represents the mechanical filter followed by an LC highpass filter. The mechanical filter is a two-resonator bar-flexure mode design (see Figure 7.34) with a center frequency of 11.333 kHz and a bandwidth of 100 Hz. The vibration amplitude is a constant 5 g in the direction of maximum microphonic output. This type of planar filter structure is flexible when vibrated in the direction of the normal bar resonator motion and therefore has a high output at 500 Hz.

In contrast to the bar-flexure mode design, the disk-wire filter (see Figure 7.1) has a stiff three-dimensional structure and therefore will have its maximum amplitude resonances at higher frequencies. This is shown in Figure 9.18 where the measured characteristic is audio modulation of an RF signal located at the center frequency of the mechanical filter. The filter is designed for single-sideband data communication use. The carrier frequency is 455 kHz but the data is at audio frequencies and therefore post filtering as in the previous example is not possible. If the modulation level creates a problem it may be necessary to relocate the filter on the board or use a mounting fixture that provides acoustic damping. Also, because the microphonics amplitude varies greatly from filter to filter, it may be necessary to vibrate each filter and sort out those which do not meet the specification. This type of 100 percent testing is done, for example, on the 48 kHz telephone channel filters shown in Figure 7.13.

In previous chapters we have seen that spurious responses are present in all mechanical filters. The amplitudes and frequencies of these responses vary from one type of filter to another, so they may not cause problems for the system designer. If they are a problem, the filter user must either modify his system or provide additional selectivity, in front of or following the mechanical filter. This selectivity, when needed, is usually no more than one or two pole-pairs of low pass, high pass, or bandpass sections composed of active, LC, or piezoelectric ceramic elements.

Selectivity Improvement

There are two ways of improving stopband selectivity without redesigning a filter. One method is to couple two or more filters in a cascade (tandem)

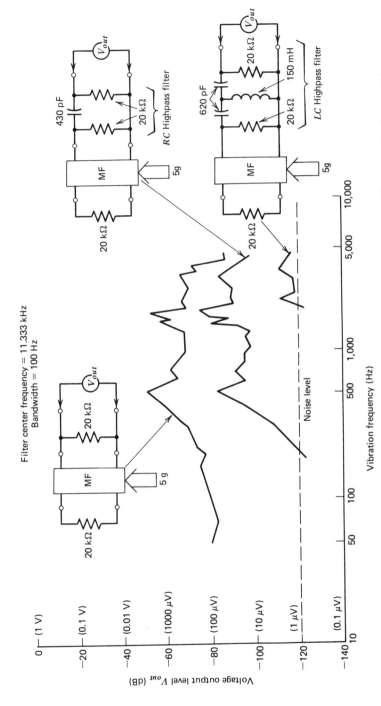

FIGURE 9.17. Microphonic responses of a bar-flexural mechanical filter to a 5g vibration input under different postfiltering conditions.

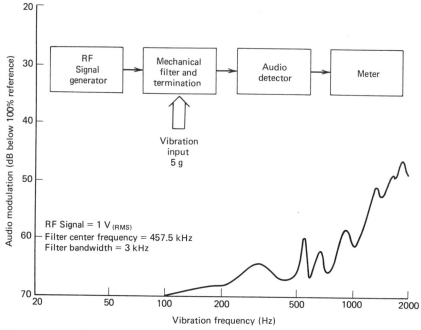

FIGURE 9.18. Audio modulation as a function of the frequency of vibration. Envelope of maximum outputs of 16 filters.

arrangement; the other way is to bridge the filter with a small value of capacitance.

Improving Selectivity by Cascading

Cascading two mechanical filters which are separated by a buffer amplifier as shown in Figure 9.19(a) results in a composite frequency response attenuation that is the decibel sum of the two responses. If the filters are identical, completely isolated, and properly terminated, the 1.5 dB frequency points become the 3 dB points and the 30 dB frequency points become the 60 dB points. In other words, the 1.5 dB bandwidth becomes the 3 dB bandwidth and the 30 dB bandwidth becomes the 60 dB bandwidth. We must remember that although we obtain a large improvement in stopband selectivity, we lose some of that benefit in the passband becoming narrower. This is a point that is often overlooked. An alternate method of isolating the filters is to use an isolation pad (a pi or tee network of resistors) in place of the amplifier. The pad also acts as output and input resistance for the two filters. A 6 dB loss pad provides roughly 4/1 isolation between the filters. In other words, a 20 percent change in the input impedance of the second filter looks like a 5 percent change in the input resistance of the pad. The amount of attenuation that is necessary in the pad is a function of the allowable variation of the passband amplitude ripple.

The filters can also be coupled with reactances as shown in Figure 9.19(*b*) [7]. This type of coupling makes it possible to obtain the selectivity of a four-resonator filter as opposed to the selectivity of two, two-resonator filters in cascade. The 60 dB to 3 dB shape factors, from Figure 8.7, are 4.50/1 (four-resonator filter) and 6.38/1 (two decoupled two-resonator filters). Note in Figure 8.7 that the two-resonator 1.5 dB bandwidth is 0.83 times that of the 3 dB bandwidth. Although there are selectivity advantages in using the reactive coupling between filters, the primary disadvantage is that the filter manufacturer must control the filter's input and output impedances, as well as the amplitude response of each individual section, in order to prevent excessive variations in passband characteristics. Because of this problem, the large majority of cascaded mechanical filter applications use isolation circuits such as the amplifier of Figure 9.19(*a*). In addition, reactive coupling is limited to two-resonator sections.

Improving Selectivity with Capacitive Bridging

In the case of two-resonator filters, capacitive bridging like that shown in Figure 9.19 (C_1 and C_2 are the bridging capacitors) is used to improve

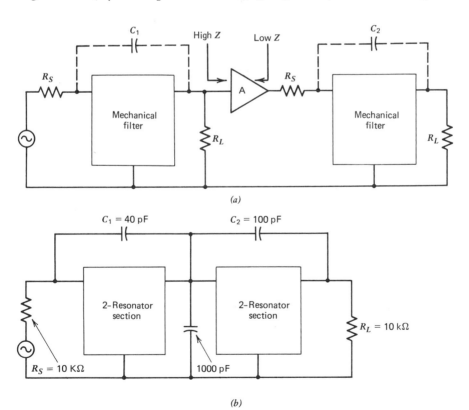

(a)

(b)

FIGURE 9.19.   Cascading mechanical filters.

stopband selectivity. Examples of this technique were shown in Figures 7.6, 7.31, and 7.37. When two-resonator sections are cascaded, the bridging capacitors should be different values in order to stagger the attenuation-pole frequencies. In this way we can maximize the stopband attenuation. If the stopband response of an individual section remains below 30 dB after bridging, the 3 dB bandwidth will be affected very little, but there will be some change in the differential delay.

REFERENCES

1. H. Albsmeier, A. E. Günther, and W. Volejnik, "Some special design considerations for a mechanical filter channel bank," *IEEE Trans. Circuits Syst.*, **CAS-21**, 511–516 (July 1974).

2. B. Birn, "A modified insertion loss theory for mechanical channel filter synthesis," in *Proc. 1976 IEEE ISCAS*, 754–757 (Apr. 1976).

3. K. Yakuwa, M. Yanagi, and K. Shirai, "DA-Series channel translating equipment using pole-type mechanical filters, Part 2: Components and devices," *Fujitsu Scientific and Tech. J.*, **15**(3), 23–45 (Sept. 1979).

4. R. A. Johnson and W. A. Winget, "FDM equipment using mechanical filters," in *Proc. 1974 IEEE ISCAS*, San Francisco, 127–131 (Apr. 1974).

5. C. F. Kurth, "Generation of single-sideband signals in multiplex communication systems," *IEEE Trans. Circuits Syst.*, **CAS-23**(1), 1–17 (Jan. 1976).

6. W. Haas, "Channeling equipment technology using electromechanical filters," *Electrical Communication*, **48**(1, 2), 16–30 (1973).

7. R. A. Johnson and W. D. Peterson, "Build stable compact narrow-band circuits," *Electron. Design*, 60–64 (Feb. 1, 1973).

INDEX